Hydrogeological and Environmental Investigations in Karst Systems

Hydrogeological and Environmental Investigations in Karst Systems

Editor

Hubert O'Kelly

scitus academics

Hydrogeological and Environmental Investigations in Karst Systems
Edited by **Hubert O'Kelly**

ISBN: 978-1-68117-422-8
Library of Congress Control Number: 2016937446

www.scitusacademics.com
Box No. 4766, 616 Corporate Way,
Suite 2, Valley Cottage,
NY 10989

Printed in United States of America on Acid Free Paper

Preface

Karst is a type of landscape, and also an aquifer type. Karst areas consist of solid but chemically soluble rock such as limestone (most important) and dolomite, but also gypsum, anhydrite and several other soluble rocks. It is characterized by underground drainage systems with sinkholes and caves. It has also been documented for weathering-resistant rocks, such as quartzite, given the right conditions. Subterranean drainage may limit surface water with few to no rivers or lakes. However, in regions where the dissolved bedrock is covered or confined by one or more superimposed non-soluble rock strata, distinctive karst surface developments might be totally missing. Karst aquifers are characterized by a network of conduits and caves formed by chemical dissolution, allowing for rapid and often turbulent water flow.Karst aquifers form by flowing water containing carbon dioxide which dissolves carbonate rocks. Therefore, there is a close relation between aquifer evolution, the formation of caves and groundwater flow. Hundreds of millions of people worldwide live in karst areas and are supplied by drinking water from karst aquifers. These aquifers include valuable freshwater resources, but are sometimes difficult to exploit and are almost always vulnerable to contamination, due to their specific hydrogeologic properties. Therefore, karst aquifers require increased protection and application of specific hydrogeologic methods for their investigation. Other problems frequently encountered in karst areas include: soil erosion and rock desertification, leakages of channels and reservoirs, collapse of underground cavities and formation of sinkholes, and flooding. Resolution of these problems requires involvement of karst hydrogeology experts. The book Hydrogeological and Environmental Investigations in Karst Systemscovers the topics understanding of cave and karst related processes and facilitates the translation of current discipline-specific research to an interdisciplinary readership by dealing with specific cave or karst systems. The book addresses a multidisciplinary approach involved in anthropology, archaeology, biology, chemistry, geography, geology, geomorphology, hydrogeology, paleontology, sedimentology, and all other disciplines related to speleology and karst terrains.

Table of Contents

1

CHAPTER

SIMULATION OF REGIONAL KARST AQUIFER SYSTEM AND ASSESSMENT OF GROUNDWATER RESOURCES IN MANATÍ-VEGA BAJA, PUERTO RICO

Balati Maihemuti[1], Reza Ghasemizadeh[1], Xue Yu[1], Ingrid Padilla[2], Akram N. Alshawabkeh[1*]

[1]Department of Civil and Environmental Engineering, Northeastern University, Boston, MA, USA

[2]Department of Civil Engineering and Surveying, University of Puerto Rico, Mayaguez, PR, USA

ABSTRACT

The North Coast karst aquifer system of Puerto Rico, the most productive aquifer of the island, is a vital water source for drinking water and local ecosystems. High freshwater demands alter the coastal groundwater system that impacts both human populations and coastal ecosystems of the island. To predict how this system might respond to rainfall events and high pumping demands, we used the equivalent porous medium (EPM) technique to develop a three-dimensional groundwater flow model to estimate hydrogeological parameters and assess groundwater resources in the Manatí-Vega Baja karst aquifer. The approach is based on the hypothesis that the simplified EPM approach will reproduce groundwater hydrodynamics in this complex karst environment. The steady-state model was calibrated with trial and error and parameter estimation methods using an observed groundwater table of 1995 ($r = 0.86$, $p < 0.0001$, $n = 39$). The large-scale simulation suggested that groundwater flow roughly follows the elevation slope (i.e. south to north). Calibrated hydraulic conductivities range from 0.5 to 86 m/d, whereas the hydro-geologic data strongly suggest higher permeability in the middle karst section of the study area. The transient model adequately estimates the observed groundwater fluctuations in response to rainfall events from 1980 until 2014. The transient results indicate that the conceptual model accuracy is more acceptable with a mean error (ME) of −0.132 m, mean absolute error (MAE) of 0.542 m and root mean square (RMSE) error of 0.365 m. The results of water budget simulation show that the total recharge satisfies the total groundwater withdrawal rate in the past, but continuous closure of more contaminated wells causes groundwater levels to increase in the future. The results indicate that the assumption of applicability of EPM approach is sustained and supported by measured data in the study area. Taking future water demands into account, this model could be applied further to predict the changes of groundwater levels and mass balance under different exploitation scenarios.

Keywords: Karst Aquifer, Groundwater Modeling, Seawater Intrusion, MODFLOW, Puerto Rico

1. INTRODUCTION

Numerical modeling is an important predictive tool for managing groundwater resources. Groundwater models can be used to test or refine different conceptual models, estimate hydraulic parameters, and most importantly for water resource management to predict how the aquifer might respond to changes in pumping demands and climate change [1] . Groundwater flow models are helpful in determining the direction of groundwater flow, distribution of hydraulic heads and flow magnitudes.

Simulating groundwater flow and accurately assessing groundwater resources are more complex in karst aquifers due to higher uncertainty in hydraulic parameters. Karst terrains are underlain by highly soluble rocks (such as limestone or dolomite) with well-developed secondary porosity, exhibiting distinctive hydrology and landforms such as sinkholes, sinking streams, springs, and caves [2] -[4] . A more complete understanding karst formation and hydrogeological characteristics would improve the simulation modeling accuracy [5] [6] .

Karst aquifers show very high spatial heterogeneity [7] [8] . The EPM method is one of simplest yet most debated approach to characterize karst aquifer systems [2] [9] -[14] . EPM assumes that a karst aquifer is equivalent to a porous medium in which karst channels and large faults are treated as areas of higher hydraulic conductivity [10] [15] . In karst systems, the concept of a representative elementary volume is used when the size of the area of interest (or the cell in a model) becomes large enough to approximate equivalent porous media [16] [17] . Although accurate simulation of transport processes is still problematic in small (local) scales, one may be able to model hydraulic heads, flow volumetric, and general flow directions by dealing with the complex aquifer as a porous media, especially in regional-scale and intermediate-scale studies [18] .

Karst aquifers contribute to a major portion of the groundwater supply in Puerto Rico (PR) as well as in many parts of the world. Karst covers between 28% [19] and 34% of Puerto Rico [20] and is most extensive in the northern karst belt, which is well documented by several USGS reports [20] -[22] .

Public-water supply (PWS) withdrawals within the Manatí quadrangle area constitute about 70% of the total water demand within the municipalities of Manatí and Vega Baja. Groundwater conservation and resource distribution in subsurface systems have elicited considerable interest from both public and private sectors. Due to environmental contaminants, the Manatí-Vega Baja area upper aquifer received more concerns recently. Some studies were conducted to estimate groundwater flow, such as Gregory (2001), which estimated its water budget using a steady state two-dimensional model, but their model did not evaluate the temporal trend and dynamics of the aquifer. The objective of this study is to use the EPM approach to establish a three-dimensional groundwater flow model and to evaluate the adequacy of EPM in reproducing the groundwater flow behavior and temporal water table fluctuations. It is questionable whether simplifying the high heterogeneity, which is often impractical to quantify, allow for accurately simulating the complex karst system. The thrust for the current study is the ability to combine limited field data with numerical simulations to generate groundwater flow estimates in higher scale as well as evaluating the spatial and temporal patterns of the groundwater flow and its influence in the environment. Insight obtained in this study will better characterize the behavior of the aquifer system, assess long-term sustainability of its groundwater, and evaluate the hydrogeological properties of the unconfined upper karst aquifer of the Manatí-Vega Baja area within the North Coast Province of Puerto Rico.

2. SITE DESCRIPTION

2.1. Location

The study area of Manatí-Vega Baja is located in north-central Puerto Rico and covers approximately 204.61 km^2 (Figure 1). This area is bordered by the Atlantic Ocean on the north and by the rugged cone karst topography in the center and to the south. The western boundary is a north-south line, just west of the Río Grande de Manatí, and the eastern boundary is a north-south line on the east of the confluence of the Río Indio and the Río Cibuco.

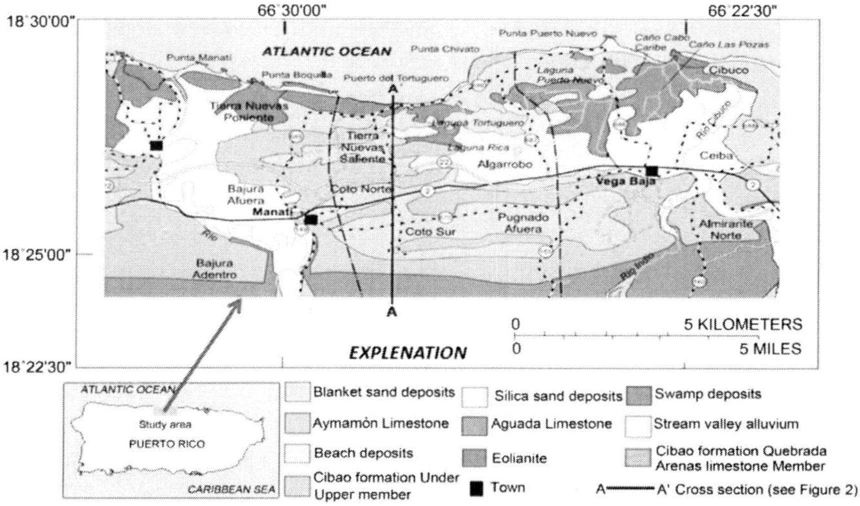

Figure 1. Location of the Manatí-Vega Baja study area and important hydrologic features and generalized surficial geology of the Manatí-Vega Baja study area. (Modified from Monroe, 1980).

2.2. Geological Setting

Tertiary limestone within the North Coast Province generally strikes to the west and dips to the north with slope of 6 to 7 degrees, near the contact with volcanic rocks south of the study area, and from about 1 to 2 degrees near the coast [23] . These rocks, grouped into six formations, are named in ascending order: the middle Oligocene San Sebastián Formation, the late Oligocene Lares Limestone, the Oligocene and Miocene Cibao Formation, the Miocene Aguada and Aymamón Limestones, and the lower Miocene Camuy Formation. Alluvial deposits occur in the river valleys of the Río Grande de Manatí and the Río Cibuco, with a sequence of "blanket sands" overlying limestone in upland areas (Figure 1).

2.3. Hydrology and Hydrogeological Setting

The North Coast karst groundwater system contains two major aquifers. The upper water-table aquifer lies within the Aymamón and Aguada limestone and alluvial deposits along the coastal areas, whereas the lower aquifer occurs within various locations of the Cibao Formation and the Lares limestone (Figure 1 and Figure 2). The upper aquifer is connected to the surface throughout most of its outcrop area. The lower aquifer is confined toward the coastal zone and outcrops to the south of the upper aquifer, where it is recharged. Groundwater in upper aquifers flows from recharge areas toward discharge areas near the coast, streams, wetlands, springs, and other surface-water features [24] . Long-term records from National Weather Service rainfall stations in the study area indicates a relative dry season from December to April, and a wet period from May to November, associated with the hurricane season [25] . The mean monthly rainfall records indicate that May is the wettest month, followed closely by November; while February and March are the driest months. Average annual rainfall in the study area ranges from less than 1.5 m near the coast to 1.8 m in the karst terrain near the southern boundary.

Although some reaches of rivers and streams are important sources of groundwater recharge, most aquifer recharge in this region is from infiltration of rainfall through the limestone outcrops via runoff to sinkholes and enclosed topographic depressions. Within these areas, recharge, which has been estimated by groundwater flow models, is in the range of 0.52 to 0.51 m/yr. The average annual recharge to the upper aquifer of the study area was estimated to be about 0.52×10^8 m^3 to 1.04×10^8 m^3 [26] -[28] .

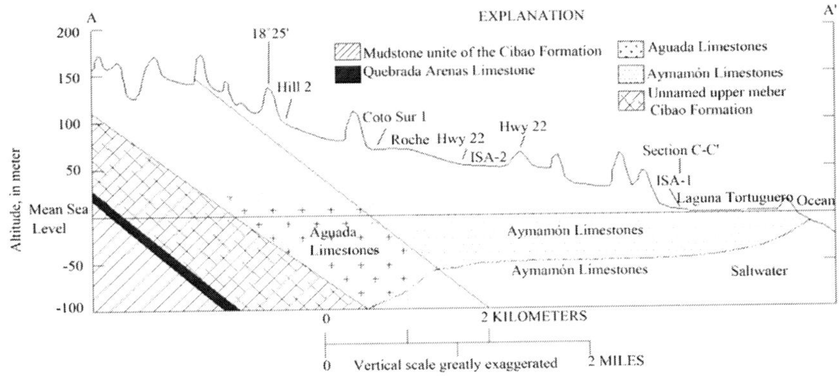

Figure 2. Generalized hydrogeologic cross sections of the Manatí-Vega Baja study area, North Coast Province, Puerto Rico. Refer to Figure 1 for cross-section location.

Prior to the 1940s, the Upper Aquifer System in the Manatí-Vega Baja region was relatively undeveloped. Drilling activities started in the early 1960s and an extensive drilling program accelerated in the 1970s. In 2014, the total number of pumping wells has reached 120, and the total annual groundwater extraction has continually increased. The locations of productive wells used to define the upper aquifers and confining unit are shown in Figure 3. The groundwater withdrawal rates were 7344, 22,974, 55,509 and 65,488 m^3/d in 1945, 1970, 1980 and 1995, respectively.

The groundwater of the Upper Aquifer System generally occurs under unconfined conditions in the Manatí-Vega Baja area. The water table is not stable and fluctuates during wet and dry seasons and also influenced by the withdrawal rates. Four observation wells (Table 1) are used to monitor the water level fluctuations of the Upper Aquifer. These wells have total depths above mean sea level (MSL) of 38.41 m, 34.75 m, 83.82 m and 76.2 m, respectively.

Besides groundwater withdrawals, the next prominent groundwater discharge feature for the upper aquifer is Laguna Tortuguero. The mean annual groundwater discharge into Laguna Tortuguero was estimated as 0.6 m^3/s by Bennett and Giusti [29] , which was comparable to the 0.62 m^3/s mean annual discharge rate through the ocean outlet channel obtained by Quiñones and Fusté [30] .

3. METHODOLOGY

To simplify the complex field system and organize available data, an approximated representation of field situation is reconstructed through MODLFOW [31] implemented in an advanced graphical preprocessing and post processing program, Groundwater Modeling System (GMS v.10, Aquaveo 2014). While simplified under Equivalent Porous Media (EPM) approach, the developed conceptual model retains enough complexity in representing hydrological conditions to sufficiently reproduce the groundwater behavior. The conceptual model for Manatí-Vega Baja region adapted by this study consists of three-dimensional hydrogeological layers, repre- senting different layers of the unconfined upper limestone aquifers. Hydrogeological maps and data were imported into GMS coverage layers and were used as input for developing the conceptual model. Coverages were also used to define flow budgets, model boundary condition, sources and sinks, observation wells, elevations, and recharge and hydraulic conductivities. The conceptual model was then converted into a 3D numerical MODFLOW Framework, where the mathematical calculations are performed, and to reproduce the groundwater heads, water flow budget, and water flow velocity and flow directions.

3.1. Model Setup

The model domain, which was divided into 100 columns and 70 rows, included 6789 grid cells. The western and eastern edges of the model area are no-flow boundaries (Figure 3). These boundaries were extended beyond the rivers to simulate the surface-water/groundwater interaction, and to minimize computational errors resulting from the effects of model boundaries. The southern surface runoff entering the aquifer was represented in the model by injection wells located in south of the model domain. The bottom altitude of the aquifer is a no-flow boundary condition at depth of −100 m above MSL. Figure 3 shows the location of productive wells, springs, filtration plants and lagoons as well as river and stream direction and longitudes in the Manatí-Vega Baja study area. North of the blue line, the lower boundary of the upper aquifer is delimited by the saltwater/freshwater interface. The southern, western and eastern edges of the model area are no-flow boundary.

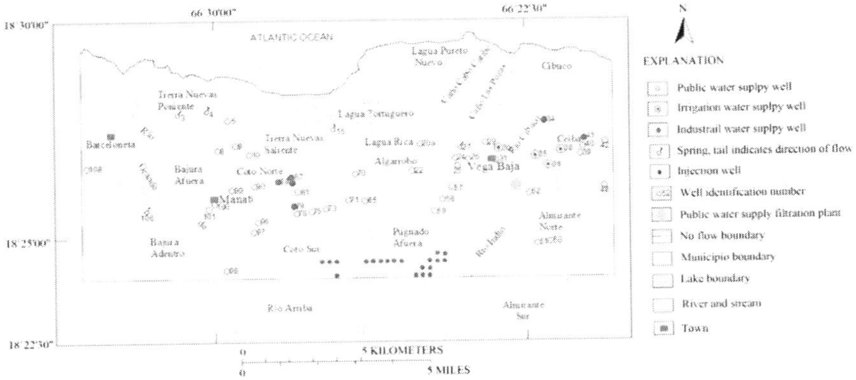

Figure 3. Conceptual model boundary map of the study region.

Table 1. USGS monitoring well water table fluctuation range and total well depths.

No.	Well Name	WT Fluctuation Range (m)	Total Depths (m)
1	Cantito Laluisa Well Manatí PR	1.2 - 2.4	38.41
2	USGS 166 Observation Well Manatí PR	2.0 - 3.4	34.75
3	Coto Sur 5 Well Manati PR	5.2 - 8.0	83.82
4	Rosario 2 Well Vega Baja PR	5.8 - 7.2	76.20

The River Package was used to simulate the interaction between the upper aquifer and Laguna Tortuguero (Figure 2). The seabed along the coast of the Atlantic Ocean was represented as a streambed with a head of 0 m. The major springs in the study area were simulated using the Drain Package within MODFLOW [31] . Using the mass-balance approach, a maximum aquifer recharge rate for the karst terrain in northern Puerto Rico was established at 0.51 to 0.64 m/yr [32] . Initially for the steady-state model, the recharge rate for areas with closed depressions or areas with poorly defined surface-water drainage was assigned a maximum value of 0.46 m/yr.

3.2. Steady State Calibration

To establish that the model can reproduce measured field data, MODLFOW was used together with Parameter Estimation [33] implemented in GMS. Calibrating groundwater flow and accurately adjusting the hydrogeologic parameter values are more complex in karst aquifers due to high uncertainty in hydraulic parameters. Model calibration is the essential to anticipate and predict the geologic uncertainties, where model inputs are modified until the resultant predictions give a reasonably good fit to the observed data. In this study, the model was calibrated in steady-state with a trial and error approach using parameter estimation methods matched to the observed groundwater table of 1995 for which sufficient spatially distributed data are available.

Steady state calibration for the flow model was achieved by comparing the hydraulic heads obtained from available groundwater level contour maps or observation wells (n = 39) of the aquifer and the calculated hydraulic heads of the MODFLOW simulation in order to simulate the flow lines of this aquifer layers [34] . Calibration of the steady state model was performed using the root mean square error (RMSE), mean absolute error (MAE), mean error (ME) and mass balance discrepancy of water into and out of the system. The overall RMSE of the steady state model is 1.86 m, MAE is 1.254 m and the ME is 0.212 m for the calibrated wells (n = 39).

Horizontal hydraulic conductivity and recharge were initially assigned based on a previous study [35] and then the PEST method was applied to estimate the optimum values for recharge and conductivity within defined specific ranges for each zone. After calibration, horizontal hydraulic conductivity values obtained were between 0.5 m/day to 86 m/day for different units, e.g. limestone, silt, sand, and gravel. In Figure 4, the red color area represents the higher transmissivity zones, which appears in the center of the study area, near Manatí, and southwest of Vega Baja. In these areas, the hydraulic conductivity values decrease both to the north and south within the limestone units, and in the part of the aquifer contained within the alluvial deposits. The water table comparisons between observed versus simulated data are shown in Figure 5. To provide information on the overall match of all the monitoring wells (n = 39).

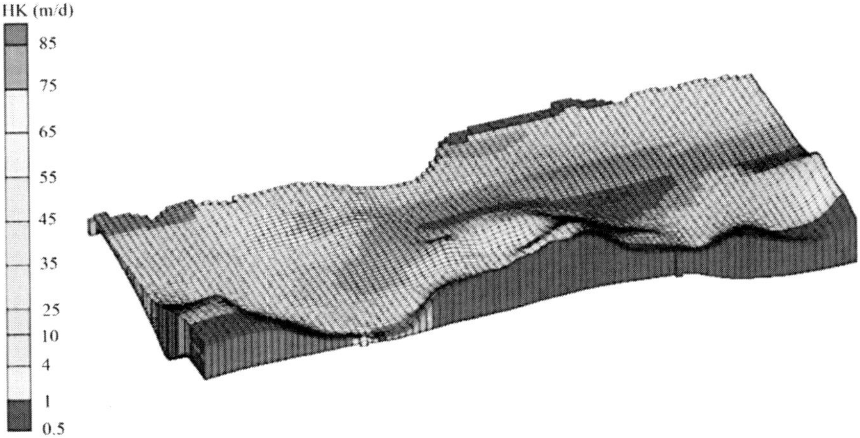

Figure 4. Regional distribution map of hydraulic conductivity.

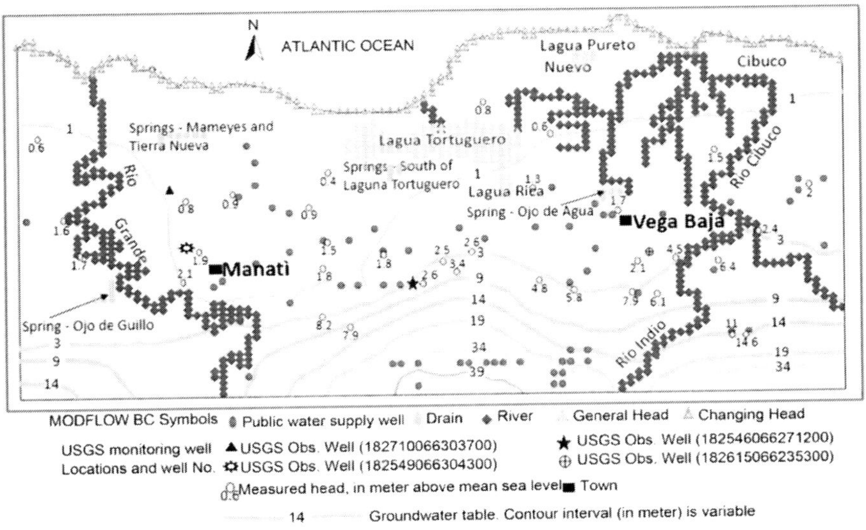

Figure 5. Comparison value of the computed head and observed head.

3.3. Transient Calibration

Upon achieving a satisfactory steady state condition, the transient groundwater model was implemented for the 34 year period between 1980 and 2014. The transient model consists of 412 monthly stress periods. Each stress period had 10 time steps and used days as a time unit. The first stress period represented the steady state period with no significant drawdown in the water level. Sufficient pumping data were available for verification from 1980 to 1995 and the model is capable of predicting behavior from 1996 to 2014. The groundwater withdrawal rates used for the transient model simulation correspond to the historical pumping rates reconstructed from PRASA public water-supply and the PRDNER records, and the United States Geological Survey (USGS) water-use database, which includes production data for industrial, public water-supply, and irrigation uses.

The data from 1980 to the end of 2014 were used for transient calibration and from 1990 to 1995 for verification. The initial transient simulation used a storage coefficient of 1×10^{-5} - 7×10^{-6} and specific yield of 0.01 - 0.2 for different zones. The first step of the calibration assigned an initial value for storage coefficients and specific yield for model layer. Calibration was performed using trial and error procedure by changing the specific yield, storage coefficient, and with very limited range in the hydraulic conductivity values. In this study area there were only four USGS monitoring wells with continues measurements. The available data were used to calibrate the transient state of the model. The main limitations of equivalent porous media approach in karst aquifers are failing to capture groundwater hydrodynamics on a local scale, and failing to simulate transient turbulent flow through conduit network and its interaction with the karst aquifer rock matrix. For this reason, the head error range ($-1 < h < 1$) has been defined for this transient model calibrations.

3.4. Sensitivity Analysis

As with any model, there is uncertainty in the measurements of head (datum errors), hydraulic conductivity, estimation of recharge, and well withdrawal rates. It is therefore important to estimate the effect of improperly simulating a specific parameter. The hydrologic parameters that have the greatest impact on simulation results are identified through sensitivity analyses. Also the non-uniqueness of the calibrated model can be evaluated using sensitivity analysis. A sensitivity analysis was conducted to show that the values of the aquifer's hydraulic properties as estimated by the model calibration procedure were not random and do in fact influence the final results. In other words, significant variations of the simulated value will drive proportional changes of the model results. This sensitivity analysis quantified the dependency of the model to calibrated aquifer parameters, and hydrogeologic stresses such as withdrawals. The effects of the main parameters governing the regional hydrodynamics (i.e. recharge, conductivity, and pumpage) on average water levels were evaluated separately by running the model under reduced or enhanced parameter values. The water levels are more sensible to a decrease in conductivity values than to a decrease in pumping conditions, but less sensitive to an increase in conductivity than an increase in pumping conditions. The sensitivity line for pumping is almost linear and therefore the aquifer is moderately vulnerable to over withdrawal activities. As the Figure 9 indicates, the aquifer is the most sensitive to changes in recharge conditions among other parameters, especially a reduction in average recharge significantly drop the average water levels.

4. RESULTS AND DISCUSSION

4.1. Model Calibrations

The hydraulic conductivities calibrated through the steady state model ranged from 0.5 to 86 m/d, whereas the hydrogeologic data strongly suggested higher permeability in the middle karsts of the study area (see Figure 4). Simulated aquifer recharge values ranged from 0 m/d in the wetland areas of Cabo Caribe, to 0.0003 m/d in the alluvial valleys of Río Grande de Manatí and Río Cibuco, and 0.0013 m/d in sinkholes and areas of enclosed depressions (Figure 6). Recharge is from infiltration of rainfall through the limestone outcrops via runoff to sinkholes and enclosed topographic depressions. While the high value of recharge appears in the center of the study region, along the north coast (wetland area) the recharge rate is insignificant. The overall RMSE of the steady state model is 1.86 m, MAE is 1.254 m and the ME is 0.212 m for the calibrated wells. The simulated steady- heads are in good agreement with the observed values ($r = 0.86$, $p < 0.0001$, $n = 39$) shown on Figure 7.

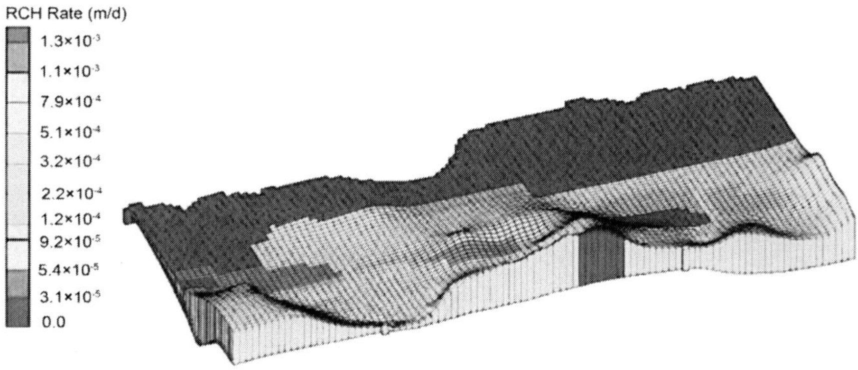

Figure 6. Regional distribution map of recharge rate.

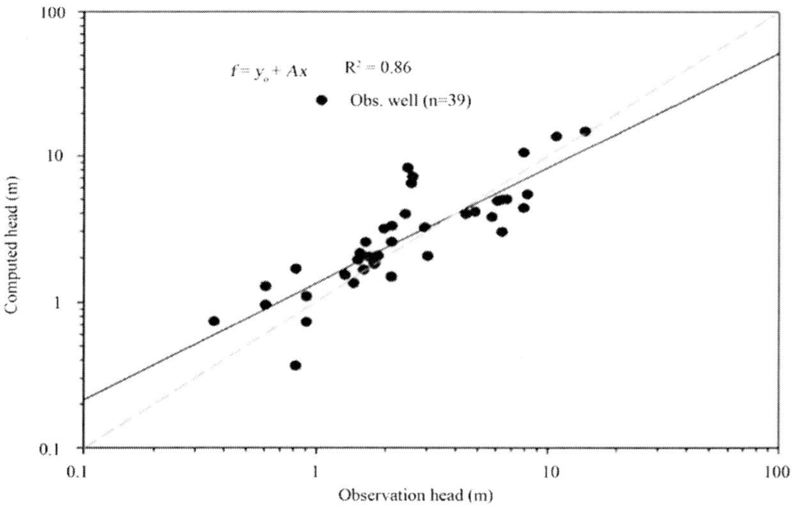

Figure 7. Comparison result of observation and calculation head.

The model's tendency to overestimate or underestimate the actual heads is attributed to the effect of possible conduit or fracture network in this karst aquifer. While pinpointing and simulating the exact location and behaviors of such highly permeable pathways are expensive and in most cases not practical, the employed equivalent porous media approach in our developed model is the easiest (but not most accurate) method available.

The steady state model results show that the computed values are in well-fit to the observed data (Figure 7), indicating that the conceptual model is reasonably calibrated and could be further applied to predict changes in groundwater levels and mass balance under different exploitation scenarios when considering the future water demands in the study area.

After steady state model calibrations achieved a satisfactory steady state, transient groundwater model was developed for a period of 34 years from 1980 to 2014. The transient calibration results were evaluated by comparing the temporal variations in simulated heads with those of observed water levels at four target observation wells. Figure 8 shows the monthly transient simulation comparison between computed and observed head value. The measured and simulated fluctuations were similar and fit each other well; the head error value is under the target value ($-1 < h < 1$ m) as shown Figure 8. The transient state calibration results are shown in Figure 8. The overall RMSE of the transient model was 0.365 m, MAE was 0.542 m, and ME was −0.132 m, which all well below calibration target value.

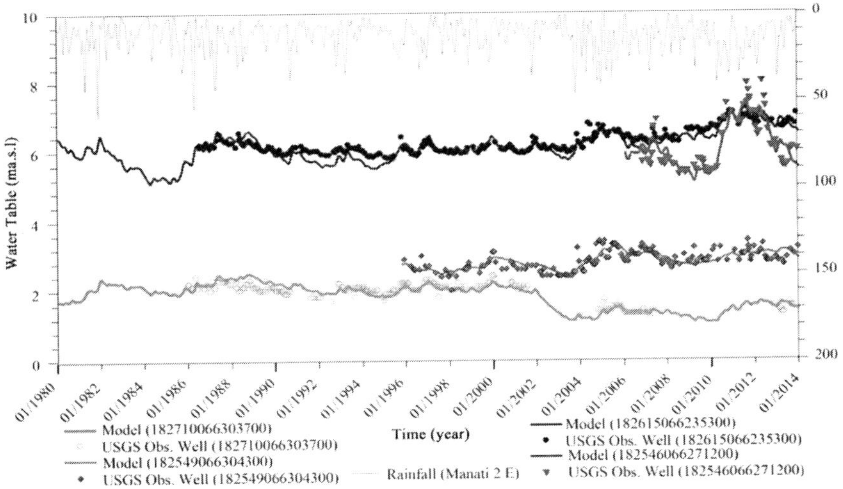

Figure 8. Monthly simulated and observed WT vs. monthly rainfall for 1980-2014. The USGS monitoring wells location showed in Figure 5.

The observed water level data fluctuations matched well with the simulated water level changes resulting from rainfall. For example, groundwater levels in 2011 in the USGS observation (well: 182546066271200) well increased up to a maximum of 2 m according to both measured and simulated data (Figure 8). The fluctuations of the water table in all wells are a function of local groundwater withdrawal trends and the precipitation intensity. As shown in Figure 8, the general magnitude and the sharp fluctuations in water level data (as well as in simulations) are in direct response to monthly rainfall and could be high pumpage in the dry season. Our modeling analyses suggested that aquifer replenishment correlated remarkably with rainfall events both in terms of magnitude and changes.

Figure 8 represents the long term variation of the water table at four USGS wells and the rise in the water table of about 0.4 m at USGS observation well (182615066235300) over the last three decades. The rise is due to the closure of some of the pumping wells over time. Intermediate-term variation in water table corresponds to wet and dry years, as well as groundwater withdrawal activities.

4.2. Sensitivity Analysis

As presented in section 3.4., this karst aquifer is extremely vulnerable to pumping activities as evident by the linear correlation between pumping intensity and the average water levels (Figure 9). A 50% increase in pumping rate would significantly lower the average water level by one meter, and would alter the regional groundwater hydrodynamics. The effect of such over-withdrawal may only be neutralized by about a 30% increase in precipitation. The coastal karst aquifer is especially vulnerable during dry seasons; since a 20% reduction in precipitation would result in lowering the water table by one meter on average. As expected and similar to other karst aquifers, the sensitivity of the water table to hydraulic conductivity is non-linear [15] [36] .

The hydrologic parameters that have the greatest impact on simulation results can also be identified through sensitivity analyses. Water levels were most sensitive to variations in recharge and hydraulic pumpage rates (see Figure 9). Simulated water levels increased with increasing recharge and with decreasing hydraulic conductivity.

4.3. Water Budget

The water inputs in this system derive from recharge, the constant head (ocean), general head (lake), and river or stream boundaries. The main outputs of water from the aquifer were drainages, river or stream segment and the Atlantic Ocean. The water budget of the entire aquifer obtained from the groundwater flow model is presented in Table 2. The balance between inflows and outflows was consistent with the steady state modeling hypothesis. The flow budget result indicates that the main recharge comes from precipitation, which represents about 76%, followed by 17.9% river, 5.1% sinkholes streamflow, 0.82% lakes and 0.11% Atlantic Ocean. The discrepancy between inflow and outflow was estimated to be 0.04% (shown Table 2), which is a significant improvement of the previous modeling study of the same area (24.27%; [35]).

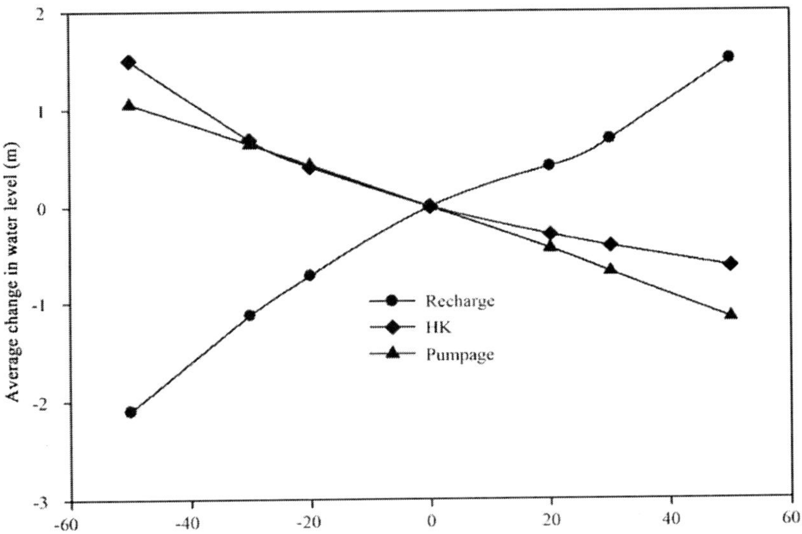

Figure 9. Sensitivity of the predicted water levels to chpanges in recharge, hydraulic conductivity and pumpage rates.

The groundwater levels increased along the cross section A-A', which is from north to south (From Atlantic Ocean to Land; Figure 10). The groundwater flow over the Cibao formation is faster and the flow is thinner due to significantly higher hydraulic gradients. This is typical of all coastal aquifers, and the longitudinal head profiles for springs also follow the same trend with higher heads farther away from the oceans. Farther from the coast (>9000 m), the Aguada formation at the bottom of the aquifer becomes shallower and thinner. Therefore, the storability and conductivity values decrease and the average groundwater level rises with higher rates for areas farther away from the ocean. We also noticed that the magnitudes of seasonal fluctuations of groundwater levels are higher on Aguada limestone. The highest hydraulic gradient occurred in areas close to the southern boundary, which coincident with the lowest values of hydraulic conductivity in this karst aquifer [15] . The water-table gradient is almost flat (25 to 40 meters per kilometer) in the Aymamon limestone. The water table increases to the south in the Aguada limestone (500 to 600 meters per kilometer) toward the model boundary (Figure 10). This hydro diagram demonstrated the groundwater heads in the aquifer of study and its region rising from north to south (from Atlantic Ocean to land).

In the dry seasons, when higher pumping periods decrease the groundwater level, the surface water (river, Atlantic Ocean, stream and lack water) infiltrates the ground and recharges to groundwater. In contrast, the outflow value increases during dry periods. Total ground water pumping ranged from 87,871 m^3/d to 58,495 m^3/d. The transient model results indicated that the groundwater level of the aquifer tended to rise in future simulated stress periods.

Table 2. Water budget of the aquifer for steady state condition (1995).

Flow Budget Term	Flow In (m^3/d)	Flow Out (m^3/d)
Constant Heads	128	−2629.74
Drains	0	−12,983.35
General Heads	939.72	−7785.47
Rivers	20,685.61	−12,508.15
Wells	5904.9	−79,726.56
Recharge	87,975.08	0
Total (Simulated)	115,633.31	−115,633.27
Total (Gregory 2001)	118,879.106	−118,903.572
Discrepancy (Simulated)	0.04%	
Discrepancy (Gregory 2001)	−24.47%	

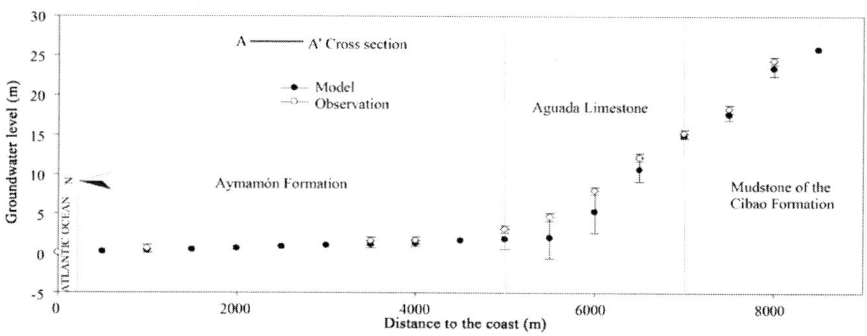

Figure 10. Comparison of simulated and observed groundwater level on cross-section A-A'.

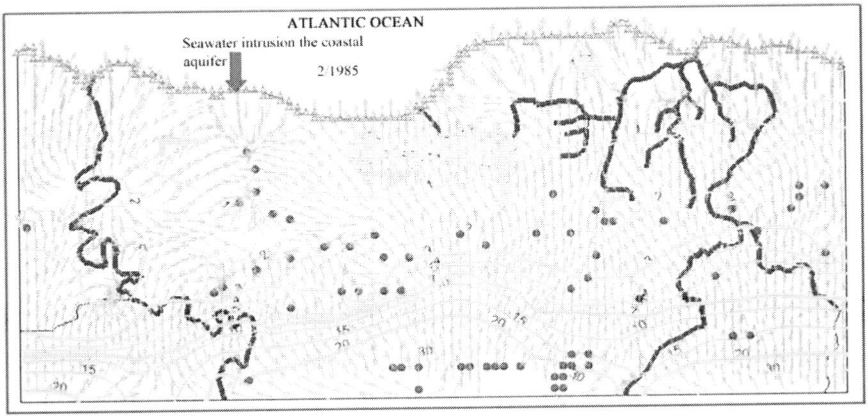

Figure 11. Groundwater level trends, flow directions and seawater intrusion.

4.4. Flow Field and Seawater Intrusion

In the coastal region of the study area, the model results show that the seawater and groundwater complement each other (Figure 11). Groundwater levels throughout most of the region have been relatively stable or have shown an increasing trend in this study region. In the coastal regions, however, this stability may be misleading as there may be large seawater intrusion since groundwater levels are below sea level throughout much of this region.

Seawater intrusion is often associated with over pumping in coastal regions, resulting in overdraft conditions and creating an inland gradient of seawater [37] . Despite the fact that the total recharge amount matches the total groundwater withdrawal rate, in the dry season and high pumpage periods some portion of recharge to the aquifer comes from seawater of Atlantic Ocean due. The seawater intrusion to this karst aquifer versus time step hydro-diagram is shown in Figure 12. The largest discharge of groundwater (13,479 m^3/d) from the karst aquifer to Atlantic Ocean occurred at February 1985. While during the same period, a total volume of 1230 m^3/d seawater flowed into the public water supply well Boquillas (PS) in Manatí (see Figure 12), as a result of over pumping in this region. The facts that water budget is well balance and groundwater was overdraft in the Boquillas (PS) pumping area of Manti region indicate the seawater intrusion into this aquifer, which potentially leads to decline of groundwater quality by raising salinity to levels exceeding acceptable drinking and irrigation water standards and endangering future exploitation of coastal aquifers.

The intrusion of seawater into groundwater systems will also impact coastal ecosystems such as marshes by changing the elevation of the freshwater-saltwater interface [38] . Generally, pumping from public supply wells in coastal aquifers underlain by saltwater can lower the water table, decreasing the depth to the freshwater-salt- water interface beneath the pumping well. This increases the potential for saltwater intrusion, and potentially limits the amount of fresh water available from the well. Therefore, our study strongly suggested the control of pumping activities in the coastal region of this study area.

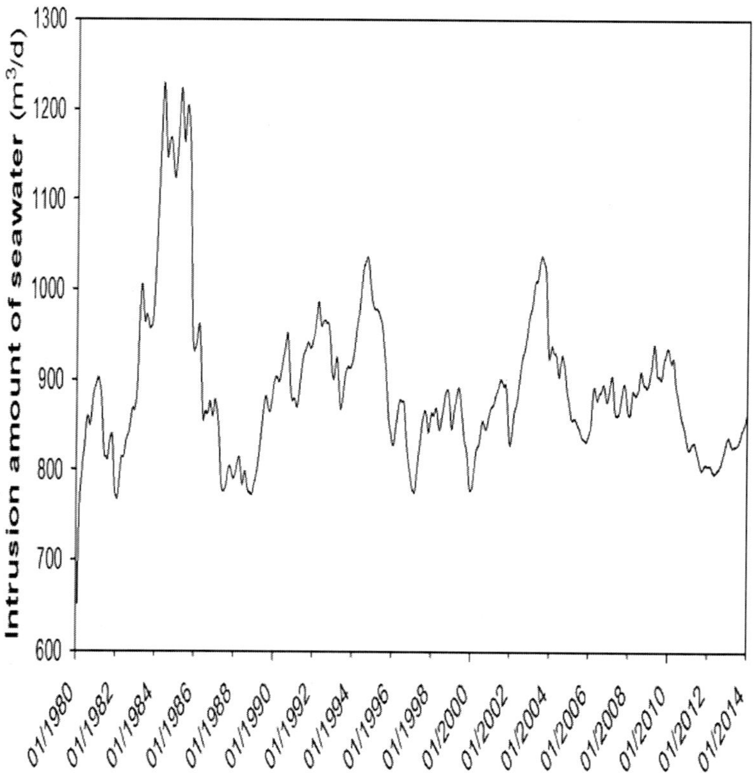

Figure 12. Seawater intrusion amount vs. time step (Transient Model).

5. CONCLUSIONS

This study applied a three-dimensional equivalent porous media (EPM) approach to simulate regional groundwater flow in a highly permeable karstic aquifer and to provide data for planning water and management purposes. The groundwater system in the Manatí-Vega Baja area was simulated in three-dimensions in steady state and transient conditions using MODFLOW to investigate the aquifer potentiality with emphasis on groundwater flow direction, head distribution and mass balance. We conclude that the EPM approach was adequate in representing the hydrodynamics of the Manatí-Vega Baja karst aquifer despite the high potential of conduit dominated flow. The developed conceptual model can be applied further to predict the changes of groundwater levels and mass balance under different exploitation scenarios to predict future water demands in the study area. The results of water budget simulation reveals that the total outflow and inflow water to be equal in the aquifer and that the groundwater level of the aquifer tends to increase in future simulations. The simulated flow field showed that the regions around water supplying Boquillas pumping area is subject to seawater intrusion and decreased water quality due to salinization.

ACKNOWLEDGEMENTS

Support for the work described in this study was provided by the National Institute of Environmental Health Sciences and the National Institute of Health to the Puerto Rico Test site for Exploring Contamination Threats (PROTECT) Superfund Research Program Center (P42ES017198). The content is solely the responsibility of the authors and does not necessarily represent the official views or policies of the National Institute of Environmental Health Sciences and the National Institute of Health. We are also thankful to the anonymous reviewers and the academic editor whose pertinent comments have greatly improved the quality of this paper.

REFERENCES

1. Regli, C., Rauber, M. and Huggenberger, P. (2003) Analysis of Aquifer Heterogeneity within a Well Capture Zone, Comparison of Model Data with Field Experiments: A Case Study from the River Wiese, Switzerland. Aquatic Sciences, 65, 111-128.

2. Angelini, P. and Dragoni, W. (1997) The Problem of Modeling Limestone Springs: The Case of Bagnara (North Apennines, Italy. Ground Water, 35, 612-618.

3. Kaufmann, G. and Braun, J. (2000) Karst Aquifer Evolution in Fractured, Porous Rocks. Water Resources Research, 36, 1381-1391.

4. Ford, D.C. and Williams, P.W. (2007) Karst Hydrogeology and Geomorphology. Wiley, Chichester, 576.

5. Király, L. (2003) Karstification and Groundwater Flow. Speleogenesis and Evolution of Karst Aquifers, 1, 1-26.

6. Hu, C.H., Hao, Y.H., Yeh, T.C.J., Pang, B. and Wu, Z.N. (2008) Simulation of Spring Flows from a Karst Aquifer with an Artificial Neural Network. Hydrological Processes, 22, 596-604.

7. Bakalowicz, M. (2005) Karst Groundwater: A Challenge for New Resources. Hydrogeology Journal, 13, 148-160.

8. Faulkner, J., Hu, B.X., Kish, S. and Hua, F. (2009) Laboratory Analog and Numerical Study of Groundwater Flow and Solute Transport in a Karst Aquifer with Conduit and Matrix Domains. Journal of Contaminant Hydrology, 110, 34-44.

9. Greene, E.A. (1992) Hydrologic Properties of the Madison Aquifer System in the Western Rapid City Area, South Dakota. USGS Water Resources Investigation, 56, 93-108.

10. Teutsch, G. (1993) An Extended Double-Porosity Concept as a Practical Modeling Approach for a Karstified Terraine. Hydrogeological Processes in Karst Terraines. Proceedings of the Antalya Symposium and Field Seminar, International Association of Hydrological Sciences, Wallingford, October 1990, 281-292.

11. Keeler, R.R. and Zhang, Y.K. (1997) Modeling of Groundwater Flow in a Fractured-Karst Aquifer in the Big Springs Basin, Iowa. GSA Abs with Programs, 29, 25.

12. Larocque, M., Banton, O., Ackerer, P. and Razack, M. (1999) Determining Karst Transmissivities with Inverse Modeling and an Equivalent Porous Media. Ground Water, 37, 897-903.

13. Bauer, S., Liedl, R. and Sauter, M. (2003) Modeling of Karst Aquifer Genesis: Influence of Exchange Flow. Water Resources Research, 39, 1285.

14. Abusaada, M. and Sauter, M. (2012) Studying the Flow Dynamics of a Karst Aquifer System with an Equivalent Porous Medium Model. Groundwater, 51, 641-650.

15. Scanlon, B.R., Mace, R.E., Barret, M.E. and Smith, B. (2003) Can We Simulate Regional Groundwater Flow in a Karst System Using Equivalent Porous Media Models? Case Study, Barton Springs Edwards Aquifer, USA. Journal of Hydrology, 276, 137-158.

16. 16. Pankow, J.F., Johnson, R.L., Hewetson, J.P. and Cherry, J.A. (1986) An Evaluation of Contaminant Migration Patterns at Two Waste Disposal Sites on Fractured Porous Media in Terms of the Equivalent Porous Medium (EPM) Model. Journal of Contaminant Hydrology, 1, 65-76. http://dx.doi.org/10.1016/0169-7722(86)90007-0

17. Neuman, S.P. (1987) Stochastic Continuum Representation of Fractured Rock Permeability as an Alternative to the REV and Fracture Network Concepts. In: Custodio, E., Gurgui, A. and Lobo-Ferreira, J.P., Eds., NATO Advanced Workshop on Advances in Analytical and Numerical Groundwater Flow and Quality Modelling, NATO ASI Series, Series C: Mathematical and Physical Sciences, Volume 224, Reidel Publications, Dordrecht, 331-362.

18. Huntoon, P.W. (1994) Is It Appropriate to Apply Porous Media Groundwater Circulation Models to Karstic Aquifers? In: El-Kadi, A.I., Ed., Groundwater Models for Resources Analysis and Management, CRC/Lewis Publishers, Boca Raton, 339-358.

19. Lugo, A.E. and Helmer, E. (2004) Emerging Forests on Abandoned Land: Puerto Rico's New Forests. Forest Ecology and Management, 190, 145-161.

20. Monroe, W.H. (1976) The Karst Landforms of Puerto Rico. US Geological Survey Professional Paper 899, 68.

21. Troester, J.W. (1992) The Northern Karst Belt of Puerto Rico: A Humid Tropical Karst. In: Back, W., Herman, J.S., and Paloc, H., Eds., Hydrogeology of Selected Karst Regions, Verlag Heinz Heise, Hannover, 475-486.

22. Lugo, A.E., Castro, L.M., Vale, A., et al. (2001) Puerto Rican Karst—A Vital Resource. US Department of Agriculture, Forest Service, General Technical Report WO-65, 100.

23. Monroe, W.H. (1980) Geology of the Middle Tertiary Formations of Puerto Rico. US Geological Survey Professional Paper 953, 93. (1 pl., Scale 1:250,000)

24. Renken, R.A., Ward, W.C., Gill, I.P., Gómez, G.F., Rodríguez, J.M., et al. (2002) Geology and Hydrogeology of the Caribbean Islands Aquifer System of the Commonwealth of Puerto Rico and the US Virgin Islands. US Geological Survey Professional Paper 1419, 1-139.

25. Yu, X., Ghasemizadeh, R., Padilla, I., Irizarry, C., Kaeli, D. and Alshawabkeh, A. (2015) Spatiotemporal Changes of CVOC Concentrations in Karst Aquifers: Analysis of Three Decades of Data from Puerto Rico. Science of The Total Environment, 511, 1-10.

26. Heisel, J.E., González, J.R. and Cruz, C. (1983) Analog Model Analysis of the North Coast Limestone Aquifer, Puerto Rico. US Geological Survey Open-File Report 82-52, 49.

27. Gomez-Gomez, F. and Torres-Sierra, H. (1988) Hydrology and Effects of Development on the Water-Table Aquifer in the Vega Alta Quadrangle, Puerto Rico. US Geological Survey Water-Resources Investigations Report 87-4105, 54.

28. Torres-González, S., Planert, M. and Rodríguez, J.M. (1996) Hydrogeology and Simulation of Ground-Water Flow in the Upper Aquifer of the Río Camuy to Río Grande de Manatí Area, Puerto Rico. US Geological Survey Water-Resources Investigations Report 95-4286, 102.

29. Bennett, G.D. and Giusti, E.V. (1972) Ground Water in the Tortuguero Area, Puerto Rico, as Related to Proposed Harbor Construction. US Geological Survey Water-Resources Bulletin 10, 25.

30. Quinones-Marquez, F. and Fuste, L.A. (1978) Limnology of Laguna Tortuguero, Puerto Rico. US Geological Survey Water-Resources Investigations Report 77-122, 86.

31. McDonald, M.G. and Harbaugh, A.W. (1988) A Modular Three-Dimensional Finite-Difference Groundwater Flow Model. US Geological Survey, Techniques of Water-Resources Investigations, Book 6, Chapter A1, USGS, Reston, 586.

32. Bennett, G.D. and Giusti, E.V. (1976) Ground Water in the Tortuguero Area, Puerto Rico, as Related to Proposed Harbor

Construction. US Geological Survey Water-Resources Bulletin 10, 25.

33. Doherty, J., Brebber, L. and Whyte, P. (1994) PEST: Model-Independent Parameter Estimation. Watermark Computing, Brisbane.

34. Weiss, M. and Gvirtzman, H. (2007) Estimating Ground Water Recharge Using Flow Models of Perched Karstic Aquifers. Ground Water, 45, 761-773.

35. Gregory, S.C. (2001) Simulation of Flow in the Upper North Coast Limestone Aquifer, Manatí-Vega Baja Area, Puerto Rico. Water-Resources Investigations Report, San Juan, Puerto Rico, 4266.

36. Ghasemizadeh, R., Hellweger, F., Butscher, C., Padilla, I., Vesper, D., Field, M. and Alshawabkeh, A. (2012) Review: Groundwater Flow and Transport Modeling of Karst Aquifers, with Particular Reference to the North Coast Limestone Aquifer System of Puerto Rico. Hydrogeology Journal, 20, 1441-1461.

37. Moujabber, M.E., Samra, B.B., Darwish, T. and Atallah, T. (2006) Comparison of Different Indicators for Groundwater Contamination by Seawater Intrusion on the Lebanese Coast. Water Resources Management, 20, 161-180.

38. El Moujabber, M., Atallah, T., Darwish, T. and Bou Samra, B. (2004) Monitoring of Groundwater Salination by Seawater Intrusion on the Lebanese Coast. Lebanese Science Journal, 5, 21-36

2

CHAPTER

COUPLING GROUNDWATER MODELING AND BIOLOGICAL INDICATORS FOR IDENTIFYING RIVER/AQUIFER EXCHANGES

Didier Graillot[1], Frédéric Paran[1*], Gudrun Bornette[2], Pierre Marmonier[2], Christophe Piscart[2,3] and Laurent Cadilhac[4]

[1]*CNRS, Ecole Nationale Supérieure des Mines de Saint-Etienne, UMR 5600 Environnement Ville et Société, 158 Cours Fauriel, F-42023 Saint-Etienne Cedex 2, France.*

[2]*CNRS, Université de Lyon, Université Lyon 1, UMR 5023 Ecologie des Hydrosystèmes Naturels et Anthropisés, 43 Boulevard du 11 novembre 1918, F-69622 Villeurbanne Cedex, France.*

[3]*Present address: CNRS, Université Rennes 1, UMR 6553 ECOBIO, Campus Beaulieu, 263 Avenue du Général Leclerc, F-35042 Rennes Cedex, France.*

[4]*Agence de l'eau Rhône-Méditerranée & Corse, 2-4 allée de Lodz, F-69363 Lyon, France.*

ABSTRACT

Future climate changes and the resulting modifications in anthropogenic activities will alter the interactions between rivers and groundwater. The quantification of these hydraulic interactions is absolutely necessary for achieving sustainable water use and requires accurate analytical methodologies. This report proposes an interdisciplinary approach to the quantitative and qualitative characterization of hydraulic interactions between rivers and shallow aquifers, wherein it outlines the advantages of coupling groundwater modeling with biological markers. As a first step, we built independent diagnostic maps of hydrological exchanges at the sector scale on the basis of hydrogeological modeling and biological indicators. In a second step, these maps were compared to provide a quantitative and qualitative understanding of exchanges between groundwater and surface water. This comparison significantly improved the calibration of groundwater models through a better assessment of boundary zones. Our approach enabled us to identify the conditions under which it could be possible to use biological indicators instead of a large set of piezometric measures. The integration of such combined tools in a future decision support system will assist governmental authorities in proposing appropriate long-term water policies for the preservation of groundwater resources, such as for supplying potable water and/or mitigating pollution risks.

Keywords: Groundwater Hydrological modeling Aquatic vegetation Aquatic invertebrates Hyporheic zone

INTRODUCTION

The importance of surface water/groundwater exchanges that are related to i) river discharges, ii) the availability of potable water, iii) river thermal regimes, iv) river productivity regarding both nutrient inputs and interstitial fauna biomass and v) the functioning and dynamics of alluvial wetlands has been frequently outlined (Bravard et al. 1997; Hayashi and Resenberry 2002; Lucassen et al. 2005; Jansson et al. 2007). The assessment of groundwater fluxes between aquifers, rivers and their floodplains thus represents a relevant issue for the long-term management of water resources. A precise delineation of the sections of aquifers that are mostly supplied by rivers and of river sections and floodplain areas that are mostly fed by groundwater upwellings (Bornette and Arens 2002) may constitute a valuable tool for watershed managers to identify strategic water reserves (Griebler et al. 2010). Although some methods exist for this type of quantification (GICC Research Group 2005), they have two main drawbacks. First, building a groundwater model based on a very large-scale grid is time consuming. Second, the amount of synchronous data that are necessary to feed and fit the model is prohibitive. In the framework of global change, the preservation of water and the functions that are associated with groundwater/surface water exchanges requires new methodologies or combinations of existing methods. Such methods should lead to more sustainable water usage and, consequently, for the preservation of groundwater in good quality and the associated biodiversity throughout large rivers.

The objective of this study was to characterize the hydraulic transfers between a large river and the lateral shallow aquifers. Traditionally, in studies of groundwater/surface water interactions on the sector scale, hydraulic models have been ubiquitously used to locate areas of interactions with aquifers (e.g., the Murray River, Lamontagne et al. 2002; the Snake River, Hortness and Vidmar Hortness and Vidmar 2004; the Jordan River, Holtzman et al. Holtzman et al. 2005 Urbano et al. 2006).

More recently, biological and ecological indicators have been used, e.g., microorganisms (Steube et al. 2009; Stein et al. 2010) and invertebrates (Schmidt et al. 2007; Korbel and Hose 2011). The findings of these studies suggest that faunal assemblages reflect hydrological exchanges.

The originality of our study lies in an interdisciplinary approach that combines scientific knowledge regarding the hydraulic control of groundwater transfers with two biological indicators: aquatic plants and groundwater fauna, for which the ecological requirements with respect to groundwater connectivity are known. This approach was applied to a large floodplain section to test its efficiency in providing the integrative mapping of surface water/groundwater interactions one a large scale (10 to 100-km).

MATERIALS AND METHODS

The originality of the proposed method lies in its combination of physical and biological methods (Bornette et al. 2008) which will be successively presented and then integrated.

Site location and description

The Rhône River runs from Switzerland in the Alps to the Mediterranean Sea along a 545-km course. The river catchment covers approximately 5220 km^2 of surface area (Figure 1). Along this river, several types of aquifers occur: free and confined sedimentary aquifers, impervious geological formations and aquifers in fissured bedrock. Many man-made structures have been built along the Rhône River, including approximately 40 dams for hydroelectric power production and flood prevention, navigational enhancement, and the management of water for irrigation (Bravard et al. 1986; Bravard 1987).

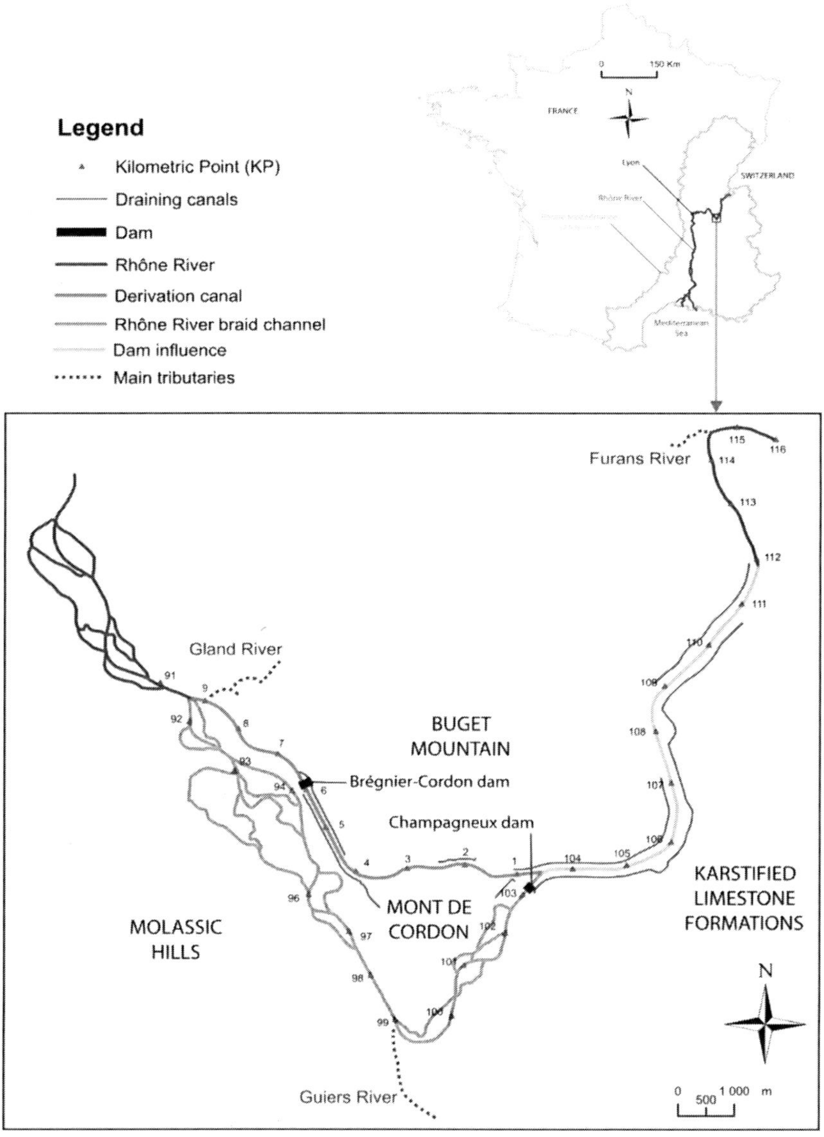

Figure 1. LOCATION OF THE RHÔNE RIVER IN FRANCE (TOP) AND MAP OF THE STUDY SITE (BOTTOM). Kilometric points (KP) decrease from upstream to downstream for the main and by-passed channels (from KP 110 to KP 91) but increase in the deviation canal from the dam (KP1) to the main channel (KP 9).

Our study focused on the Brégnier-Cordon river stretch (Ain, France). This site was chosen based on three main criteria: i) a low level of human activities; ii) the existence of hydraulic and biological data (Creuzé des Châtelliers 1991; Bornette et al. 1998a) and iii) the existence of previous studies (Michal 1988; Bornette et al. 1994). This river stretch is located 100-km upstream from the city of Lyon from KP116 to KP91 (KP: kilometric distance from a reference point, the Rhône-Saône confluence, thus decreasing from upstream to downstream; Figure 1). The Rhône River flows from the north-east to the south-west, and the slope ranges between 0.7 and 1‰. The average river discharge is approximately 423 m^3/s, varies from 160 m^3/s to 1900 m^3/s, and is governed by a pluvio-nival type regime. The strong slope is associated with plentiful sediment transport from the tributaries of the Pré-Alps Mountains, and a wide range of braided-style flow regimes has developed (Creuzé des Châtelliers 1991; Bravard et al. 1986). The geology of the area is a complex mix of karstic limestone mountains and molassic hills that surrounds a large alluvial floodplain. The left bank of the river consists of partially karstified limestone formations in the eastern part (KP110 to KP100) and a large area molassic hills in the western part from KP99 to the downstream end of the sector. The right bank consists of karstic limestone and marls of the Jurassic period, which is known as Bugey Mountain and stretches from KP105 to the downstream end of the sector. A small karstic Mountain, Mont de Cordon, is isolated between two channels of the river between KP2 and KP3 (Figure 1). The floodplain around the river is a glacial valley that has been refilled with quaternary alluviums. The hydraulic conductivity of these alluviums decreases from 1 × 10–3 to 1 × 10–5 m/s and the thickness close to the Rhône River is approximately 10 m (Chauve 2006).

The river and its floodplain have been considerably modified by a hydroelectric power station that was constructed between 1982 and 1984 (Figure 1). A derivation dam divides the river channel into two branches from KP104 to KP91 (KP103, Champagneux dam). Upstream from this dam, the channel has been transformed into a 12-km long reservoir (KP112 to KP103). Downstream of this dam, the former braided channel (KP103 to KP91) is now a by-passed channel with a regulated discharge that ranges from 80 m^3/s in the winter to 150 m^3/s in the summer (Bornette et al. 1994). This by-passed channel is also fed by the Guiers River (KP99). Most of the Rhône River water is diverted by the upstream dam to an 8-km derivation canal (KP1 to KP9) that feeds a hydroelectric facility (KP6) with a maximum fall of approximately 14.3 m ($Q = 780\ m^3/s$). To collect water that exfiltrates along the banks of the derivation canal and the reservoir, two drainage canals have been dug at the foot of the reservoir and two at the foot of the derivation canal to reduce the potential rise in the floodplain groundwater level (Claret et al. 1999).

Hydraulic diagnosis

Hydraulic interactions between groundwater and the river were preliminarily estimated through gradients that were calculated from digitalized piezometric maps with GIS spatial analysis tools (ArcGIS 9.2, ArcInfo, ESRI, http://www.esri.com/). The method was founded on the properties of the triangulated irregular network (TIN) model, which is very useful for generating surfaces. Each triangle contains three values as defined in ArcInfo: elevation, slope (or gradient) and aspect (or angle) (Bornette et al. 2008; Paran et al. 2008).

The proposed method permits the characterization of exchanges between any aquifer and river along a linear interface (bank) that can be represented as a polyline consisting of several segments. The exchange characterization implies: i) the quantification of the rate of flow (Q in m^3/j) and ii) the identification of the direction of hydraulic interactions, wherein fluxes from the aquifer to the river are characterized by positive values (+Q) and exchanges from the river to the aquifer by negative values (−Q). Infiltration (Q: m^3/s) through the bank was calculated using Darcy's law (Equation 1) considering gradients (i), the cross-section of the river (A: m^2) and the permeability of the aquifer (K: m/s).

$$Q(m^3/s) = K.A.i \tag{1}$$

The gradient values (i) at the aquifer/river interface were obtained from digitalized piezometric maps (low water level in February of 1990), which were converted in the TIN. Cross-section (A) values were estimated by subtracting groundwater elevation values from riverbed elevation values. Permeability (K) values were obtained from the literature (Michal 1988; Rampnoux 1992; Hydratec Design Office 2002).

The exchange directions were given by the sign (+ or -) of the sinus of the angle formed by the groundwater flow and river flow. This angle (D) was calculated by subtracting the angle values given by the TIN of the piezometric levels and the TIN of the river levels. The infiltration estimation (Qe, Equation 2) was weighted by the value of the sinus of angle D to take into account the influence of the angle between the groundwater and superficial flows on the exchanges. The groundwater flow intensity was usually too weak when Qe was provided in m^3/s; hence, Qe was given in m^3/d:

$$Qe(m^3/d) = Q.\sin(D).86400 \tag{2}$$

Qe was calculated for each segment of the Rhône river bank polyline. The final infiltration estimation (Qf) was given by Equation 3. Qf is the sum of each Qe calculated for a segment of the polyline multiplied by the length (L) of this segment:

$$Qf\left(m^3/d\right) = \Sigma_{(1\ \text{to}\ n)}(Qe.L) \tag{3}$$

This algorithm facilitated a selection of the scale for which the exchange flows had to be calculated. It was then possible to aggregate values for several 1-km-long unitary segments along a predetermined river Section.

Interstitial invertebrates

The interstitial fauna was sampled within the sediment using the Bou-Rouch technique (Bou and Rouch 1967; Bou 1974), which involves the use of a perforated metal pipe. This mobile pipe was driven 50 cm deep at each sampling point, where 10 L of a mix of interstitial water and sediment was pumped prior to filtering through a 200-µm-size mesh net. Samples were preserved in the field in a 5% formaldehyde solution, stained with Bengal pink, and sorted in the laboratory. A total of 41 sites were sampled in the main active channel of the river, the drainage canals and the wetlands located in the floodplain. They were sampled from 1987 to 2009 with varying periodicity: 16 stations were sampled in the main river channel (sampled twice in 1988 and 1989; Creuzé des Châtelliers 1991) and three were sampled again in 2008 and 2009 for the present study. A total of 16 stations were investigated in the drainage canals [three stations were sampled during the summer in 1987 and 1995 (Claret et al. 1999), six others during the summer in 1995, 1996 and 1997 (Marmonier et al. 2000), and ten others in 2008 (unpublished data)]. Finally, five stations were sampled in abandoned river channels: two stations were sampled in April, July and November in 1987 and 1995, which were years before and after the restoration of the wetland, respectively [fine sediment was removed by dredging in 1993 (Claret et al. 1999), and three others were sampled in 2007 (unpublished data)].

To locate groundwater upwelling zones, we used the absolute and relative taxonomic richness (S and S%) and the abundances (Q) of the stygobite fauna. We used abundances and richness because the stygobite fauna is more abundant and diverse in upwelling zones than in downwelling zones (Marmonier and Dole 1986; Dole-Olivier and Marmonier 1992). We also considered relative richness values because in areas fed by groundwater inputs, the ratio between stygobite and epigean species is expected to increase in interstitial assemblages and should consequently represent a relevant indicator for hydrological exchanges.

Aquatic vegetation

Aquatic vegetation is strongly governed by the groundwater/surface water connectivity in aquatic ecosystems (Bornette and Amoros 1991; Bornette and Large 1995; Bornette et al. 1998b). The distribution of aquatic plant species is related to the nutrient levels (Carbiener et al. 19901995; Onaindia et al. 1996; Britto et al. 2001) and seasonal thermal variation in the water (Bornette and Puijalon 20092011). In aquatic habitats, there is a strong correlation between the intensity of groundwater exfiltrations and seasonal temperature variability (Sakura 1993; Bornette and Large 1995). The various water sources that supply aquatic habitats usually strongly contrast in terms of nutrient levels (Bornette et al. 1998a; Bornette and Arens 2002). Knowledge of the ecological requirements of species that occur in a given habitat allows the determination of the origin and turnover rate of the water that supplies the habitat (Amoros et al. 2000). For this purpose we had to i) identify the ecological range of aquatic species in terms of trophic levels in the water and temperature, ii) map the occurrence of aquatic species in the aquatic habitats of the floodplain and iii) determine the weight of groundwater supplies in each of these aquatic habitats and their potential origin, as estimated using the abundance and ecological requirements of the aquatic species.

Identification of the ecological range of aquatic species

The ecological range of each species was based on the occurrence of a significant relationship between species occurrence and relative abundance and i) the thermal regime and ii) the water trophic levels of wetlands. For this purpose, we used floristic relevés that were collected from 1993 and 1999 in 36 wetlands of the Rhône and Ain Rivers. The physico-chemical characteristics of the wetlands were simultaneously measured monthly for at least a year in the same set of wetlands (i.e., 12 replicates in the Rhône River wetlands and 24 for the Ain wetlands in terms of ammonia nitrogen, phosphates and temperature). This sampling was performed in the same year as (Rhône River) or the same year and year after (Ain River) vegetation sampling. At the same time, three to six substrate samples were collected in all of the wetlands to determine the phosphorous level of the substrate. Ecological profiles were produced for all species occurring in at least five sites among the 36.

Cool, well-oxygenated groundwater supplies limit plant growth and competition processes but also allow evergreen species to grow throughout the year (Bornette and Large 1995; Greulich and Bornette 2003). Consequently, the distribution of aquatic plants is partly ruled by water temperature. To quantify the groundwater supplies to the wetlands, the relative standard deviation of temperatures was calculated for each wetland, and the wetlands were divided into five equilibrated classes according to this coefficient. In a second step, the average abundance of each species for all of the wetlands belonging to a given class was calculated. The difference between the classes was then tested for each species. First, the equality of variances between classes was tested using a Levene test. Second, either parametric or non-parametric tests were performed to determine whether the abundance of species significantly differed between classes.

Combined with knowledge about groundwater supplies, the nutrient content of the water in the wetlands provides information about the origin of groundwater, that is, the seepage water is nutrient rich in the Brégnier-Cordon area, whereas the karstic lateral aquifers should be less eutrophic (Bornette et al. 19941998a). First, the species affinity for phosphorous (both in the water [PO_4^{3-}] and the substrate total P content) was assessed. For this purpose, monthly [PO_4^{3-}] values were log transformed and averaged per site, and the total phosphorous values were standardized and log transformed. Then, both values were summed for each site, and the same method that was used for temperature was applied: wetlands were categorized into five classes according to their total phosphorous contents, and the non-random character of the species distribution along the phosphorous gradient was tested. The same procedure was applied to the ammonia nitrogen contents of water [$N\text{-}NH_3$]; however, wetlands were separated into four instead of five equilibrated classes. The species affinity for nitrate nitrogen is not assessed because it has been demonstrated to rarely rely to eutrophication. Phosphorous is frequently the major limiting nutrient for aquatic plants (Amoros et al. 2000; Bornette and Puijalon 20092011). The affinities of aquatic species for phosphorus and ammonia nitrogen were then qualitatively combined in a trophic profile of four classes by evaluating the relative affinity of the species for eutrophication. The analyses were carried out with Jump (JMP® Statistical Discovery Software, http://www.jmp.com/).

The mapping of the aquatic species occurrence

The aquatic species contents and relative abundances were then roughly determined for each aquatic habitat of the Brégnier-Cordon reach. For this purpose, at least one floristic relevé was carried out in each wetland in an area that was considered to be representative of the floristic contents of each wetland. For rivers, canals and riverine wetlands, several relevés were performed to take into account changes in water quality along the upstream/downstream gradients. For each relevé, the relative abundance of each species was estimated using the Braun-Blanquet (1932) index. Then, the trophic levels were estimated by i) multiplying the abundance of each species by its ecological profile and ii) summing the values obtained for all species of the relevé (Amoros et al. 2000). We thus obtained the affinity of the wetland for each trophic class. We considered the class presenting the highest values to indicate the most probable trophic level (Additional file 1). For groundwater inputs, each species was classified according to its affinity for stenothermic water. The difference between the number of species that were intolerant to stenothermic waters and those preferentially occurring in groundwater-fed ecosystems was determined. Then, the wetlands were mapped according to four groundwater influence classes (null: index value = 0; low: index value < 4; intermediate: index values ranging from 5 to 9; high: index value > 9) and four trophic level classes (oligotrophic, mesotrophic, eutrophic and hyper-eutrophic). A comparison of the two maps then facilitated the identification of several geographic areas with contrasting groundwater influences.

Map crossed analysis

The three types of describers resulted in the production of three separate sets of maps that all located the shallow aquifer recharge areas by the river and sections of the river fed by groundwater. These maps cannot be directly compared because of their characteristics; however, it was possible to analyze the global patterns that resulted from the three different describers using overlay mapping.

A this stage of our work, we can only analyze the global patterns because the available data were not collected simultaneously. However, the crossed analysis is possible because data were sampled during comparable hydrologic conditions (low water level), during the same decade (February 1990 for piezometric values; summers 1987, 1995, 1996, 1997 and 2008 for interstitial invertebrates data; summers 1993 to 1999 for aquatic vegetation and chemical data) and after the modifications inferred by the hydroelectric power station. Furthermore, we will look if significant correlations can be found between estimated infiltration, stygobite richness and abundance and macrophyte index.

RESULTS

The diagnoses of the surface water/groundwater exchanges that were made using the three describers were first independently considered for the three types of metrics, and the subsequent three maps were then superimposed.

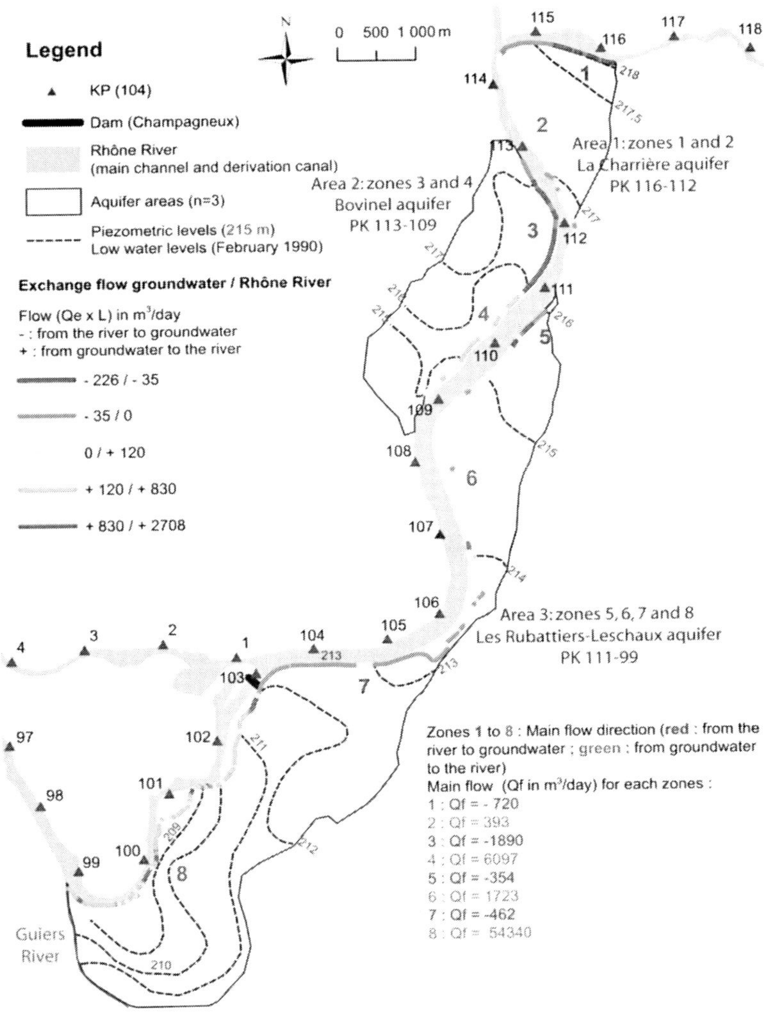

Figure 2 . Mapping of the piezometric levels in the alluvial aquifer and the hydraulic exchanges between the river and the alluvial aquifer.

Hydraulic modeling

Gradients (i) were obtained from the piezometric map TIN (low water level in February of 1990). The average cross-section of the river (A) was approximately 4.2 ± 1.2 m (min = 1.8, max = 8.1). The following permeability values (K) were obtained: Area 1, 1.3 × 10−3 m/s; Area 2, 5 × 10−3 m/s; Area 3, KP111 to KP106, K = 5 × 10−3 m/s, KP106 to KP102.5, K = 3.9 × 10−4 m/s and KP102.5 to KP99, K = 8.5 × 10−3 m/s (Figure 2). The infiltration estimation (Qe) was calculated for each area. For each of the zones with the same global exchange behavior, we calculated the Qf and the unit flow (Qu) for 1-km (m^3/day/km).

From a hydrogeological point of view, the GIS spatial analysis method was tested on three distinct aquifers to calculate groundwater/river exchanges (Figure 2 and Table 1):

- First, the polyline of Area 1 (the La Charrière zone, approximately 5600 m long, KP116 to KP112) consisted of 62 segments. The results provide evidence that the left bank of the Rhône River can be divided into two zones (Figure 2): zone 1 (KP116 to KP114.5), which is associated with the river supplying the aquifer and a Qf of −668 m^3/day, and zone 2 (KP114.5 to KP112), which consists of an aquifer drained by the Rhône River and a Qf of 393 m^3/day.
- Second, the polyline of Area 2 (the Bovinel zone, approximately 4850 m long, KP113 to KP109) was comprised of 74 segments. The results provide evidence that the left bank of the Rhône River to can divided into two zones (Figure 2): zone 3 (KP113 to KP111), with the river supplying the aquifer and a Qf of −1890 m^3/day, and zone 4 (KP111 to KP109), with an aquifer drained by the river and a Qf of 6097 m^3/day.
- Third, the polyline of Area 3 (the Les Rubattiers-Leschaux zone, approximately 12400 m, KP111 to KP99) consisted of 312 segments. These results provide evidence that the left bank of the Rhône River can be divided into four zones (Figure 2): zone 5 (KP111 to KP110), with the river supplying the aquifer and a Qf of −354 m^3/day; zone 6 (KP110 to KP106), with the river draining the aquifer and a Qf of 1723 m^3/day; zone 7 (KP106 to KP102.5), with the river supplying the aquifer and a Qf of −462 m^3/day; and, finally, zone 8 (KP102.5 to KP99), with the river draining the aquifer and a Qf of 54340 m^3/day.

Table 1 synthesis of the exchange flow between groundwater and rhône river in the brégnier-cordon sector

Name	Stretch / KP	Flow direction	Qf m^3/day	Bank length m	Qu m^3/j/km
Area 1 (A1)	Zone 1 (Z1) 116 to114.5	-	-668	1547	-432
La Charrière	114.5 to 112	-	-52	531	-98
	Zone 2 (Z2)	+	393	3545	111
	Total A1 (Z1 + Z2)				
	-		*-720*	*2079*	*-347*
	+		*393*	*3545*	*111*
Area 2 (A2)	Zone 3 (Z3) 113 to 111	-	-1890	2040	-927
Bovinel	Zone 4 (Z4) 111 to 109	+	6097	2809	2170
Area 3 (A3)	Zone 5 (Z5)	-	-354	587	-603
	110.5 to 110	+	5	76	64
	Zone 6 (Z6)	-	-77	431	-178
	110 to 106	+	1723	3709	465
Les Rubattiers-Leschaux	Zone 7 (Z7)	-	-462	3057	-151
	106 to 102.5	+	8	349	24
	Zone 8 (Z8)	-	-35	21	-1693
	102.5 to 99	+	54340	4190	12968
Total A3 (Z5 + Z6 + Z7 + Z8)					
-			*-928*	*4096*	*-226*
+			*56077*	*8323*	*6737*
Total A1 + A2 + A3					
-			**-3538**	**8214**	**-431**
+			**62566**	**14678**	**4263**

The zones (Z) and stretches (S) are shown in Figure 2.

Overall, the Rhône River in the Brégnier-Cordon sector gains 62566 m^3/day of water (~4263 m^3/day/km) and loses 3538 m^3/day of water (~431 m^3/day/km). A general finding which is not explicitly mentioned is that the river gained much more water from the aquifer than it lost to the aquifer within the stretch investigated. Indeed, the groundwater discharge into the river was very low (i.e., ~0.72 m^3/s) in comparison to the Rhône River discharge (i.e., ~160 m^3/s during the season of low discharge).

Stygobite fauna

A total of 106 taxa were sampled, of which nine were stygobite species (Table 2). The local richness, abundance and composition of the stygofauna greatly varied with the channel type and location (Figure 3).

Table 2. List of stygobite species sampled in the brégnier-cordon sector between 1988 and 2008

Order	Species	KP of occurrence
Mesogastropoda	*Islamia moquiniana*	3 – 6; 97 – 103
Ostracoda	*Fabaeformiscandona wegelini*	97 – 103
Bathynellacea	*Parabathynella cf stygia.*	96 – 97
Amphipoda	*Crangonyx subterraneus*	96 – 97
	Nipgargopsis casparyi	4 – 5; 96 – 103; 109 – 110
	Niphargus fontanus	4 – 5; 96 – 98; 109 – 110
	Niphargus kochianus	4 – 6, 97
	Niphargus rhenorhodanensis	96 – 103; 109 – 110
	Salentinelia juberthiae	96 – 97; 109 – 110

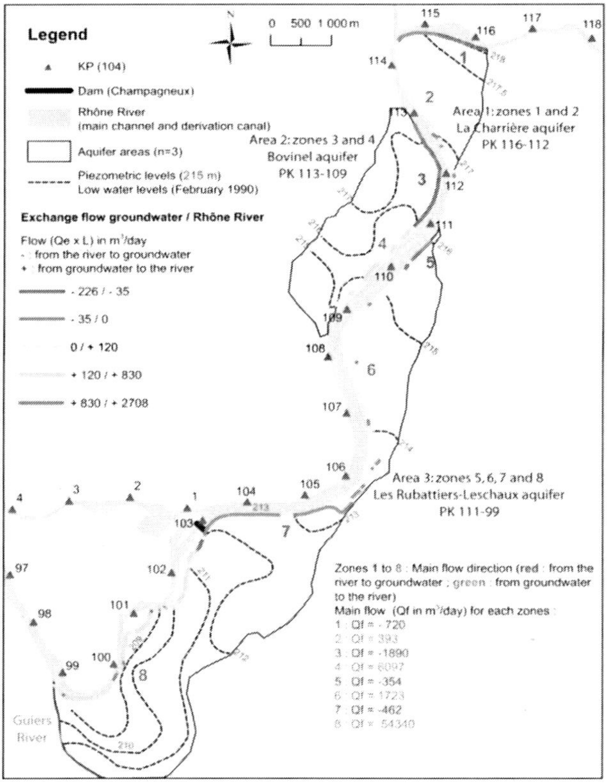

Figure 3. Maps of the potential exchanges between surface water and groundwater based on the stygobite taxonomic richness (s and s%) and abundances (q). The five river sections with contrasting hydrological exchanges are shown for the Rhône River (dashed lines) and for the drainage canals (solid lines).

In the by-passed channel, two contrasting sections were distinguished. From the dam to the confluence with the Guiers River and to the beginning of the anastomosed sector (KP103 to KP96), seven stations harbored stygobite species in their interstitial assemblages (Section a, Figure 3). The abundance and richness of stygobite species greatly varied, with locally significant values being obtained, especially for the varied abundance (15 individuals at KP101 and 13 individuals at KP99), suggesting punctual inputs of groundwater into the by-passed channel. Here, the stygobite species consisted of shallow interstitial fauna (*Fabaeformiscandona wegelini, Islamia moquiniana, Niphargopsis casparyi* and *Niphargus rhenorhodanensis*). Downstream of this section (KP91 to KP96), no stygobite fauna was sampled in the by-passed channel, suggesting little or no groundwater inputs there (Section b, Figure 3). This result was consistent with the geomorphological characteristics of this section, which is associated with sediment deposition and clogging.

Additionally, contrasting results were obtained in the four drainage canals (Figure 3). Upstream of the dam (KP112 to KP109), the abundance and richness were high at the upstream end of the right and left side drainage canals, suggesting important groundwater inputs (Section c, Figure 3). Here, the stygobite fauna consisted of two crustacean amphipod species living in shallow interstitial habitats of the main channel (*N. casparyi, N. rhenorhodanensis*) and two species living deeper in the sediment (*Salentilella juberthiae*), or frequently found far from the main channel (*Niphargus fontanus*). This last species may indicate inputs of groundwater from the nearby karstic system on the left bank. From KP109 to KP103, in the area influenced by the reservoir, no stygobites were sampled in the drainage canals, suggesting a weak influence of groundwater in comparison to important inputs of surface water that filtrate through the river banks (Section d, Figure 3). In contrast, in the left and right side drainage canals that are located downstream of the dam (KP4 to KP6), stygobite species were sampled in the entire length of the left canal and at the upstream end of the right one, but with low richness and low abundances (Section e, Figure 3). Here, the stygobite fauna consisted of two amphipod species living in shallow interstitial habitats of the main channel (*N. casparyi, N. rhenorhodanensis*) and four species living deep in the sediment far from the main channel (*N. fontanus, Niphargus kochianus, Crangonyx subterraneus* and *Parabathynella* cf *stygia*), suggesting a mix of surface water infiltrating through the bank and deep groundwater coming from the nearby karstic systems (i.e., Bugey for the right-side canal and Mont Cordon for the left one).

Aquatic vegetation

The observed aquatic vegetation facilitated the discrimination of four areas. In the first upstream area, the Rhône River flowed as a unique channel along a karstic massif. All of the wetlands were oligotrophic or mesotrophic, with the groundwater influence ranging from low to high. No eutrophic wetlands occurred along the right side of the river, suggesting that the wetlands were not supplied by seepage water but, instead, by a drained hillslope aquifer in this area. Along the left bank of the river, some hyper-eutrophic wetlands occurred along the karstic massif, suggesting either that the wetlands did not drain the hillslope aquifer but, instead, were supplied by river seepage, or that the aquifer was eutrophic in this area. Downstream of this reach, a zone where the wetlands appeared to be eutrophic occurred. Because the river seepage is hyper-eutrophic in this area (Bornette et al. 1994), this pattern could indicate either the drainage of a eutrophic aquifer by the wetlands, or that the wetlands are supplied by a mixture of both oligotrophic groundwater and hyper-eutrophic seepage. This second interpretation was reinforced by the occurrence of oligotrophic wetlands along the floodplain margins. A third zone was discriminated by the braided zone of the river, wherein the sampled vegetation indicated that most wetlands were poorly supplied by groundwater and were hyper-eutrophic, indicating that the river was aggrading in comparison to its floodplain, and that only a few seepage-supplied wetlands occurred. Along the river margin, the wetlands were highly supplied by groundwater and less eutrophic, suggesting some influence of a hillslope aquifer.

Superimposition of the maps

The superimposition of the three maps (Figure 4 and Table 3) highlighted the convergences and divergences among the describers:

- In most cases, the results of the three methods were consistent. In area 1, at La Charrière (left bank, KP114.5 to KP112), there was a lack of groundwater invertebrates; however, hydraulic modeling and the macrophyte index both indicated poor to low (Qu = 111 m^3/day/km) inputs of groundwater to the river. From KP110 to KP106 (left bank), the macrophyte index and the hydraulic method converged and indicated low intensity drainage of the aquifer by the river (Qu = 465 m^3/day/km). From KP106 to KP102.5 (on the left bank), the three methods provided similar results, i.e., a low supply of river water to the aquifer (Qu = −151 m^3/day/km). Finally, the results were also consistent for the KP102.5 to KP99 section (left bank) and indicated medium to high intensity drainage of the aquifer by the river

(Qu = 12968 m^3/day/km). For the right bank, all of the methods provided similar results for the reach located from KP110.75 to KP109, i.e., medium to high intensity drainage of the aquifer by the river (Qu = 2170 m^3/day/km).

- Discrepancies only occurred between the different methods in two cases. First, in zone 2, at Bovinel (right bank, KP113 to KP111), the macrophyte index did not detect a flow from the river to the aquifer, whereas the hydraulic method did (Qu = −927 m^3/day/km). Second, in zone 3, at Les Rubattiers-Leschaux (left bank, KP111 to KP110), the three methods diverged over a small distance: the macrophyte index did not highlight groundwater inputs, whereas the hydraulic method indicated a weak flow from the river to the aquifer (Qu = −603 m^3/day/km) and the invertebrate index indicated an important mixing of surface water and groundwater in the drainage canal.
- Finally, in the zones where no hydraulic diagnosis was available (downstream of KP99 and along the derivation canal), the macrophyte and groundwater fauna indices gave rather similar results. From KP4 to KP9 (right bank, derivation canal) and around Mont de Cordon (left bank, derivation canal), both biological indices indicated groundwater upwelling to the drainage canals and to the river. Downstream of KP99 (left bank) and KP9 (right bank) to the end of the study area, both macrophyte and invertebrate indices indicated a lack of groundwater influences.

Significant correlations were observed between the three methods (i.e., estimated infiltration (Qu), stygobite richness and abundance and macrophyte index, n = 11 zones). Thus, we found a significant correlation either between the macrophyte index and invertebrate richness ($R^2 = 0.407$, $p = 0.034$) and invertebrate abundance ($R^2 = 0.382$, $p = 0.042$) or between the macrophyte index and the infiltration estimation ($R^2 = 0.503$, $p = 0.014$). No significant correlation was found between the hydraulic infiltration estimation and the groundwater invertebrates.

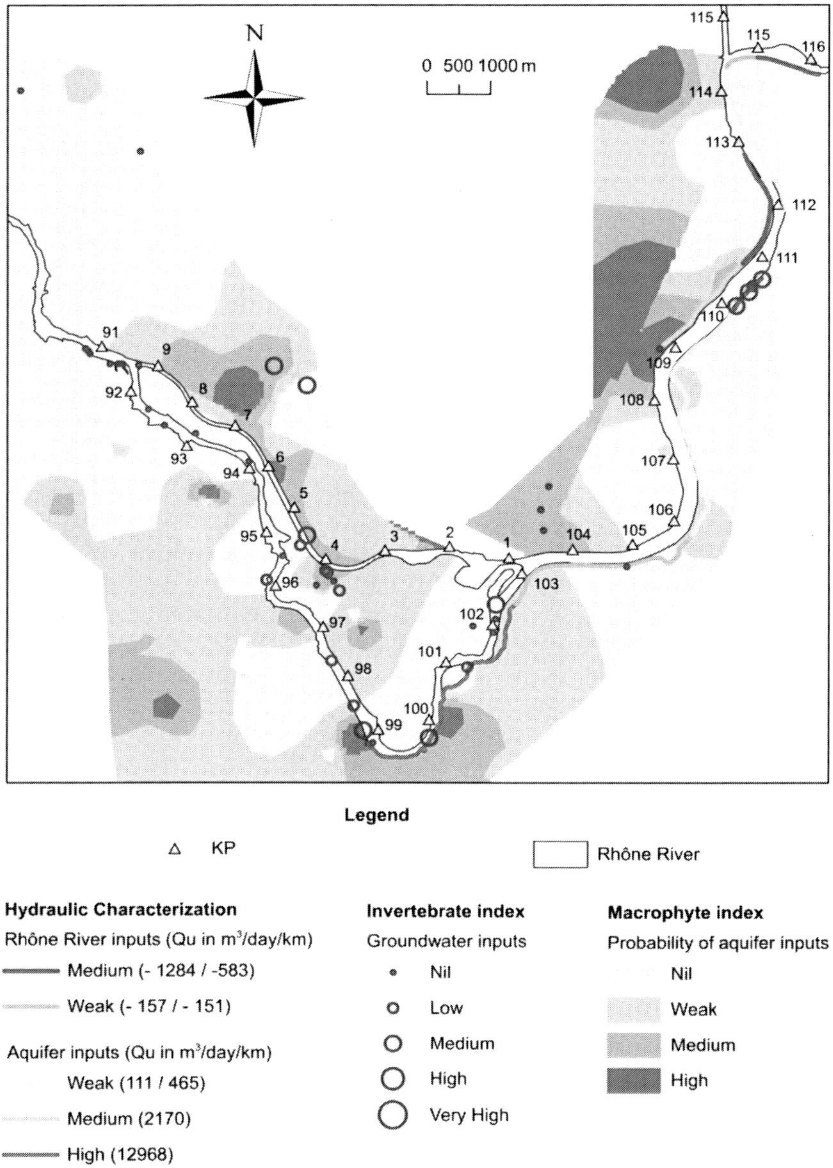

Figure 4. Comparison map that combines the results from the hydraulic, invertebrate and macrophyte indicators.

Table 3. Comparison of the results of the hydraulic and biological methods

KP	Bank	Type	Hydraulic Qf m³/day	 Qu m³/day/km	Macrophytes	Invertebrates
115_114.5	Right	Rhône River	-	-	Weak, karst	-
114.5_113	Right	Rhône River	-	-	Nil	-
113_111	Right	Rhône River	-1541	-871	Nil	-
111_110.75	Right	Dam influence	-349	-1294	Medium to high, karst	-
110.75_109	Right	Dam influence	6097	2170	Medium to high, karst	-
109_108	Right	Dam influence	-	-	Medium to high, karst	-
108_105	Right	Dam influence	-	-	Nil to weak, karst	-
105_103	Right	Dam influence	-	-	Weak to medium, karst	-
1_4	Right	Derivation canal	-	-	-	-
4_9	Right	Derivation canal Restitution	-	-	Medium to high, aquifer of the Gland River	Deep groundwater flow
9_dowstream	Right	Restitution, Rhône River	-	-	Weak to nil	Weak groundwater flow
116_115	Left	Rhône River	-582	-583	-	-
115_114.5	Left	Rhône River	-86	-157	Nil	-
114.5_112	Left	Rhône River	393	111	Nil	-
112_111	Left	Rhône River	-	-	Nil	-
111_110	Left	Dam influence	-354	-603	Nil, pollution from karst	Important exchanges, mixing
110_106	Left	Dam influence	1723	465	Nil to weak	-
106_102.5	Left	Dam influence	-462	-151	Nil to weak, flow from the Rhône River to the aquifer	Flow from the Rhône River to the aquifer
102.5_99	Left	Rhône River	54340	12968	Medium to high, groundwater	Flow from the aquifer to the Rhône River
99_dowstream	Left	Rhône River	-	-	Weak, preferential corridor, groundwater	preferential corridor, groundwater
103_99 / 1_5	Mont Cordon	Rhône River, Derivation canal	-	-	Karstic reappearance	Karstic reappearance

DISCUSSION

Validation elements for hydraulic exchanges

Hydraulic studies have been previously conducted to estimate exchanges between the Rhône River and groundwater in the same area of Brégnier-Cordon based on hydraulic modeling and different piezometric measurements (Rampnoux 1992; Michal 1988; Košmelj 1982) but without the quantification of fluxes. These studies provided hydraulic gradient values for five zones along the right and left banks of the river. The directions of these exchanges were confirmed by the results of the present study. The quantifications of the fluxes that were performed in the present analyses highlighted very weak values that could be associated with groundwater fluxes. This suggests that the Brégnier-Cordon aquifer represents a limited reserve of groundwater and that its use for irrigation or as a source of potable water must be restricted. The multidisciplinary strategy employed in this study could lead to a new modeling approach that can be applied on a broad scale in instances where it would be too costly and time consuming to implement methods that use gridded datasets.

Multi-metric comparison

Convergence of the results

Despite the absence of significant correlation between hydrology and groundwater fauna, we observed a surprisingly good convergence of the results that were obtained using hydraulic and biological indicators. This was especially observable on the left bank of the river from KP106 to KP102.5. In this section, which was influenced by the dam, the estimated flow value was 460 m^3/day from the river to the aquifer. The value of the macrophyte index was very low and indicated a lack of water exchange between groundwater and surface water. The invertebrate index confirmed the direction of the flow and its intensity in comparison to other sections of the river. Between the dam and the confluence of the Guiers River at KP99, the flow rate value significantly increased until reaching 45.367 m^3/day from the aquifer to the river. This result was confirmed by both the macrophyte and invertebrate indices.

Divergence needing complementary approach

Interestingly, we did not find this convergence at all of the sites. For example, between KP111 and KP110, the hydraulic gradient and macrophyte index demonstrated that the river fed the aquifer; however, the invertebrate indicators highlighted important water exchanges. In this case, further investigations through geochemical and isotopic analyses may facilitate the identification of the origin of the water. The use of environmental isotopic indicators would be a valuable method for validating the results of the present study (Fette et al. 2005). For example, the ratios of isotopic hydrogen (δD) or oxygen ($\delta^{18}O$) are usually used for identifying river/aquifer exchanges (Gonfiantini et al. 1995; Walker and Krabbenhoft 1995). This approach has been used on other sections of the Rhône River to identify the origin of surface-water and groundwater (Paran et al. 2012).

Conclusion about multi-metric comparison

Moreover, a combination of these three methods provides complementary information (Bornette et al. 1998a; Bornette and Arens 2002; Paran et al.

2005). The hydraulic method precisely quantifies the flow and direction of the exchange, whereas the macrophyte and groundwater invertebrate indices may reveal the relative importance of groundwater inputs in comparison to surface water flows and can help to distinguish the origin of groundwater.

Method limitations and accuracy

The accuracy of the results provided by the method discussed herein is different for each metric. For hydrogeology, we find the traditional sources of error, such as the depth of the aquifer, discontinuities in the alluvial aquifer and measurement errors related to groundwater level and hydraulic conductivity. For hydraulics, the depth of the channels has been estimated to be up to 5 m based on the quality of the data provided by the National Company of Rhône. The possible clogging of the banks has been estimated from the results given by the models when they were available.

For invertebrates, the limitations of the method lie in: i) the time required for sampling, ii) the identification of species, which is also time consuming and requires a large panel of specialists; and finally, iii) the lack of knowledge about the ecological preferences of many groundwater species. For the Brégnier-Cordon sector and for most of the northern part of Europe and North America, another limit related to the use of groundwater fauna is their low diversity in biogeographic zones that were frozen during the last glaciations (Ferreira et al. 2007). Most populations of groundwater species disappeared under quaternary glaciers and recolonized these areas during the last 10 ky (Ferreira et al. 2007). This low species richness certainly reduces the efficiency of the groundwater invertebrate index. Future research must test this index in areas located outside of the quaternary glacial limit.

Aquatic vegetation surveys are not time consuming in comparison to analyses of aquatic invertebrates, as most species can be identified in the field. Furthermore, the ecological requirements of plants are usually rather narrow because of their sessile character, improving their indicative value. However, there are some limitations related to i) the paucity of wetlands in the floodplain and ii) the existence of poorly informed species, either because they are rare and not documented, or because they exhibit a broad ecological niche. Finally, it is still necessary to test whether these ecological profiles are usable for other river systems.

At this stage of our work, we know that our method works but remains perfectible. To improve the method and its accuracy, our future works will focus on synchronous data to reduce the risk of errors in the interpretation of results and to make the statistical analysis efficient. Furthermore, we have to better consider the scale of interpretation: the spatial resolution of the estimated infiltration and the macrophyte index which is about 1-km versus 100 m for the stygobite richness and abundance.

CONCLUSIONS

The numerical modeling of groundwater/surface water interactions is not always efficient for large-scale river basins. In very well-instrumented areas, the exchanges between groundwater and surface water can be calculated from a spatial analysis of piezometric gradients, and they can be validated by the results given by models that have been implemented on a local scale. Biological markers that are or are not influenced by phreatic conditions and discharge fluctuations can also confirm these surface water/groundwater interactions. This novel approach provides accurate maps of surface water/groundwater exchanges and demonstrates that hydraulics and biological metrics provide complementary information, despite the absence of significant correlation between hydrology and groundwater fauna, both in terms of spatial scale and the quality of information obtained. The results of this method will provide relevant hydrological and biological information on the complex interactions between groundwater and surface waters on the scale of the entire course of large rivers for decisions makers.

In terms of water resource management, the correspondence between hydraulics and biological indicators can help to determine sensitive areas where groundwater should be preserved. This is the case for the pumping well located in the southern part of the study area close to the Guiers River on the left bank of the floodplain. The maximum groundwater flow has been estimated to be 10.774 m^3/day over a 1-km distance on this left bank. Considering that the influence radius of the well is equal to 300 m, the extraction rate of the well (1.355 m^3/day) for supplying drinking water cannot be increased more than twice. In the case of groundwater overextraction, the exchange will be reversed, and the flow of the river can reach the pumping well.

Future research will be required to validate the spatial correlation found between the metrics in another case study to take into account contrasting situations regarding latitude, climate, geological context, the biogeography of organisms and anthropogenic conditions. Similarly, transient conditions remain to be studied because surface water/groundwater interactions can be modified during the fall low-flow period. Another field of research should be developed to predict the modification of these hydraulic exchanges in the framework of climate changes. Indeed, the thermal modifications that are induced by global warming could have harsh repercussions on the water demand of human populations. For example, nuclear power plants would require increasing amounts of stenothermal groundwater, and the groundwater demand for agriculture during summer periods would significantly increase.

DECLARATIONS

Acknowledgements

This research program was supported by the French Water Agency Rhône-Meditérranée-Corse with the help of the Compagnie National du Rhône in providing hydraulic data. This work has been carried out under the aegis of the Rhone Basin Long Term Environmental Research Observatory (Zone Atelier Bassin du Rhône).

Competing interests

The authors declare that they have no competing interests.

Authors' contribution

DG created the design of the study and defined the scientific objectives. FP carried out the hydraulic diagnosis and coordinated the map crossed analysis. GB carried out the aquatic vegetation diagnosis. PM carried out the interstitial invertebrate diagnosis. CP carried out the interstitial invertebrate diagnosis and performed the statistical analysis. LC participed in the design of the study and focused on the operational objectives. All authors read and approved the final manuscript.

REFERENCES

1. Amoros C, Bornette G, Henry CP: Environmental auditing-A vegetation-based method for ecological diagnosis of riverine wetlands. Environ Manage 2000, 25: 211-227.

2. Bornette G, Amoros C: Aquatic vegetation and hydrology of a braided river floodplain. J Veg Sci 1991, 2: 497-512.

3. Bornette G, Arens MF: Charophyte communities in cut-off river channels the role of connectivity. Aquat Bot 2002, 73: 149-162.

4. Bornette G, Large ARG: Groundwater-surface water ecotones at the upstream part of confluences in former river channels. Hydrobiologia 1995, 310: 123-137.

5. Bornette G, Puijalon S: Macrophytes: ecology of aquatic plants. In Encyclopedia of Life Sciences. Chichester: Wiley J & Sons; 2009:1-9.

6. Bornette G, Puijalon S: Abiotic controls on aquatic plants: a review. Aquat Sci 2011, 73: 1-14.

7. Bornette G, Amoros C, Collilieux G: Role of seepage supply in aquatic vegetation dynamics in former river channels-Prediction testing using a hydroelectric construction. Environ Manage 1994, 18: 223-234.

8. Bornette G, Amoros C, Lamouroux N: Aquatic plant diversity in riverine wetlands: the role of connectivity. Freshwat Biol 1998, 39: 267-283.

9. Bornette G, Amoros C, Piégay H, Tachet J, Hein T: Ecological complexity of wetlands within a river landscape. Biol Conserv 1998, 85: 35-45.

10. Bornette G, Dechomets R, Dufour M, Ferreira D, Germain A, Gibert J, Graillot D, Jezequel C, Lafont M, Marmonier P, Paran F, Piscart C, Puijalon S, Simon L: Characterization of Exchanges between Rhône River and Groundwater. 2008. Research report ZABR Phase 2, AERMC

11. Bou C: Methods to sample in interstitial groundwaters (Les méthodes de récolte dans les eaux souterraines interstitielles). Annls Spéléol 1974, 29: 611-619.

12. Bou C, Rouch R: New researches in subterranean aquatic fauna. C R Acad Sci 1967, 265: 369-370.

13. Braun-Blanquet J: Plant Sociology: the Study of Plant Communities. New York: Transl. Fuller GD, Conard HS, Mc Graw Hill Book Co; 1932.

14. Bravard JP: The Rhône River from the Leman Lake to Lyon City. Lyon: La Manufacture; 1987.

15. Bravard JP, Amoros C, Pautou G: Impact of civil engineering works on the successions of communities in a fluvial system. Oikos 1986, 47: 92-111.

16. Bravard JP, Amoros C, Pautou G, Bornette G, Bournaud M, Creuzé des Châtelliers M, Gibert J, Peiry JL, Perrin F, Tachet H: Stream incision in southeast France: morphological phenomena and impacts upon biocenoses. Regul Rivers 1997, 13: 75-90.

17. Britto DT, Siddiqi MY, Glass ADM, Kronzucker HJ: Futile transmembrane NH4+ cycling: a cellular hypothesis to explain ammonium toxicity in plants. Proc Natl Acad Sci U S A 2001, 98: 4255-4258.

18. Carbiener R, Trémolières M, Mercier JL, Ortscheit A: Aquatic macrophyte communities as bioindicators of eutrophication in calcareous oligosaprobe stream waters (upper Rhine plain, Alsace). Vegetatio 1990, 86: 71-88.

19. Carbiener D, Trémolières M, Miuller S: Vegetation of running waters and water quality: thesis, debates and prospects. Acta Bot Gallica 1995, 142: 489-531.

20. Chauve P: Jura. In Aquifers and Groundwaters (Aquifères et eaux souterraines), Collection Scientifique et Technique, vol 2. Edited by: BRGM. Orléans: BRGM; 2006:561-590.

21. Claret C, Marmonier P, Dole-Olivier MJ, Castella E: Effects of management works on the interstitial fauna of floodplain aquatic systems (River Rhône, France). Biodiv Conserv 1999, 8: 1179-1204.

22. Creuzé des Châtelliers M: Dynamics of Distribution of River Rhône Interstitial Biocenosis with Regard to Geomorphological Characteristics (Dynamiques de répartition des biocénoses interstitielles du Rhône en relation avec des caractéristiques géomorphologiques (Brégnier-Cordon, Miribel-Jonage et Donzère-Mondragon). 1991. PhD Thesis, University Lyon 1

23. Dole-Olivier MJ, Marmonier P: Patch distribution on interstitial communities: prevailing factors. Freshwat Biol 1992, 27: 177-191.

24. Ferreira D, Malard F, Dole-Olivier MJ, Gibert J: Obligate groundwater fauna of France: diversity patterns and conservation implications. Biodiv Conserv 2007, 16: 567-596.

25. Fette MW, Kipfer R, Schubert CJ, Hoehn E, Wehrli B: Assessing river-groundwater exchange in the regulated Rhone River (Switzerland) using stable isotopes and geochemical tracers. Appl Geochem 2005, 20: 701-712.

26. GICC Research Group: Study of Potential Impacts of Global Climate Change in the River Rhone catchment (Etude des impacts potentiels du changement climatique sur le bassin versant du Rhône en vue de leur gestion). 2005. Gicc-Rhône Project, Final revised report final, short version, ENSMP, BRGM, CEMAGREF, CNRS, EDF, MéteoFrance

27. Gonfiantini R, Fröhlich K, Araguás-Araguás L, Rozanski K: Isotopes in groundwater hydrology. In Isotope Tracers in Catchment Hydrology. Edited by: Kendall C, McDonnell JJ. Amsterdam: Elsevier science; 1995:203-246.

28. Greulich S, Bornette G: Being evergreen in an aquatic habitat with attenuated seasonal contrasts-a major competitive advantage? Plant Ecol 2003, 167: 9-18.

29. Griebler C, Stein H, Kellermann C, Berkhoff S, Brielmann H, Schmidt S, Selesi D, Steube C, Fuchs A, Hahn HJ: Ecological assessment of groundwater ecosystems-vision or illusion? Ecol Eng 2010, 36: 1174-1190.

30. Hayashi M, Resenberry DO: Effects of ground water exchange on the hydrology and ecology of surface water. Groundwater 2002, 40: 309-316.

31. Holtzman R, Shavit U, Segal-Rozenhaimer O, Gavrieli I, Marei A, Farber E, Vengosh A: Quantifying ground water inputs along the lower Jordan River. J Environ Qual 2005, 34: 897-906.

32. Hortness JE, Vidmar P Geological Survey, Scientific Investigations Report 2004–5115. Surface-water/ground-water Interaction along Reaches of the Snake River and Henrys Fork, Idaho, U.S 2004.

33. Hydratec Design Office EPTB territoire Rhône, CNR, DIREN Rhône-Alpes, AE-RMC. Global Study for a River Rhône Flood Risk Decreasing Strategy (Etude globale pour une stratégie de réduction des risques dus aux crues du Rhône) 2002.

34. Jansson R, Laudon H, Johansson E: The importance of groundwater discharge for plant species number in riparian zones. Ecology 2007, 88: 31-139.

35. Korbel KL, Hose GC: A tiered framework for assessing groundwater ecosystem health. Hydrobiologia 2011, 661: 329-349.

36. Košmelj K: A Methodology for Hydraulic Data Processing (Méthodologie pour le traitement des variables hydriques du Haut-Rhône français). 1982. PhD thesis, Grenoble 2 University, France

37. Lamontagne S, Leaney F, Herczeg A: Streamwater-groundwater Interaction: the River Murray at Hattah-Kulkyne Park, Victoria: Summary of Results. 2002. CSIRO Land and Water Australia, Technical Report 27/02

38. Lucassen ECHET, Smolders AJP, Lamers LPM, Roelofs JGM: Water table fluctuations and groundwater supply are important in preventing phosphate-eutrophication in sulphate-rich fens: consequences for wetland restoration. Plant Soil 2005, 269: 109-115.

39. Marmonier P, Dole MJ: Sediment amphipods of a river Rhône bypassed channel-distribution according to flood reaction. Rev Sci Eau 1986, 5: 461-486.

40. Marmonier P, Claret C, Dole-Olivier MJ: Interstitial fauna in newly created floodplain canals of a large regulated river. Regul Rivers 2000, 16: 23-36.

41. Michal P: Recharge of a Rhône river valley alluvial aquifer (Modes d'alimentation d'une nappe alluviale : nappe de la vallée du Rhône entre La Balme et Saint-Genix-sur-Guiers-Savoie). 1988. PhD Thesis, Scientific, technologic and medical University of Grenoble

42. Onaindia M, de Bikuna BG, Benito I: Aquatic plants in relation to environmental factors in northern Spain. J Environ Manage 1996, 47: 123-137.

43. Paran F, Malard F, Mathieu J, Lafont M, Galassi D, Marmonier P: Distribution of groundwater invertebrates along an environmental gradient in a shallow water-table aquifer. In Proceedings of the Symposium on World Subterranean Biodiversity. Edited by: Gibert J. Villeurbanne: CNRS-UCBL; 2005:99-106.

44. Paran F, Dechomets R, Graillot D: Spatial analysis characterization of river/groundwater exchanges (Identification et quantification des échanges nappes/rivière par analyse spatiale). Géomatique Expert 2008, 64: 73-79.

45. Paran F, Arthaud F, Bornette G, Graillot D, Lalot E, Marmonier P, Novel M, Piscart C: Characterization of Exchanges between Rhône River and Groundwater. 2012. Research report ZABR Phase 4, AERMC

46. Rampnoux JP: Aquifers n°12, 12bis, 12ter et 13, Geologic and Hydrogeological Characteristics of the Rhône Valley and the Guiers

Valley-Synthesis of Existing Data. Chambery: CG Savoie, DDAF Savoie; 1992.

47. Sakura Y: Groundwater flow estimated from temperatures in the Yonezawa basin, northeast Japan. IAH Publications 1993, 215: 161-170.

48. Schmidt SI, Hahn HJ, Hatton TJ, Humphreys WF: Do faunal assemblages reflect the exchange intensity in groundwater zones? Hydrobiologia 2007, 583: 1-19.

49. Stein H, Kellermann C, Schmidt S, Brielmann H, Steube C, Berkhoff SE, Fuchs A, Hahn HJ, Thulin B, Griebler C: The potential use of fauna and bacteria as ecological indicators for the assessment of groundwater quality. J Environ Monit 2010, 12: 242-254.

50. Steube C, Richter S, Griebler C: First attempts towards an integrative concept for the ecological assessment of groundwater ecosystems. Hydrogeol J 2009, 17: 23-35.

51. Urbano L, Waldron B, Larsen D, Shook H: Groundwater-surfacewater interactions at the transition of an aquifer from unconfined to confined. J Hydrol 2006, 321: 200-212.

52. Walker JF, Krabbenhoft DP: Groundwater and surface-water interactions in riparian and lake dominated systems. In Isotope Tracers in Catchment Hydrology. Edited by: Kendall C, McDonnell JJ. Amsterdam: Elsevier science; 1995:203-246.

3
CHAPTER

SPELEOGENESIS AND EVIDENCES TECTONIC INTO LAKHSSAS KARST SYSTEM (WESTERN ANTI-ATLAS, MOROCCO)

Abderrahmane Wanaim*, Mhamed Alaeddine Belfoul, Sophia Bouzid, Farid Faik

GESCAT Laboratory, Faculty of Science, Ibn Zohr University, Agadir, Morocco

ABSTRACT

This study aims to highlight evidences of tectonic and geodynamic features within Lakhssas high plateau karst area (Moroccan Western Anti-Atlas), using an original approach based on geospeleothems analysis. This work is organized by the speleogenesis process, controlled by tectonics and interference of karst intrinsic parameters. Regarding the first issue, a multi-layered study compiling inventoried speleological spaces with tectonic field data analysis then with lithological, hydrogeological and geomorphological features, has demonstrated that speleogenesis processes are deeply controlled by tectonic framework so described as "speleotects". Subsequent studies in terms of sedimentology, geomorphology, hydrogeology and dating deposits, are planned to ground those conclusions.

Keywords: Geospeleothems Superposed Tectonic, Lakhssas Karst, Anti-Atlas, Uplift

1. Introduction

Caves are natural archives that preserve sediments, tectonic motions and paleo-earthquakes for a considerable period [1] and [2] . This work concerns geospeleological approach as an original thematic study used rarely in Morocco, probably in whole Africa to highlight tectonic motions through an endokarstic device.

2. GEOMORPHOLOGICAL AND GEOLOGICAL FRAMEWORK OF LAKHSSASS KARSTIC DEVICE

2.1. Geomorphological Context

Morocco takes the northwestern corner of the Africain continent and the country represents a mosaic of various geodynamic events from the geological annals of Africa, from the Pre-Cambrian until the present. The favorable conditions here help the preservation of unique paleo and neo karst [3] . Lakhssas area (Moroccan western AntiAtlas) is a flattening surface (plateau) rising up to 2000 m of altitude overlapping a hercynian complex compression structure [4] [5] . From satellite view, it appears in the shape of a bat: narrow in the middle with about 30 Km width and elongated over 47 Km length from north to south. It builds a natural bridge between two inliers: Ifni in the west and Kerdous in the east separating the Tiznit basin at north (southern extension of Souss basin), from that of Bouizakarn at south (Figure 1).

Lakhssas plateau is an interesting karstic device that provides a signified diversity of speleothems that refers to all forms of endokarstic environment. Its fascinating underground network constitutes a geological patrimony. It develops groundwater, sink-holes, lapiez and dry conditioning systems; typical forms of karstic areas that are excellent markers of neotectonic and seismic activities but also useful to interpret paleoclimate.

It is considered as prehamadian lands surrounded by wide dolomitic and calcareous adoudounian outcrops [6] .

2.2. Geological Setting

Lakhssas outcrops belong to the western Anti-Atlas. It exhibits a Neoproterozoic and Palaeozoic sedimentary cover [7] , dominated by Adoudounian (terminal Proterozoic) and Georgian (lower Cambrian)

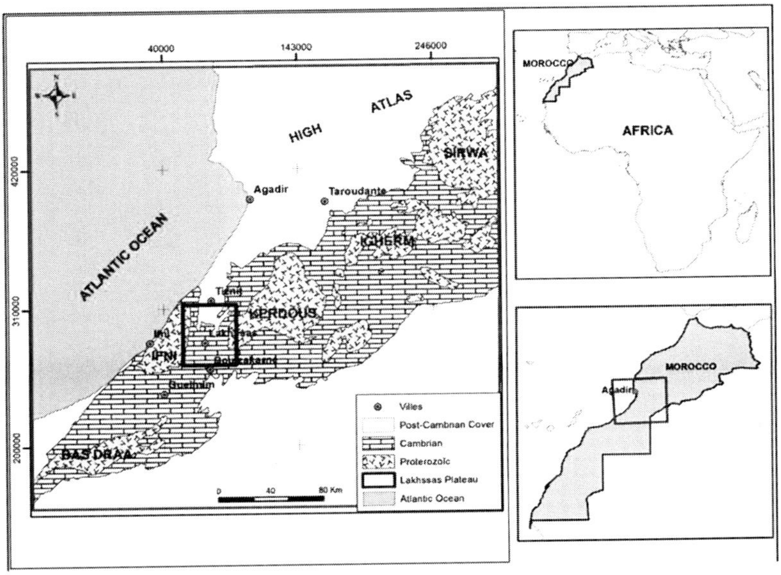

Figure 1. Geographical and geological map of the Anti-Atlas showing the location of Lakhssas plateau.

limestones and dolomites (Figure 2). Those teguments surround Precambrian inliers with the most closers are from SW to NE: Ifni, Kerdous, Ighrem. The main structures are of hercynian orogeny network: submeridionale verticale faults associated to wide folds such as those of Jbel Inter north and south of Id Yahia. During this time, Lakhssass pla teau has been up lifted by a shortening WNW-ESE reworking in the cambrian by listric Faults (Figure 3(a)), into reverse to strike-slips [8] , where the Ifni and Kerdous blocs motions squeezed the central part of the plateau (Figure 3(b)).

3. LAKHSSASS SPELEOLOGICAL SPACES GENESIS

3.1. The Underground Network: Geometry and Features

This issue concerned a detailed field investigation that involved more than 140 underground sites dispatched over all the studied area, with various inside trends and where 60% are sinkholes and 40% are caves [9] (Figure 4(a)).

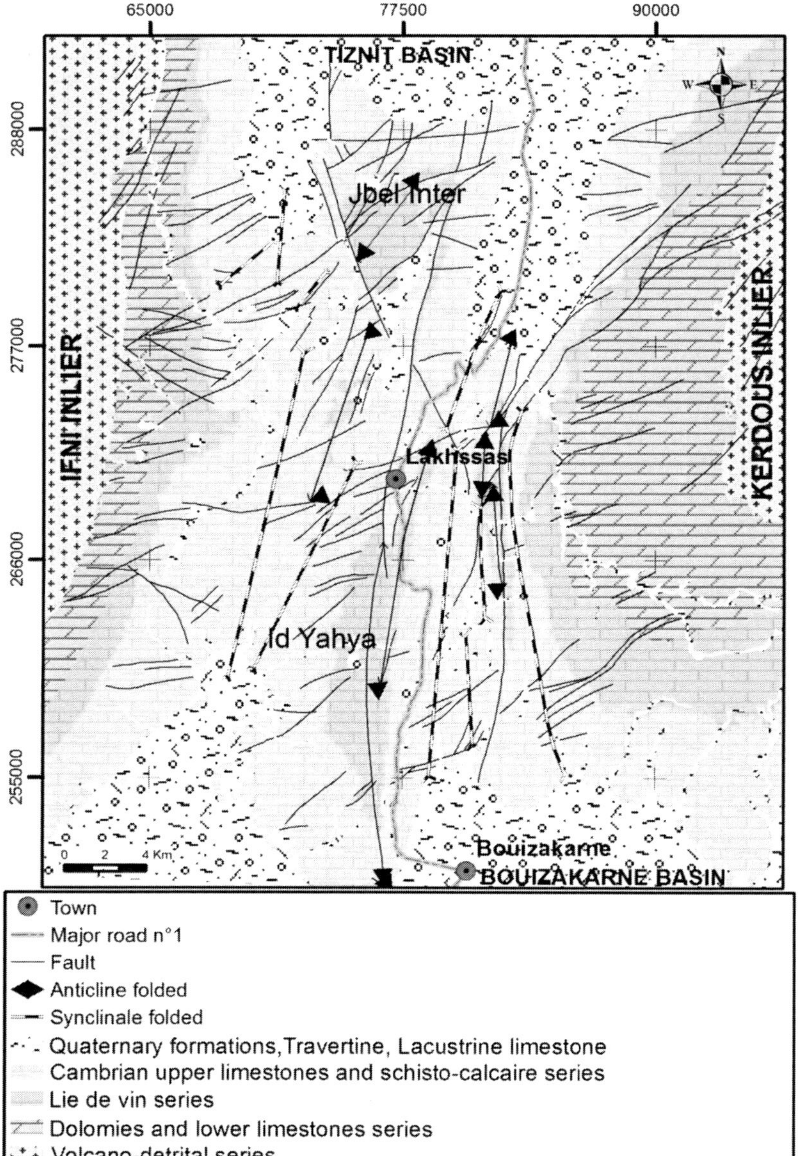

Figure 2. Schematic geological map of Lakhssas plateau 1/100,000.

The geometry of the underground network is treated through trend measurements. More than 500 values are plotted using stereographic direction diagram (stereonet software).

According to this diagram, the ENE-WSW direction is clearly the major trend followed by the speleogenesis processes. N-S trend is also a preferred section beside NE-SW. NW-SE to WNW-ESE trend does exist but seems to be less used.

The explanation of this network will be exposed later in this paper, after giving, in the following paragraphs, evidences of several parameters impacts on the underground network genesis.

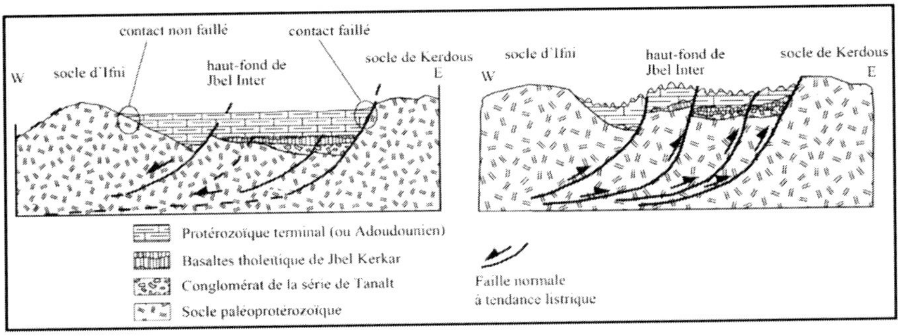

(a) (b)

Figure 3. Adoudounian deposits corridor in terms of Lakhssas hemigraben open to the west with listric faults system (Belfoul, 2005).

3.2. Evidences of Tectonic Control

To achieve this purpose, several stereographic diagrams are established to represent the directions the tectonic structures (major fractures, joints, strike slips and fold axis), then compiled with the underground network geometry to fixe if there is any kind of relationship. To quantify the measurement over the study area, both field and map data [10] -[13] are used (Figure 4).

3.2.1. Faults and Joints

Joints and faults play different roles in controlling karst distribution, depending on their relative spatial dimensions and characteristics (length, width, surface area and roughness, opening, etc.), on their frequency and on their intersections with other joints sets [14] -[17] .

The stereographic analysis of faults from map and field data using Wulffdiagram, shows three main directions (Figure 4(b)): ENE-WSW, N-S, and E-W.

The geometric analysis of joints reveals NE-SW, E-W and another direction NW-SE.

According to those diagrams, fractures trends (Figure 4(b) 1) are widely well represented in the underground network (Figure 4(b) 3) with ENE-WSW as the main direction and the N-S as a secondary one.

Joints geometry is relatively less represented into the underground network. N-S and NE-SW trend exist but the NW-SE seems not a preferred underground trend.

3.2.2. Folded and Foliated Structures

Lakhssas plateau is mainly structured by more or less wide anticlines that spread from Jbel Inter at north and the Id Yahia area at south, with the narrow synclines at its sides.

These folds are often with (Figure 4(c)): a lengths from a few kilometers to a few cms, N-S direction, subvertical axial planes, low axial plunge (to the north or to the south), isopach folds type 1b [18] , sometimes evolving to type 2 isoclinal folds, hinges highly stretched in the foliation planes (total transposition S0 (stratification plane) into S1 (cleavage plane)).

These structures (Figure 4) concentrated generally along major submeridional faults so that cleavage is itself submeridional with subvertical to vertical dip. The directional concordance between the cleavage and the folds axis is due to the almost vertical plans flattening.

The compilation of these structures with the cartographic underground spaces shows a remarkable concordance. According to the map in (Figure 4), it seems that the best link coincides with the synclinal structures. Figure 4 provides some additional information, folds and cleavages trends are not enough combined with NWSE underground geometry.

3.2.3. Interpretation and Synthesis

Statistical analysis of the elongation and alignment of the underground forms (caves, sinkholes, lapiez, ⋯)

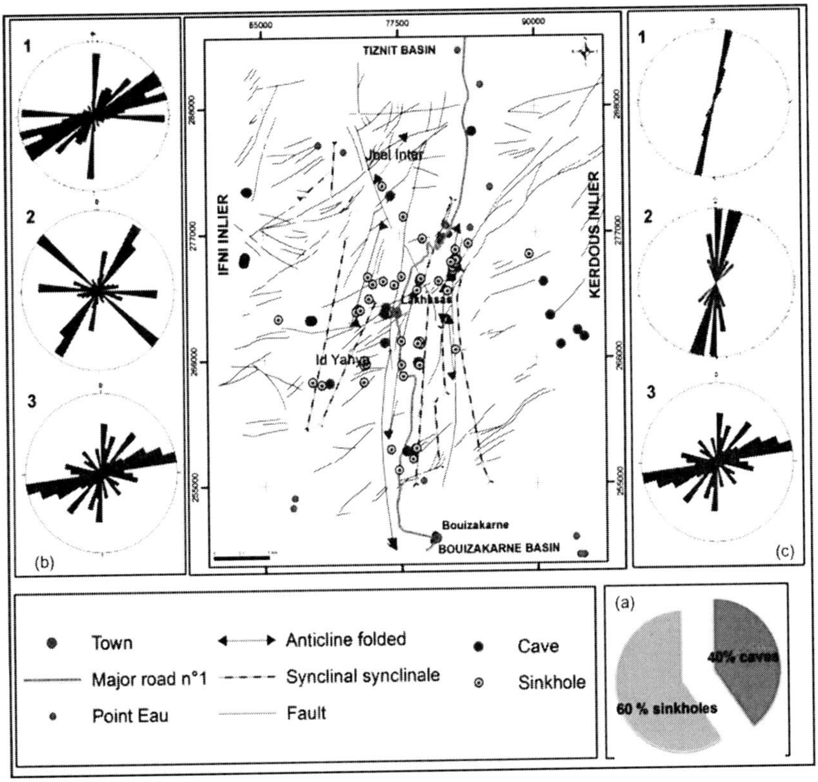

Figure 4. Map representing the various parameters controlling the distribution of underground spaces in the speleological Lakhssas plateau (extracted from Bouizakarn geological map 1/100000). (a) Rates of different underground component. (b) Comparative diagram of main fractures directions from different sources: exokarstic fractures (1) and joints (2) and endokarstic lineaments (3). (c) Comparative diagram of main folded (1) and foliated (2) structures and endokarstic lineaments (3).

listed, shows that the direction of speleological architecture goes with the interference of the faults direction, joints, fold axes and cleavage planes that are: NS, NE-SW, EW and NW-SE but in different rates.

The main direction of underground trend is N-S; this may be explained by the major direction of faults, folds axis and cleavage. The NE-SW and NW-SE directions are due to the conjugate joins that are developed with the N-S major folding. The E-W trend must be connected to the fractures and opening spaces induced by the equatorial main hercynian shortening.

The sinkholes deserve a particular attention. It seems that, in the major cases, their distribution coincides with intersection of fractures or joins so that they present a vertical geometry.

Actually, the density of speleological sites is proportional to that of tectonic structures. Faults, folds and foliation structures are rather favorable to the speleogenesis process.

3.2.4. Role of Lithologic Composition

The adoudounian and lower Cambrian age the sedimentary pile is largely dominated by calcareous deposits in various forms (limestone, dolomitic limestone, and dolomite) with some intercalations of sandstones to pelitic nature (Figure 5(a)).

The Quaternary formations are almost outcropping toward the lowland. Slightly thinner in the east but thicker toward the west (Figure 5(b)) [12] .

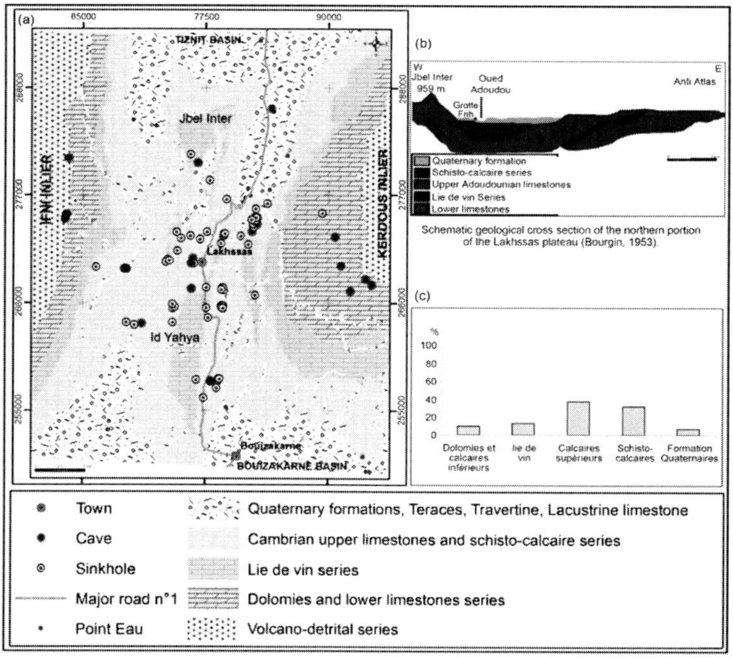

Figure 5. Displays the underground sites in Lakhssass plateau over the lithologic composition and the rate of those speleological units by type of rock.

At the Lakhssas plateau, recent calcareous crusts are very common. They are expressed either into slabs form sonflatsor more dramatically by concretions at the ancient water springs or at the location of river network.

The upshot of this figure is that the Cambrian upper limestones and schisto-calcaire series are the most elected facies to host speleological spaces. Dolomies and lower limestones as well as "lie de vin" series contribute less to this issue. If the first group of limestone rocks is logical to be connected to the speleological network spreading, the second one raises up some troubles (Figure 5(c)). Why dolomies and lower limestones allow a poor part of speleological spaces? This may be explained by the fact that those rocks are stratigraphically, more or less deep so that the serie lie de vin, as it is impermeable, can be a shield that stopped water infiltration to attain lower limestones and dolomies. In this situation, speleological spaces are not well developed at those levels. The rate shown in the (Figure 5) reflects those that are developed when those rocks are outcropping.

Karsts forms can also be found in the villafranchian lakes calcareous of Tiznit lowlands, such as the cave of Frih douar in south of Tadayght. These sites don't show many forms of concretions.

4. OTHER PROCESS AND INTERFERENCE OF DIFFERENT PARAMETERS

4.1. Topographic and Hydraulic Processes

To achieve this purpose, we use arcscene as a software application that allows GIS data 3D visualization. The on-working document is giving by "MNT" ("Field Numerical Model").

3D analysis extension leads to lay images or vectorial data upon a surface then to extract vectorial entities of a surface to create lines, walls and solids.

3D symbols can be used to give realism in displaying of GIS data and to create animation of high quality. Scenes can be displayed from different angles with vision option. It is also possible to change 3D layers features in order to use transparency or shadows.

This figure show how the hydrographic network is deeply controlled by the fracture field. This can be seen by the main ENE-WSW trend recorded both in the two diagrams (Figure 6(a), Figure 6(b)). Folds axis and foliation trends (Figure 6(c)) are also similar to those of the hydrographic network which means that folds and foliations are guiding the hydrographic flows.

Confronting the hydrographic network to the underground architecture diagrams (Figure 6(d)), it reveals that those trends are closely connected especially with ENE-WSW direction. This led us to input the importance of the hydrographic network in the underground space genesis processes.

Concerning the interaction between hydrographic and underground systems, most of sinkholes and caves are supplied either directly from their entry or by some of their tributaries or even from associated fractures. They therefore, represent spaces where water is infiltrated. This is the case of Ait Taleb Brahim lapiaz in Sidi Hssain district. At this level, around a dozen of deep sinkholes are developed in a surface of about 3 km^2. Meteoric waters runoffs are seeping into the caves to feed directly neighbourhood springs and water-sheets as the case of Ifri spring, Tiznit and Bouizakarn plains.

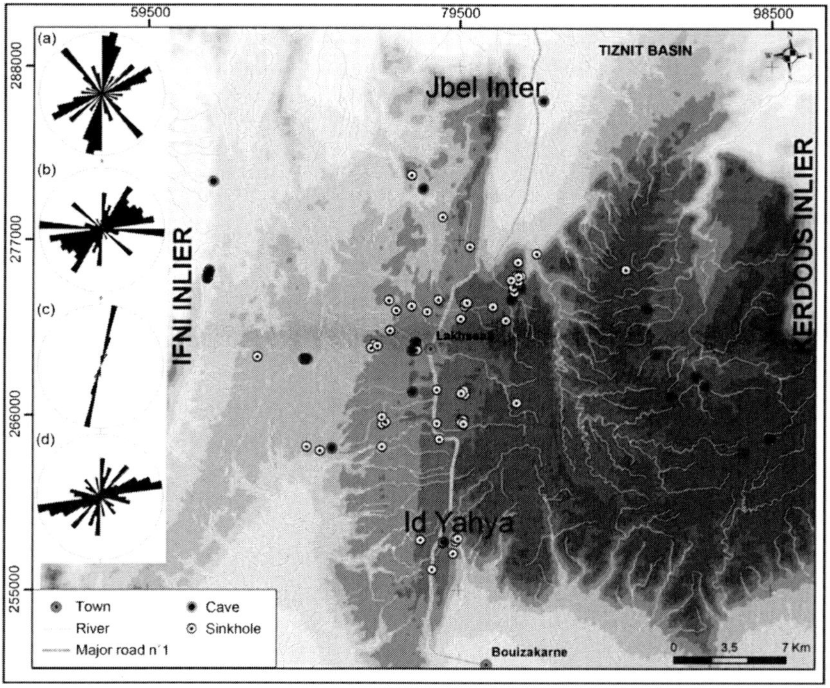

Figure 6. Simplified distribution of speleologic spaces in Lakhssas plateau using MNT and hydrographic network (from topographic map of Bouizakarn, [19]). (a) Trend diagram of hydrographic network cumulative length. (b) Trend diagram of major fractures. (c) Trend diagram of folds and foliations plans. (d) Trend diagram of underround architectural system.

Hydrographic flow is expressing topographic level changes. The attention was also to check the possible relationship between the speleological network and the altitude. Most of available underground topographic maps show that the altitude gets no impact on the speleology spaces distribution; so caves and sinkholes can be found in Lakhssas plateau whatever the topographic level is (Figure 7).

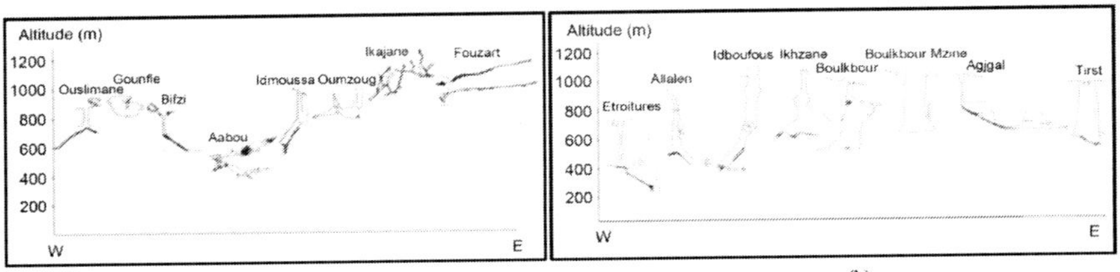

(a) (b)

Figure 7. Sample distribution of Lakhssas Karstic spaces in a function of altitude. (a) Caves Distribution according to altitude in Lakhssas plateau. (b) Sinkhole Distribution according to altitude in Lakhssas plateau.

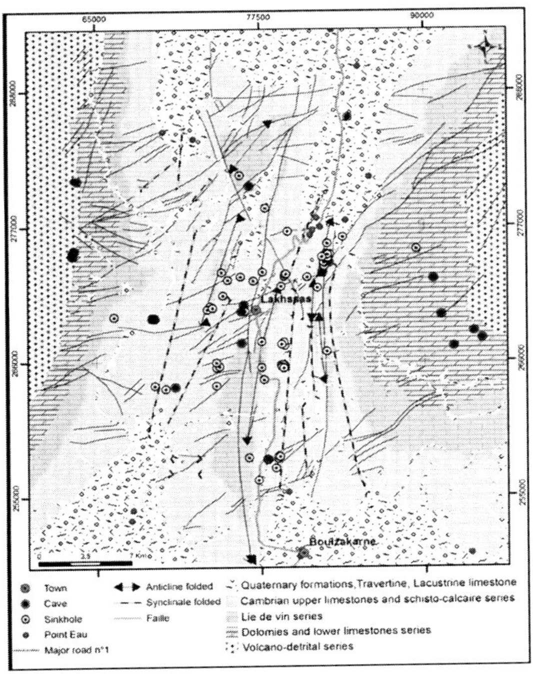

Figure 8. Map representing the various parameters controlling the distribution of underground spaces in the speleological Lakhssas plateau (extracted from Bouizakarn geological map 1/100,000).

4.1.1. Climate Processes

The climate is the main factor affecting the intensity of the karst event. Karst corrosion depends deeply on biological and chemical processes that occur in soil and vegetation covering the rock. In turn, soil and vegetation depend on the amounts of precipitation and temperature fluctuations.

In the case of Lakhssass area and during the Tertiary, the climatic conditions were probably much wetter to promoting the karstogenesis then it became dryer during the Quaternary age [7] . Currently, the arid climate prevailing in those regions doesn't allow the karstogenesis processes except in very limited cases.

4.1.2. Interference of Various Parameters

The underground architecture of spaces listed here shows unambiguous connexion with all presented parameters acting individually and in a different rates. Figure 8 shows the distribution of these parameters and the dispersion of speleology devices. It appears clearly that the tectonic lineaments ensure the main control on the underground spaces density [19] . This density is significantly higher when the rock is of calcareous composition. In other words, with limestone composition, tectonic structures prevail as "speleotect" (speleogenesis guidance). This feature is amplified by water circulation either at the surface or at depth. The intensity of fracturing, density and crushing of limestone and dolomite by tectonics, meteoric waters runoffs and groundwater movements, represent mechanically the basic parameters in the karstogenesis in this area.

5. CONCLUSIONS

The Lakhssas plateau karst developed in limestones and dolomites of the Lower Cambrian and structured by the Hercynian orogeny is an environment favorable to study several scientific disciplines. The structural study of inventoried cavities in the plateau of Lakhssas allows us to highlight the relationship between these karst forms and their genesis and controlling processes and the usefulness of speleothems as markers of neotectonics. Lithology rocks' nature and behavior are also a component of those processes.

The collection of structural data is therefore very important to understand deep karsts systems geometry, particularly where bedding and jointing are able to condition the underground water circulation [20] .

The detailed structural analysis allowed defining the relations linking the geometric arrangement of the endokarst system with bedding, jointing, folds and foliations plans. In Lakhssass plateau, the endokarst architecture is miming the exokarstic network with ENE-WSW and N-S to NNE-SSW trends as the main trends controlling the speleogenesis.

The calcareous dolomite composition is also a critical component in the speleogenesis process, beside the meteoric water circulation both at the surface and/or in the depth.

REFERENCES

1. Becker, A., Ferry, M., Monecke, K., Schnellmann, M. and Giardini, D. (2005) Multiarchive Paleoseismic Record of Late Pleistocene and Holocene Strong Earthquakes in Switzerland. Tectonophysics, 400, 153-177.
2. Gilli, é. (2005) Review on the Use of Natural Cave Speleothems as Palaeoseismic or Neotectonics Indicators. Geoscience, Tectonics p1.8.
3. Angelova, D., Belfoul, M.A., Bouzid, S. and Faik, F. (2005) Karst and Cave Systems in Bosnek Region (Vitosha Mountain, Bulgaria) and Wintimdouine (High-Atlas Mountain, Morocco). Palaeosismic Phenomena in Karst Terrains in Bulgaria and Morocco. Acta Carsologica, 34, 87-111.
4. Soulaimani, A. and Bouabdelli, M. (2005) Le plateau de Lakhssas (Anti Atlas Occidental, Maroc): Un graben finiprécambrien réactivé à l'hercynien. Annales de la société géologique du Nord t., II (2ème série), 177-180.
5. Bouchaou, L., Soulaimai, A., Boutaleb, S., Hsissou, Y. and Tagma, T. (2008) Caractéristiques hydrostructurales du plateau karstique de Lakhssas (Anti-Atlas Occidental, Maroc). 1er colloque international de spéléologie Scientifique et touristique, Résumé, Agadir (Maroc) 2008, 21.
6. Oliva, P. (1977) Karst et structures dans le plateau des Akhssas, L'Anti Atlas occidental-Maroc. Réunion de karstologie, Grenoble, Revue geogr. Alpin XVI 2 et 3.

7. Hassenforder, B. (1987) La tectonique panafricaine et varisque de l'Anti-Atlas dans le massif de Kerdous (Maroc). Thèse d'état, Université Louis Pasteur, Strasbourg, 249 p.

8. Belfoul, A. (2005) Cinématique de la déformation hercynienne et géodynamique de la marge NW Gondwana (Anti-Atlas Sud-occidental, Sahara Marocain: Zemmour, Ouled Dhlim et Mauritanides septentrionales). Thèse d'Etat, Université Ibn Zohr Agadir, Morocco.

9. Dubois, A. and Wanaim, A. (2006-2007-2008) Spéléo sur le plateau des Akhssas (Anti-Atlas occidental Maroc). Documents Inédits.

10. Carte géologique de Bouizakarn 1/100,000. Service de la Cartographie Géologique, MEM, Rabat, Morocco.

11. Carte géologique de Tiznit 1/100,000. Service de la Cartographie Géologique, MEM, Rabat, Morocco.

12. Carte topographique de Bouizakarn 1/100,000. Direction de la conservation foncière, Rabat, Morocco.

13. Belfoul, M.A., Qurtobi, M. and Faik, F. (2001) Impact de la structuration atlasique sur l'architecture interne de la grotte de Win-Timdouine (Haut-Atlas Occidental Marocain). Revue d'Analyse Spatiale, 47-50.

14. Kiraly, L. (1968) Les élements structuraux et alignement de phénomènes karstiques. Bulletin de la société Neuchateloise de Science Naturale, 127-146.

15. Kiraly, L. (1969) Anisotropie et hétèrogénéité de la perméabilité dans les calcaires fissurès. Eclogae Geologicae Helvetiae, 62, 613-619.

16. Kiraly, L. (1975) Rapport sur l'état actuel des conaissances dans le domaine de caractères physiques des roches karstiques. International Union of Geological Sciences, Serie B, Hydrologeology of Karstic Terraines, L.A.H., Paris, 53-67.

17. Boegli, A. (1980) Endokarst and Karst Hydrology. In: Boegli, A., Ed., Karst Hydrology and Physical Speleology, Springer-Verlag, Berlin, 73-76.

18. Ramsay, J.G. and Huber, M. (1987) The Techniques of Modern Structural Geology, Vol. 2: Folds and Fractures. Academic Press, London, 445-473.

19. Wanaim, A., Ezaidi, A., Belfoul, M.A., Bouzid, S., Faik, F., Oudra, M. and Dubois, A. (2008) R?le des linéaments structuraux dans la distribution cartographique des grottes du plateau de Lakhssass (Anti-Atlas occidental, Maroc). 1er Colloque International de Spéléologie Scientifique et Touristique, Agadir (Maroc) 23-26 Avril 2008, Résumé, 43.

20. Tognini, P. and Alfrido, B. (2001) Effects of the Structural Setting on Endokarst System Geometry in the Valle del Nosê (Como Lake, Northern Italy). Geolologica Belgica, 4, 197-211.

4

CHAPTER

SPATIO-TEMPORAL VARIATION AND IMPACT FACTORS FOR VEGETATION CARBON SEQUESTRATION AND OXYGEN PRODUCTION BASED ON ROCKY DESERTIFICATION CONTROL IN THE KARST REGION OF SOUTHWEST CHINA

Mingyang Zhang [1,2], Kelin Wang [1,2,*], Huiyu Liu [3,4], Jing Wang [1,2], Chunhua Zhang [1,2], Yuemin Yue [1,2] and Xiangkun Qi [1,2]

[1] *Key Laboratory of Agro-ecological Processes in Subtropical Region, Institute of Subtropical Agriculture, Chinese Academy of Sciences, No. 644, Yuanda 2nd Road, Furong District, Changsha 410125, China*
[2] *Huanjiang Observation and Research Station for Karst Ecosystems, Chinese Academy of Sciences, Huanjiang 547100, China*
[3] *College of Geography Science, Nanjing Normal University, No.1, Wenyuan Road, Xianlin University District, Nanjing 210046, China*
[4] *Jiangsu Center for Collaborative Innovation in Geographical Information Resource Development and Application, Nanjing 210046, China*

ABSTRACT

The Grain to Green Program (GTGP) and eco-environmental emigration have been employed to alleviate poverty and control rocky desertification in the Southwest China Karst region. Carbon sequestration and oxygen production (CSOP) is used to indicate major ecological changes, because they involve complex processes of material circulation and energy flow. Using remote sensing images and weather records, the spatiotemporal variation of CSOP was analyzed in a typical karst region of northwest Guangxi, China, during 2000–2010 to determine the effects of the Chinese government's ecological rehabilitation initiatives implemented in 1999. An increase with substantial annual change and a significant increase (20.94%, $p < 0.05$) in variation were found from 2000 to 2010. CSOP had a highly clustered distribution in 2010 and was correlated with precipitation and temperature (9.18% and 8.96%, respectively, $p < 0.05$). CSOP was significantly suppressed by human activities ($p < 0.01$, $r = -0.102$) but was consistent with the intensity of GTGP (43.80% positive). The power spectrum of CSOP was consistent with that of the gross domestic product. These results indicate that ecological services were improved by rocky desertification control in a typical karst region. The results may provide information to evaluate the efficiency of ecological reconstruction projects.

Keywords: rocky desertification control; carbon sequestration and oxygen production; northwest Guangxi; China; karst; vegetation

1. INTRODUCTION

Carbon sequestration and oxygen production (CSOP), which is an ecosystem regulation system that regulates the interaction between the terrestrial biosphere and the atmosphere, refers to the sequestration of carbon and release of oxygen by plants through photosynthesis, and is used to indicate major ecological changes that involve complex processes of material circulation and energy flow [1]. The Kyoto Protocol and Copenhagen conferences have focused worldwide attention on the need to reduce carbon emissions [2,3]. Moderate-resolution imaging spectroradiometry (MODIS) is widely used to monitor the spatial and temporal variation of ecosystem conditions because of its frequent and global coverage [4,5]. The normalized difference vegetation index (NDVI) is calculated as the difference between the reflectance in the red band (610–680 nm) and near infrared band (780–890 nm). NDVI is sensitive to the presence, density, and dynamics of vegetation. Consequently, the MODIS NDVI product has been applied to quantify vegetation cover and green biomass. Numerous studies have noted the effects of vegetation dynamics and climatic factors on vegetation in different ecosystems [5,6], and these studies showed that vegetation conditions were positively correlated with precipitation in most relatively dry areas, but negatively correlated in relatively humid areas [7,8].

Karst terrain in carbonate-dominated areas accounts for about 15% of the world's land area and is the underlying terrain for approximately 17% of the world's population. Karst regions are characterized by thin soil depth (generally not more than 10 cm), and poor surface water retention (infiltration coefficient of 0.3–0.6 or even 0.8) [9,10,11]. A degraded karst ecosystem may recover far more slowly than other ecosystems. One of the most harmful consequences of ecosystem degradation is rocky desertification, characterized by the formation of a desert-like landscape with a high percentage of bedrock [12]. As the bedrock is exposed, land productivity is reduced, and the distribution of cropland becomes more scattered. Because of these consequences, the karst region in southwest China is associated with severe poverty and environmental degradation [13].

In order to alleviate poverty and control rocky desertification, countermeasures have been employed, including the Grain to Green Program (GTGP) and eco-environmental emigration [14]. The objective of the GTGP is to increase vegetation coverage by planting trees or sowing grass on former farmland [15]. The focus of eco-environmental migration is to help farmers in karst areas to move to non-karst areas where ecosystem conditions were deemed to be considerably better in the 1990s. During this campaign, 49,133 families (232,705 persons) were relocated in northwest Guangxi, China. The total area of the GTGP was up to 1278.67 km^2 in Baise in 2001–2004. This included 605.33 km^2 of farmland restored to woodland and 673.33 km^2 of barren hills subjected to reforestation [16]. An increasing number of studies have focused on karst ecosystems [17,18], whereas only a few studies have investigated the efficiency of these projects. Most previous research regarding CSOP for the study area mainly focused on microanalysis through the in situ method at an ecosystem scale. The differences between anthropogenic impacts on CSOP and those of climatic variations at the landscape scale are uncertain. Residual analysis not only identifies human factors but also characterizes the separate influences of climate factors and human activities [19,20].

In this study, the ecological efficiency was evaluated with regard to rocky desertification since the implementation of GTGP and eco-environmental migration at a regional scale. Remote sensing images and climate data from 2000 to 2010 were used to extract distribution and dynamic information for CSOP in the typical karst area of northwest Guangxi, China, in which ecological projects have been implemented. The results from the present study may improve our understanding of the interactions among climate, human activities, and CSOP, and may provide auxiliary information to evaluate the efficiency of ecological reconstruction projects. Our results could also assist in ecosystem management to ensure sustainable development in fragile environments.

2. MATERIALS AND METHODS

2.1. Study Area

The study region was located in northwest Guangxi, China (104°29′–109°09′E, 23°33′–25°37′N; Figure 1), including 16 counties covering an area of 50,818.20 km^2, with a subtropical wet monsoon climate (annual temperature: 19.5 °C; annual precipitation: 1000 mm). The elevation of this region ranged from 100 to 2000 m, with a decrease in elevation from the northwest to the southeast. The main land use types in the region in 2010 included forest (28,058.83 km^2, 55.12% of the total area), shrub land (12,349.16 km^2, 24.29% of the total area), farm land (8919.58 km^2, 17.54% of the total area), and other types of land use (including building land, grass, wetland, and bare land 1513.96 km^2, 2.98% of the total area) (Figure 1). The farmland is mainly located in the industrial and economic development zones of the central region. Karst landforms are typical, and croplands are generally not fertile because of the geological conditions in this mountainous area. Most of the farmland is found in the flat areas in karst valleys or closed depressions, where major human settlements are also situated. Therefore, the demands for agricultural development and ecosystem conservation often conflict.

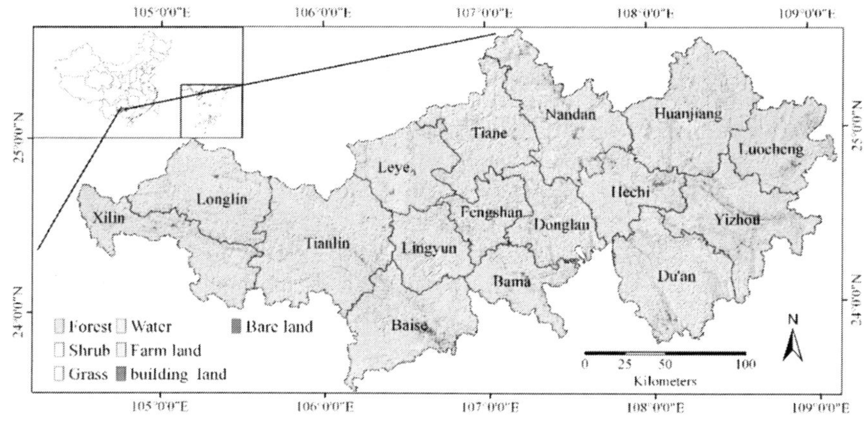

Figure 1. The location of the study region and main land uses in 2010.

2.2. Data Acquisition and Preprocessing

MODIS NDVI data (with a spatial resolution of 0.0089285714) from 2000 to 2010 were downloaded from NASA's Land Processes Distributed Active Archive Center (http://ladsweb.nascom.nasa.gov/). A digital elevation model (DEM, with a spatial resolution of 100 m × 100 m) was downloaded from the Data-sharing Network of Earth System Science in China (www.geodata.cn). Weather and radiation datasets from 97 stations within and near the study area including annual rainfall, average temperature, extreme temperature, and total and net radiation were collected for 2000–2010 from the China Meteorological Data-sharing Service System (http://cdc.nmic.cn/). Land cover in 2000 and 2010 was created by visual interpretation from Landsat-5 Thematic Mapper (TM) images with a resolution of 30 m × 30 m.

ArcGIS 10.0 (Environment Systems Research Institute, Redlands, CA, USA), ERDAS IMAGINE 9.1 (Leica, Stockholm, Sweden), and MATLAB (the MathWorks, Natick, Ma, USA) were used to compile data and perform spatial analysis. All data were projected or re-projected to the same projection (Albers Conical Equal Area projection, Krasovsky Spheroid) and re-sampled to 100 m × 100 m pixel spacing.

2.3. Measurements of CSOP

To quantitatively measure the CSOP, the terrestrial Carnegie-Ames-Stanford Approach (CASA) ecosystem model was used to estimate net primary productivity (NPP) from satellite data [21,22]. Models of the NPP calculation are as follows:

$$NPP(x,t) = APAR(x,t) \times \varepsilon(x,t) \tag{1}$$

$$APAR(x,t) = SOL(x,t) \times FPAR(x,t) \times 0.5 \tag{2}$$

$$\varepsilon(x,t) = T_{\varepsilon 1}(x,t) \times W\varepsilon(x,t) \times \varepsilon_{\max} \tag{3}$$

$$FPAR(x,t) = \frac{(NDVI(x,t) - NDVI_{i_{\min}}) \times (FPAR_{\max} - FPAR_{\min})}{(NDVI_{i_{\max}} - NDVI_{i_{\min}})} + FPAR_{\min} \tag{4}$$

$$T_{\varepsilon 1} = 0.8 + 0.02 \times T_{opt}(x,t) - 0.0005 \times \left[T_{opt}(x,t)\right]^2 \tag{5}$$

$$W_{\varepsilon}(x,t) = 0.5 + 0.5 \times \frac{E(x,t)}{E_p(x,t)} \tag{6}$$

$$E(x,t) = \frac{P(x,t) \times R_n(x,t) \times \left[(P(x,t))^2 + (R_n(x,t))^2 + P(x,t) \times R_n(x,t)\right]}{[P(x,t) + R_n(x,t)] \times \left[(P(x,t))^2 + (R_n(x,t))^2\right]} \quad (7)$$

$$E_\rho(x,t) = \left(E(x,t) + E_{\rho 0}(x,t)\right)/2 \quad (8)$$

where APAR(x,t) is the radiation available after absorption by photosynthesis; ε(x,t) is light energy utilization; FPAR(x,t) is the fraction of photosynthesis active radiation, with $FPAR_{max}$ = 0.950, $FPAR_{min}$ = 0.001 [23]; SOL(x,t) is solar radiation; $T_{\varepsilon 1}(x,t)$ and $W_\varepsilon(x,t)$ are the effects of temperature and water on light energy utilization, respectively; ε_{max} is the maximum light energy utilization, which has a different value for different vegetation types and is determined through the ecophysiological process model (BIOME-BGC) [24]; $T_{opt}(x,t)$, p(x,t), and $R_n(x,t)$ are the annual average temperature, rainfall, and radiation; E(x,t) and $E_\rho(x,t)$ are the actual and potential evapotranspiration; and $E_{\rho 0}(x,t)$ is the local available evapotranspiration, calculated using the vegetation-climate model of Thornthwaite [25].

According to the photosynthesis equation: $6CO_2 + 6H_2O \rightarrow 6O_2 + C_6H_{12}O_6$ (CH_2O), the ratio of organic matter produced by photosynthesis, carbon sequestered by photosynthesis, and oxygen released by photosynthesis is 1:1.47:1.07. The CSOP is calculated from NPP using this ratio [26].

2.4. Method of Residual Analysis

Residual analysis was used to separate the impacts of human activities from those of climatic factors. This method includes a regression analysis and is based on general observation values that are closely correlated with climate variations (mainly precipitation and temperature). This model was used to predict the value for each pixel. If the inter-annual residual change was approximately zero, human activity had no significant impacts on CSOP. Otherwise, positive residual values suggest that CSOP values had increased, which may be attributed to conservation and restoration efforts. Negative values indicate a decreasing trend that shows human-induced reduction [20].

We adopted a fitting method within the n-times power terms of precipitation, temperature, and CSOP based on R^2 set to determine the impacts of both temperature and precipitation on CSOP. Relationships between the NDVI and climatic factors for each pixel were examined using a regression analysis. Using the goodness-of-fit test (R^2 = 0.675), the best relationship was obtained with a quadratic fit of precipitation and temperature to CSOP. The formula was applied on each grid pixel:

$$CSOP = a \times P^2 + b \times P + c \times T + d \quad (9)$$

where P is precipitation (mm); T is temperature (°C); and a, b, c, and d are the multiple regression coefficients, adjusted for each variable and for each pixel.

2.5. Methods for the Human Activity Index (HAI), Intensity of the GTGP (IGTGP), and Spectrum Analysis

The HAI was calculated by integrating the gross domestic product (GDP), production of the primary industry (PRO1), production of a secondary industry (PRO2), production of a third industry (PRO3), human population (PEOPLE), and construction area (CONSTRU). The intensity of GTGP (IGTGP) index was composed of the implementation time and area for the counties in the study area. The equation for IGTGP was as follows:

$$HAI = \sum_{g=1}^{k} \left(\lambda_g / \sum_{g=1}^{m} \lambda_g \right) F_g \quad (10)$$

$$IGTGP = A/S \quad (11)$$

where λ is the characteristic root and F is the main component, A is the implementation time of the GTGP (year), and S is the area of GTGP (ha).

Spectrum analysis is the technical process of decomposing a complex signal into simpler parts [27]. The resulting plot is referred to as a power spectrum, and frequency spectra are generated by applying a fast Fourier transform. In this analysis, CSOP and the other factors (i.e., GDP, PRO1, PRO2, PRO3, PEOPLE, and CONSTRU) for every county were treated as variables.

3. RESULTS

3.1. Spatiotemporal Variations of CSOP

3.1.1. Inter-Annual Changes of CSOP

CSOP tended to increase in the study area, with substantial fluctuations and values increasing from 153.63 in 2000 to 156.41 t·ha^{-1} in 2010 (Figure 2).

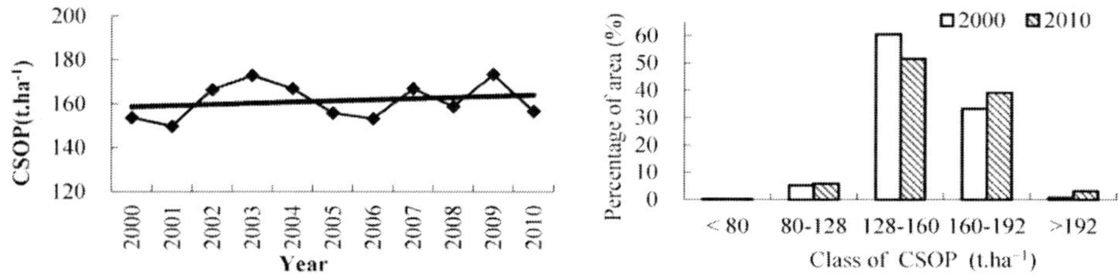

Figure 2. The annual variation of CSOP and percentage of different classes of CSOP.

Table 1. Variation of CSOP in different counties from 2000 to 2010.

County Name	Area		Mean (t.ha^{-1})			Sum (10^6 t)		
	km^{-2}	%	2000	2010	00–10	2000	2010	00–10
Nandan	3920.81	7.72	142.49	142.31	−0.18	55.87	55.80	−0.07
Huanjiang	4506.19	8.87	147.55	155.33	7.78	66.49	70.00	3.51
Tiane	3188.38	6.27	148.03	149.85	1.82	47.20	47.78	0.58
Luocheng	2633.81	5.18	158.72	160.88	2.16	41.80	42.37	0.57
Leye	2614.00	5.14	152.46	157.24	4.79	39.85	41.10	1.25
Longlin	3541.94	6.97	148.22	148.24	0.02	52.50	52.51	0.01
Hechi	2338.00	4.60	148.61	158.58	9.97	34.74	37.08	2.33
Yizhou	3815.19	7.51	155.35	159.88	4.52	59.27	61.00	1.72
Donglan	2398.94	4.72	143.32	150.99	7.67	34.38	36.22	1.84
Fengshan	1740.75	3.43	140.67	151.70	11.03	24.49	26.41	1.92
Tianlin	5532.31	10.89	162.46	164.85	2.39	89.88	91.20	1.32
Xilin	2969.56	5.84	169.11	152.14	−16.97	50.22	45.18	−5.04
Lingyun	2041.06	4.02	149.10	161.44	12.33	30.43	32.95	2.52
Du'an	3989.69	7.85	173.43	153.72	−19.71	69.19	61.33	−7.86
Bama	1906.13	3.75	127.86	137.58	9.72	24.37	26.22	1.85
Baise	3681.44	7.24	163.11	183.91	20.80	60.05	67.70	7.66
Total	50818.20	100.00	153.63	156.41	2.78	780.73	794.84	14.11

The total CSOP in the study area increased from 780.73 × 10^6 t to 794.84 × 10^6 t during the study period (Table 1). From an administrative point of view, the most rapid growth was observed in Baise county (CSOP growth density and total CSOP: 20.80 t·ha^{-1} and 7.66 × 10^6 t, respectively), followed by Huanjiang county (CSOP growth density and total CSOP: 7.78 t·ha^{-1} and 3.51 × 10^6 t, respectively). Obvious decreases were observed for CSOP in non-karst regions, such as Xilin and Du'an counties (decreases of 5.4 × 10^6 t and 7.86 × 10^6 t, respectively; Table 1, Figure 3). Results for the frequencies of CSOP values of the study area indicate that there were no significant annual vegetation changes. There was only a slight increase of 0.25 t·ha^{-1}·yr^{-1} for the annual CSOP. The CSOP had one very high peak around 160 t·ha^{-1} every year, and over 50% of the study area had CSOP values of approximately 150 t·ha^{-1} because of the widespread forest cover (79.41% of the total area) in the study area. The location of the CSOP peak ranged from 149.71 t·ha^{-1} to 173.13 t·ha^{-1}. The highest CSOP value for the study area appeared in 2000 and the lowest was in 2000. A detailed explanation for the fluctuations will be given in the following sections in terms of climate fluctuations and human activities.

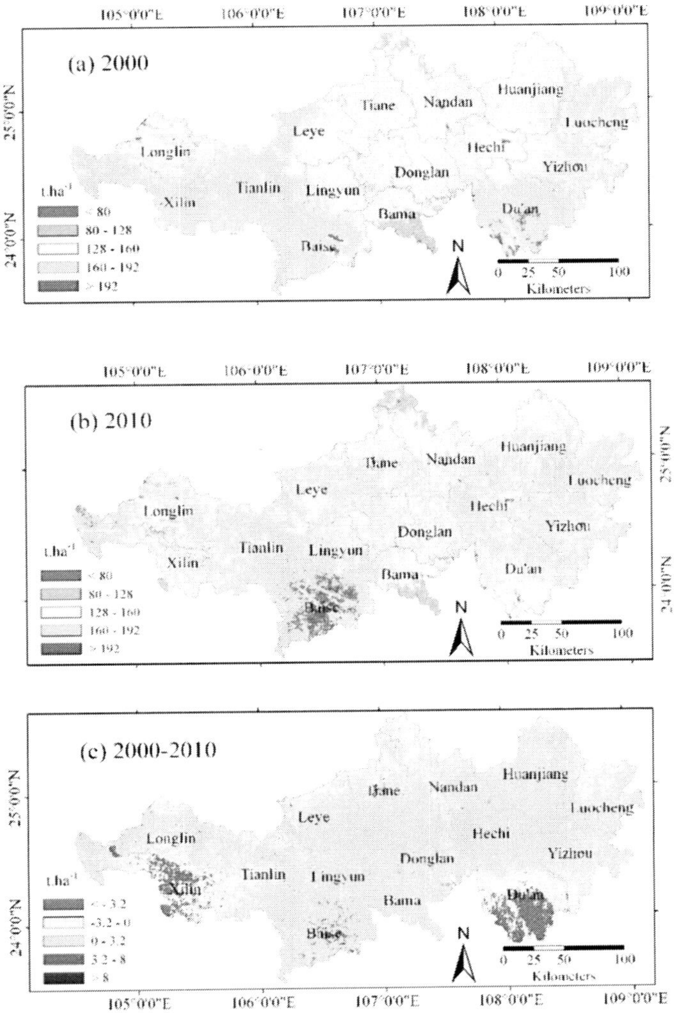

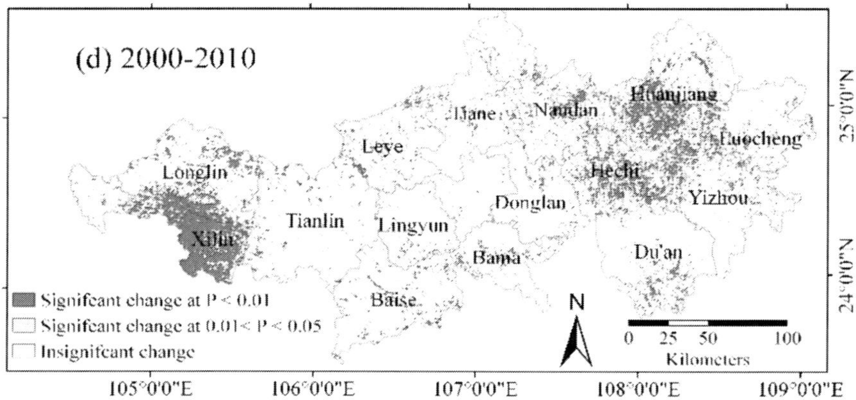

Figure 3. The distribution of (**a**) in 2000 and (**b**) in 2010; (**c**) variation; and (**d**) significant change in CSOP from 2000 to 2010 in the study area. The figure was generated using ARCGIS 10.0.

3.1.2. Spatial Variation of CSOP

The mid-west and southern regions exhibited high CSOP, while the central and northern regions exhibited low CSOP (Figure 3). The value of CSOP in most of the study regions in 2000 was 128–160 t·ha^{-1} and the proportion of land was 60.42%. A total of 90.67% of all CSOP in 2010 was concentrated into the two grades of 128–160 and 160–192 t·ha^{-1}. CSOP increased in about 64% of the study area and decreased in only 36% of the area. Such dramatic variation was mainly accounted for by slight increases (0–3.2 t·ha^{-1}), and the variation ratio was 62.11%. In contrast, the variation ratio of severe decreases (i.e., >−3.2 t·ha^{-1}) was only 4.10% (Figure 3). Significant changes were observed in 21.14% of the study area (11.23%, $p < 0.01$; 9.91%, $p < 0.05$).

CSOP increased significantly in the central part of the study region, including Huanjiang, Baise, Hechi, Donglan, Bama, and Yizhou Counties. For example, the densities of CSOP in Baise and Huanjiang counties increased 20.80 and 7.78 t·ha^{-1}, respectively, and the total CSOP increased 7.66×10^6 t and 3.51×10^6 t, respectively. Changes were significantly concentrated in the central and west regions of the study area, including Huanjiang, Hechi, Nandan, Xilin, and Du'an counties.

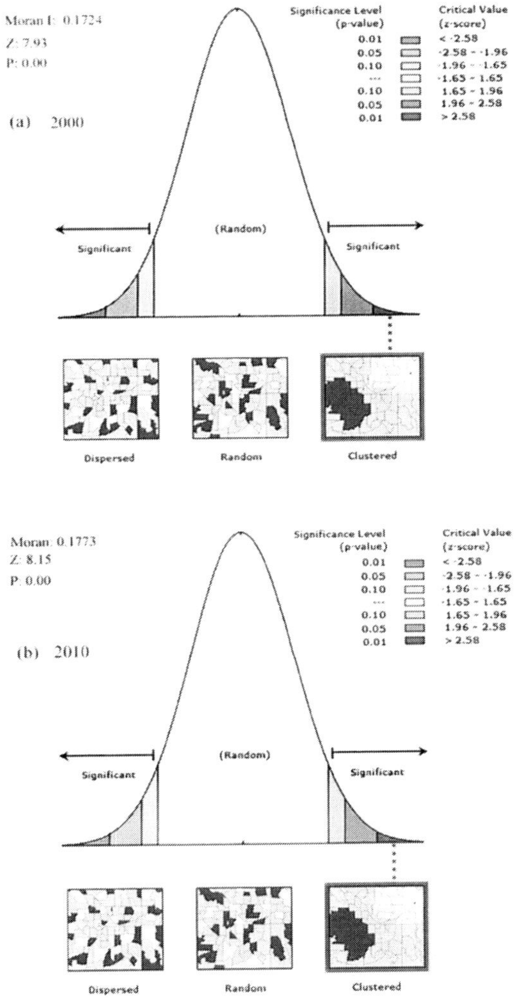

Figure 4. The self-spatial correlation index of Moran I for CSOP in 2000 and 2010. The figure was generated using ARCGIS 10.0. (a) a random pattern of distribution in 2000; (b) a higher self-spatial correlation in 2010.

3.1.3. Self-Spatial Correlation Pattern of CSOP

The resultant Z scores indicated whether features were tightly clustered or widely scattered in the self-spatial correlation index of Moran I. Features with high values that were surrounded by other features with high values were considered to have a statistically significant self-spatial correlation. Large Z scores indicated intense clustering of high self-spatial correlations. Small Z scores indicated intense clustering of low self-spatial correlations. When the difference was too large to be the result of random chance, a statistically significant Z score was generated.

Moran I of CSOP indicated a random pattern of distribution in 2000 and 2010 and a highly clustered pattern in 2010 (Figure 4). The Z score was 7.93 in 2000 and 8.15 in 2010, which indicates that the intense clustering had a higher self-spatial correlation in 2010; therefore, the distribution pattern of CSOP could have been affected by measures to control rocky desertification.

3.2. Main Factors of Influencing CSOP

3.2.1. Variation and Correlation of Natural Factors

A warming and drying trend in the study area was quite apparent over the 11 years of the study period. Average temperature increased slightly at a rate of 0.056 $°C \cdot yr^{-1}$ from 2000 to 2010 (Figure 5a). The mean annual temperature ranged from 15–18 °C in the west and north to 20–22 °C in the southeast. Annual precipitation showed a tendency to decline, and the variation rate was −2.59 $mm \cdot yr^{-1}$ (Figure 5b). There were large inter-annual variations in climatic variables in the study area (60.36% of the pixels showed positive inter-annual variation rates in temperature). In particular, in the middle parts of the study region, the inter-annual variation rates were as high as 0.12 $°C \cdot yr^{-1}$. In the western and northern parts of the study area, the inter-annual variation rates of temperature were negative. Except for the southeastern and mid-western part of the study area, the inter-annual variations rates of precipitation were negative, and the variation rates could be up to −18.24 $mm \cdot yr^{-1}$. A total of 49.91% of all the pixels showed positive rates of precipitation change.

Both positive and negative correlations existed between climatic variables and annual CSOP from 2000 to 2010 at the pixel scale in the study area (Figure 6). There was a significant correlation between CSOP and precipitation in 9.18% of the study area ($p < 0.05$), with a positive correlation within 8.69% of the area, distributed mainly in the eastern part of the study region. There was a significant correlation between temperature and CSOP in 8.96% of the study area ($p < 0.05$), and a negative correlation within 8.40% of the area. There was a strong positive correlation between temperature and CSOP in the central part of this study region, where the main vegetation types were mixed evergreen forests and shrubs. In the northeastern part of the region, the vegetation type was mainly mixed deciduous forests. Therefore, the increased temperature could lengthen the growing season of shrub and forests and strength photosynthesis, and thus promote vegetation activities. However, in the western part of the region, the correlation coefficients were negative. Specifically, the correlation coefficients were significantly negative in the western part of this study region and in most of the karst area. Most of the karst region lacked surface water. Consequently, temperature increases could increase evapotranspiration and result in vegetation stress; therefore, continuous drought in the summer may have caused the decrease in CSOP.

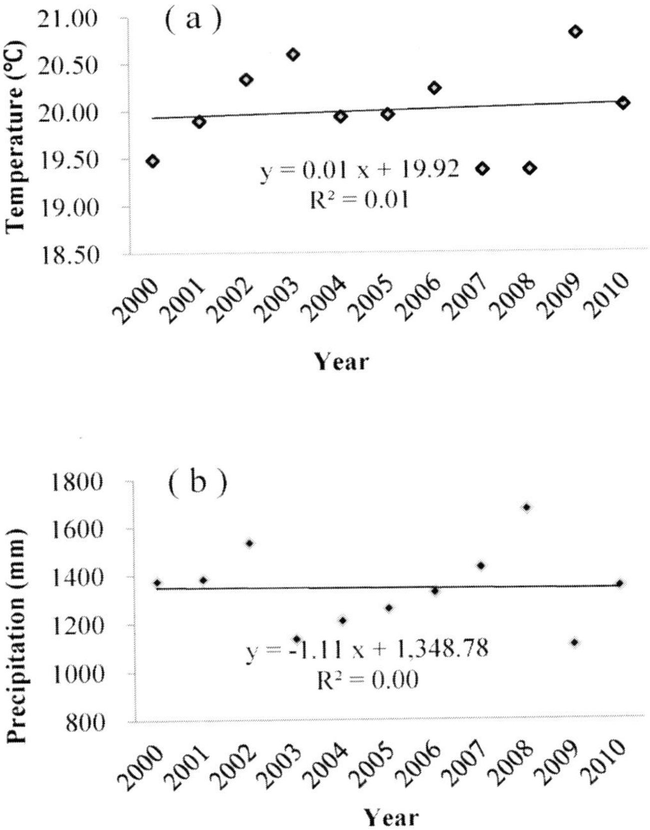

Figure 5. Climate fluctuations of (**a**) temperature and (**b**) precipitation in 2000–2010.

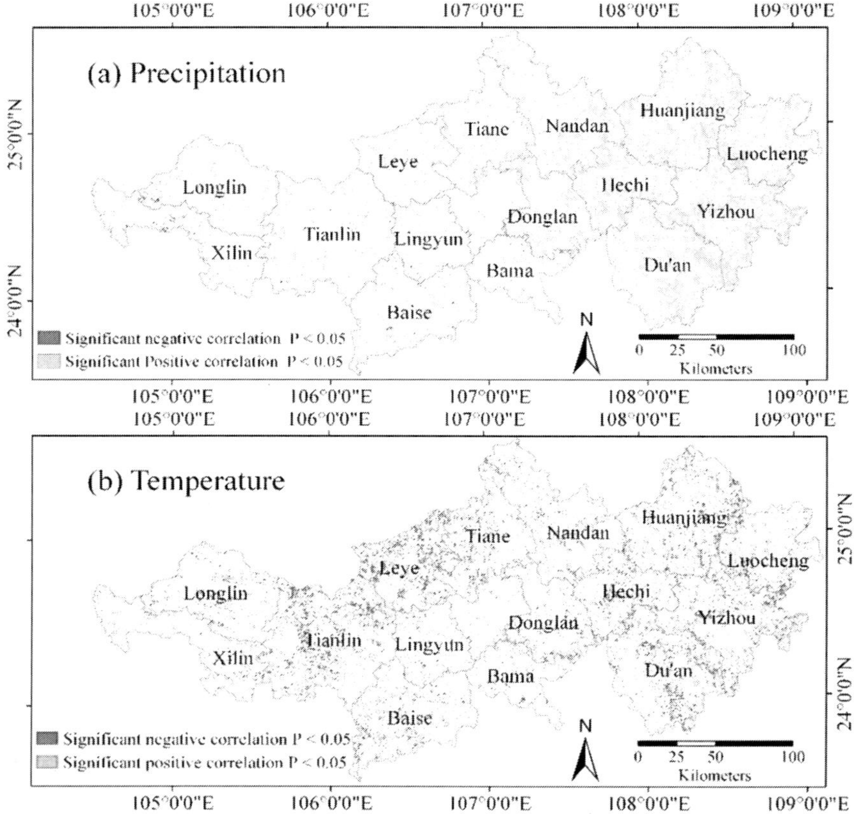

Figure 6. Distribution of correlation between CSOP and (**a**) precipitation and (**b**) temperature (0.05 and 0.01 significance levels). The figure was generated using ARCGIS 10.0.

3.2.2. Human Factors that Affected CSOP

(1) The Characteristics of Land Use Change

Land use within the study area changed over the 11 years of the study period (Table 2). The areas characterized by forest, water, farms, and buildings increased significantly, including an increase of 156.56 km² in forested land between 2000 and 2010. There were three main types of land transformations (Table 2): from farm to forest, from grass to forest, and from shrublands to forest. The shrubland area decreased the most significantly, followed by grassland and then bare land. A total of 274.36

km² of shrublands was converted to other forms of land uses between 2000 and 2010, of which 58.97% was converted into forest. A total of 215.69 km² of shrub/grassland was converted to forest, while 40.78 km² of forest became shrub/grassland. These changes were closely associated with recent vegetation reconstruction practices. Application of the GTGP and hill closure for reforestation was the primary method used to restore vegetation in the study area. A total of 21.48 km² of other types of land uses was converted to buildings. The area for building was derived mainly from shrublands and farmland, accounting for 80.77% of recently developed building land. The conversion from shrublands to buildings was one of the most significant transformations during the 11 years of the study period. It is reasonable to conclude that the implementation of vegetation restoration programs was a critical factor in increased CSOP. Meanwhile, the rapidly developing economy and accelerating urbanization process led to a massive transfer of other land uses to building land.

Table 2. Conversion and variation of land use in the study area from 2000 to 2010 (km²).

	Forest	Shrub	Grass	Water	Farm	Building	Bareness land
Forest	27,820.17	35.10	5.68	2.76	35.16	3.40	0.00
Shrub	161.78	12,290.45	8.57	99.71	50.72	11.93	0.35
Grass	53.91	15.48	406.20	0.18	11.18	0.41	0.10
Water	1.85	0.42	0.15	543.06	4.42	0.32	0.00
Farm	14.49	4.83	0.11	9.05	8813.39	5.42	0.00
Building	0.02	0.01	0.00	0.00	0.16	414.51	0.00
Bare land	6.62	2.88	0.37	0.10	4.56	0.19	1.39
Variation	156.56	−274.36	−66.39	104.64	72.31	21.48	−14.25

(2) Correlation between Human Activities and CSOP

There is a high degree of spatial heterogeneity in the relationship between human activities and CSOP (Figure 7). The mean residual slopes were between −0.0489 and 0.0263. A total of 43.80% of the residual slopes were positive, particularly in the eastern and middle parts of the study region. The areas with negative residual slopes were in the western and south-central parts of the study area. In these areas, a large number of farms were converted to building land, and the absolute values of the negative residual slopes were higher than those in other regions. Meanwhile, the strong negative trends in some areas may be the result of the intense southwest drought in 2009 and 2010. The severe drought limited vegetation growth and consequently caused a significant drop in CSOP. The negative effects caused by extreme weather events may have alleviated or offset the positive effects of the increased forest area and ecological restoration program.

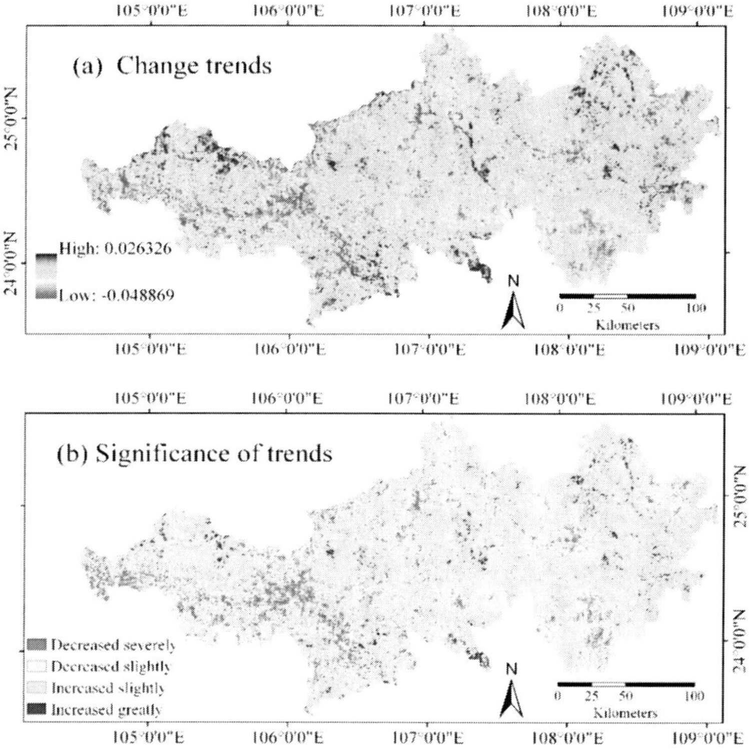

Figure 7. (**a**) Interannual change trends and (**b**) results of significance test in the study area region from 2000 to 2010. The grade of decreased severely refers to the significance $p < 0.05$ and trend < 0, and the grade of increased greatly refers to the significance $p < 0.05$ and trend > 0. The figure was generated using ARCGIS 10.0.

Although climate fluctuation is an important factor that influences the spatiotemporal variations of CSOP, the impacts of human activities should not be neglected. There were correlations among the socioeconomic statistics indexes ($p < 0.05$), with differences for varying levels of CSOP (Table 3). Obvious negative correlations were observed between CSOP and HAI ($p < 0.01$, $r = -0.102$). Significant positive correlations were observed between CSOP and PRO1 or CONSTRU ($p < 0.01$). The radial basis function network model, based on 13 environmental factors of 1377 samples, showed that human activities affected the spatial distribution of vegetation carbon in typical karst areas [14].

Population growth, urbanization, economic development, and technological advances have changed the land use and land cover [28]. Spectrum analysis showed that although the power spectral density of human activity indexes is different, CSOP was concurrent with changes in the trends of GDP, HAI, and CONSTRU, albeit in the opposite direction. In contrast, changes of power frequency density in CSOP were quite different from those of the other four indicators (i.e., PROD1, PROD2, PROD3, and PEOPLE) (Figure 8).

(3) Influence of Ecological Projects on CSOP

As one of the most serious geo-ecological problems in the karst areas in Southwestern China, karst rocky desertification had become a popular research topic. The ecological programs and rocky desertification controlling project resulted in a positive effect of the control of karst rocky desertification, thereby leading to changes of land use and land cover. In addition to the impact of climatic factors, the influence of human activities on CSOP displayed significant spatial and temporal variations. In the study region, the change in CSOP was significantly influenced by the implementation of ecological projects, such as gains in reforestation and afforestation by hill closures and comprehensive control of karst rock desertification. The implementation of the environmental emigration program had profound impacts on ecological restoration. Changes in CSOP were consistent with those of the IGTGP in typical karst counties (Figure 9): the greater the IGTGP, the more substantial the increase in CSOP, particularly in Hechi county, where the IGTGP was only 1.76 a. ha^{-1}, whereas the increase in CSOP reached 9.97 t. ha^{-1}. Previous studies in this region showed that net ecosystem productivity in typical karst areas has fluctuated substantially since 2000, and this fluctuation is closely related to the GTGP and eco-migration [12].

Table 3. Correlation of CSOP and social economic statistics.

	CSOP	HAI	GDP	PRO1	PRO2	PRO3	CONSTRU
HAI	−0.102 **						
GDP	0.052 *	−0.197 **					
PRO1	0.092 **	−0.467 **	0.333 **				
PRO2	−0.012	−0.373 **	0.621 **	0.553 **			
PRO3	0.056 *	−0.474 **	0.561 **	0.764 **	0.859 **		
CONSTRU	0.108 **	−0.425 **	0.671 **	0.386 **	0.405 **	0.591 **	
PEOPLE	−0.027	−0.110 **	0.457 **	0.378 **	0.111 **	0.455 **	0.638 **

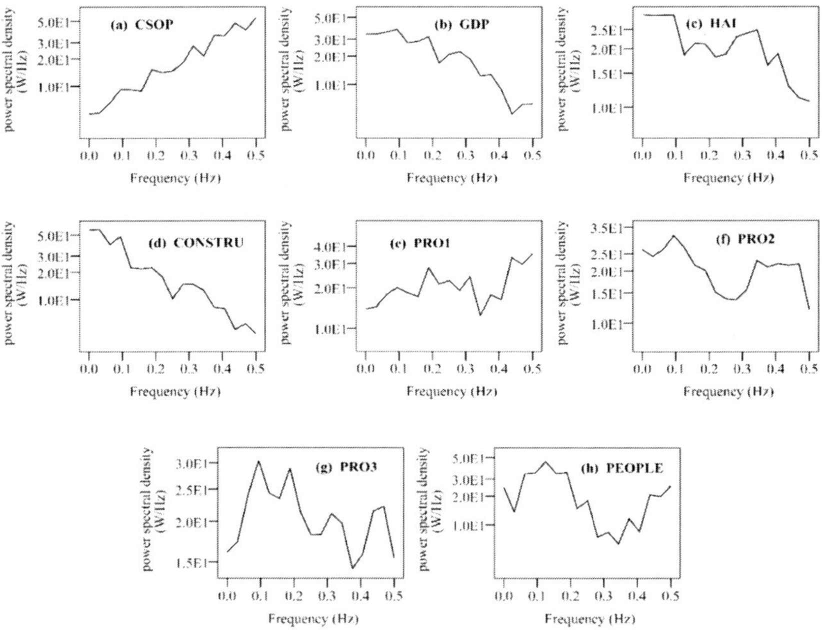

Figure 8. The spectrum trends of (**a**) CSOP (**b**) GDP; (**c**) HAI; (**d**) CONSTRU; (**e**) PRO1; (**f**) PRO2; (**g**) PRO3; and (**h**) PEOPLE. The figure was generated using SPSS (Statistical Product and Service Solutions) 10.0.

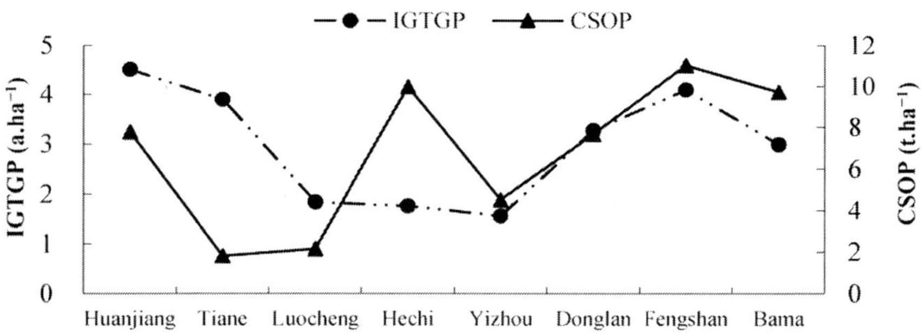

Figure 9. The change trend of CSOP and IGTGP in typical counties in the study area.

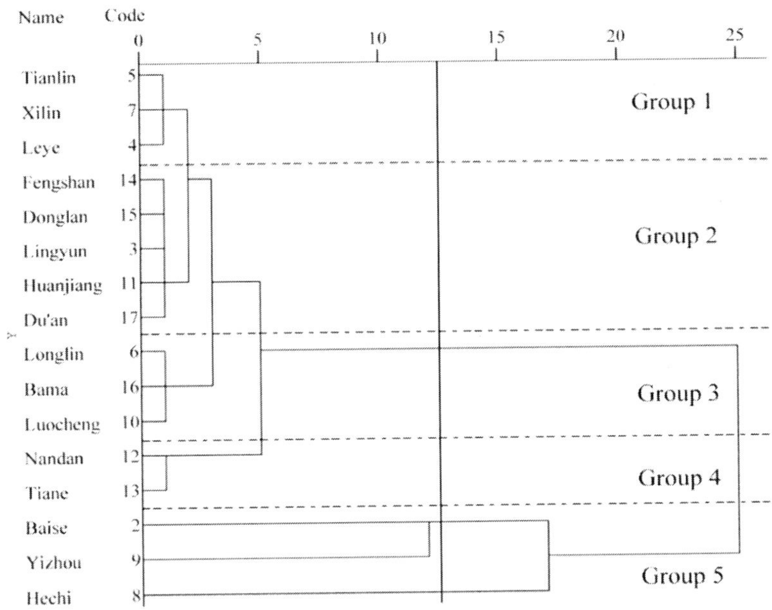

Figure 10. Clustering dendrogram of counties, based on hierarchical cluster analysis using average linkage (between groups) and rescaled distance clustering. The figure was generated using SPSS (Statistical Package for the Social Sciences) 10.0.

Table 4. CSOP and human activity in different groups of counties.

Clusters	CSOP ($t \cdot ha^{-1}$)	GDP ($\times 10^4$ $RMB¥ \cdot m^{-2}$)	PRO1 ($\times 10^4$ $RMB¥ \cdot km^{-2}$)	PRO2 ($\times 10^4$ $RMB¥ \cdot km^{-2}$)	PRO3 ($\times 10^4$ $RMB¥ \cdot km^{-2}$)	PEOPLE ($Person \cdot km^{-2}$)
Group 1	161.88	117.34	42.43	33.01	41.91	146.08
Group 2	153.29	248.85	73.88	85.29	89.67	381.04
Group 3	146.84	391.45	97.67	178.60	115.18	433.48
Group 4	144.97	734.36	34.26	217.91	61.70	290.32
Group 5	156.65	996.99	131.49	257.43	190.17	449.29

Consequently, the ecological restoration programs were one of the main drivers of the increasing trend in CSOP in the study region. Additionally, human interference, such as increased urbanization, resulted in the conversion of farmland and forest to urban land. Taken together, the results of previous research and the current study show that implementing control measures to limit rocky desertification, such as GTGP and ecological emigration, had obvious effects on biomass, carbon, and CSOP of vegetation, with potential ecological benefits. The program has made great progress in easing the impacts of humans on the natural world, increasing the use efficiency of farmland and effectively guaranteeing the implementation of policies to restore farmlands to forested areas [14].

The counties in this study region were clustered into five groups according to their landscape types (Figure 10). Group 1, the ecological shelterbelt, included Leye, Tianlin, and Xilin counties. Because large populations emigrated to more urban areas, human activities in these areas were very low (146.08 persons·km^{-2} PEOPLE), and the areas showed the highest CSOP (161.88, Table 4). Group 2, the degraded ecological shelterbelt, included Huanjiang, Donglan, Fengshan, Lingyun, and Du'an counties. They were characterized by relatively low levels of human activity (381.04 persons·km^{-2} PEOPLE) but a relatively high CSOP (153.29 t·ha^{-1}). Group 3, the agricultural area, included Luocheng, Longlin, and Bama counties. This area had low CSOP (146.84 t·ha^{-1}) and was characterized by large areas of farmland (97.67 × 10^4 RMB·km^{-2} PRO1) and small areas of permanent vegetation. Group 4, the suburban area, included Nandan and Tiane counties. This area was characterized by a low CSOP (144.97 t·ha^{-1}) and high levels of human activity (734.36 × 10^4 RMB·km^{-2} GDP, 217.91 × 10^4 RMB·km^{-2} PRO2; Table 4). Group 5, the urban area, included Hechi, Yizhou, and Baise counties. This group was the most artificially disturbed area, with high-intensity human activity (996.99 × 10^4 RMB·km^{-2} GDP, 257.43 × 10^4 RMB·km^{-2} PRO2, 449.29 persons·km^{-2} PEOPLE).

4. DISCUSSION

This study evaluated the spatiotemporal variation of CSOP and its main factors in a typical karst region of northwest Guangxi, China. This study's results suggest that factors affected by humans have important effects on the pattern of CSOP in a typical karst region, and that rocky desertification control measures also had an important impact on the pattern of CSOP, which are similar to previous results for small scale vegetation in karst obtained by sampling [8,9,14,29]. Most of the previous studies showed that spatial variation in vegetation carbon density generally increased in a typical area between 2005 and 2010, and factors associated with human activities had relatively high impacts on the distribution of vegetation carbon density. The ecosystem conditions were improved by the implementation of policies for rocky desertification in 2000–2010 [12,30,31].

The estimated results have some degree of uncertainty because of the complexity and dynamics of the ecosystems [5,32,33,34]. Several factors affected the accuracy of CSOP estimation at the regional scale. First, spatial scale is an important factor, and a multi-scale assessment of CSOP is necessary. The spatial scale at which vegetation type is measured significantly influences both the ecosystem service extent and its valuation [5,32,35]. Karst ecosystem areas are considerably fragile and heterogeneous [11], and the spatial scale needs to be taken into account in future evaluation of CSOP of this or similar regions. Second, the complex terrain in the karst area presented a challenge for deriving the spatial distribution of annual precipitation that was interpolated from climate records at 97 weather stations within and near the study area of northwest Guangxi, China. The large spatial and temporal variability in precipitation and temperature thus made accurate mapping distribution difficult at the 100-m resolution. Third, the CSOP assessment was undertaken through application of the terrestrial CASA together with remote sensing. The accuracy of the results from the CASA model is largely dependent on the resolution of remote sensing data. The remote sensing images of MODIS (1 km) and TM (30 m) were the main data sources in this study. The classifications of vegetation type were validated by referring to 1:47,000 color infrared aerial photographs. The overall accuracies for these classifications were 80.54%, with Kappa coefficients of 0.63. These errors or uncertainties can be reduced with more data, higher-resolution remote sensing data, and localized model parameters.

The analysis in this paper provides a preliminary basis for future studies of CSOP in the context of rocky desertification in karst regions. However, we did not consider the soil ecosystem in this study area. Although we compared the results with those of in situ data in part of the study area, further studies are required to determine the optimal approach for influencing the dynamic transformation of CSOP in regions with specific geographical environments and climatic backgrounds. Therefore, the present study has some limitations that should be addressed in future studies. First, the soil carbon should be considered for the whole karst system. Second, the degree of uncertainty associated with the CASA method requires further research. Finally, more comprehensive investigation of the driving mechanisms of the distribution of CSOP in karst regions is needed.

5. CONCLUSIONS

In the present study, the spatio-temporal variations of CSOP and its main factors were analyzed in a typical karst region of northwest Guangxi, China, using remote sensing data from 2000 to 2010. The results showed that the CSOP of vegetation was increasing, with substantial fluctuations (from 153.63 to 156.41 $t \cdot ha^{-1}$, reaching 173.13 $t \cdot ha^{-1}$ in 2009). Precipitation (9.18%, $p < 0.05$), temperature (8.96%, $p < 0.05$), and human activities substantially altered CSOP ($p < 0.01$, $r = -0.102^{**}$), 43.80% of the residual slopes were positive, and the increases in CSOP were consistent with the IGTGP. The power spectrum density of CSOP followed the opposite trend from that of GDP, HAI, and CONSTRU. Using these data, the counties of the study area were clustered into five groups according to their landscape types: the ecological shelterbelt, the degraded ecological shelterbelt, the agricultural area, the suburban area, and the urban area. Our results indicated that CSOP was significantly increased by the implementation of control measures to limit rocky desertification, which had an important impact on the pattern of CSOP, and that these measures resulted in an improvement in ecological services.

The conflict between economic development and environmental protection is a common issue worldwide that has been previously noted in rocky desertification in karst areas [17,32]. Poor environmental quality is known to be an important restricted condition in karst areas. Therefore, conservation of karst areas should take priority over restricting the uncontrolled reclamation of these areas for economic purposes in future land use practices. Controlling karst rocky desertification requires the optimization of land use structure. Discontinuing all reclamation activities in karst areas might not be possible, but future land reclamation projects need to be controlled and should be implemented after rigorous environmental impact assessment. More comprehensive and detailed exploration of the impacts of karst reclamation projects on ecosystem services provided in southwest China is necessary.

ACKNOWLEDGMENTS

This study was supported by National Natural Science Foundation of China (31370484, 41471445, 31470519), National Key Basic Research Program of China (2015CB452703), and the Chinese Academy of Sciences Action Plan for the Development of Western China (KZCX2-XB3-10).

AUTHOR CONTRIBUTIONS

Kelin Wang and Mingyang Zhang designed the study. Huiyu Liu, Jing Wang, Chunhua Zhang, Yuemin Yue and Xiangkun Qi performed data collection and analysis. Mingyang Zhang interpreted the results and analyzed the data.

CONFLICTS OF INTEREST

The authors declare no conflict of interest.

REFERENCES

1. Pan, Y.D.; Birdsey, R.A.; Fang, J.Y.; Houghton, R.; Kauppi, P.E.; Kurz, W.A.; Phillips, O.L.; Shvidenko, A.; Lewis, S.L.; Canadell, J.G.; et al. A large and persistent carbon sink in the world's forests. Science **2011**, 333, 988–993.
2. Wang, S.L.; Yeager, K.M.; Wan, G.J.; Liu, C.Q.; Wang, Y.C.; Lu, Y.C. Carbon export and HCO_3^- fate in carbonate catchments: A case study in the karst plateau of southwestern china. Appl. Geochem. **2012**, 27, 64–72.
3. Dixon, R.K.; Brown, S.; Houghton, R.A.; Solomon, A.M.; Trexler, M.C.; Wisniewski, J. Carbon pools and flux of global forest ecosystems. Science **1994**, 263, 185–190.
4. Pinzon, J.E.; Tucker, C.J. A non-stationary 1981–2012 AVHRR $NDVI_{3g}$ time series. Remote Sens. **2014**, 6, 6929–6960.
5. Wang, X.; Cheng, G.; Li, X.; Lu, L.; Ma, M. An algorithm for gross primary production (GPP) and net ecosystem production (NEP) estimations in the midstream of the Heihe River Basin, China. Remote Sens. **2015**, 7, 3651–3669.
6. Thevs, N.; Wucherer, W.; Buras, A. Spatial distribution and carbon stock of the Saxaul vegetation of the winter-cold deserts of Middle Asia. J. Arid Environ. **2013**, 90, 29–35.
7. Ni, J. Carbon storage in terrestrial ecosystems of china: Estimates at different spatial resolutions and their responses to climate change. Clim. Change **2001**, 49, 339–358.

8. Hou, W.; Gao, J.; Wu, S.; Dai, E. Interannual variations in growing-season NDVI and its correlation with climate variables in the southwestern karst region of china. Remote Sens. **2015**, 7, 11105–11124.

9. Cai, H.; Yang, X.; Wang, K.; Xiao, L. Is forest restoration in the southwest china karst promoted mainly by climate change or human-induced factors? Remote Sens. **2014**, 6, 9895–9910.

10. Parise, M.; de Waele, J.; Gutierrez, F. Engineering and environmental problems in Karst—An introduction. Eng. Geol. **2008**, 99, 91–94.

11. Zhao, X.; Wei, H.; Liang, S.L.; Zhou, T.; He, B.; Tang, B.J.; Wu, D.H. Responses of natural vegetation to different stages of extreme drought during 2009–2010 in southwestern China. Remote Sens. **2015**, 7, 14039–14054.

12. Zhang, M.; Zhang, C.; Wang, K.; Yue, Y.; Qi, X.; Fan, F. Spatiotemporal variation of karst ecosystem service values and its correlation with environmental factors in northwest Guangxi, China. Environ. Manag. **2011**, 48, 933–944.

13. Yang, Q.-Q.; Wang, K.-L.; Zhang, C.; Yue, Y.-M.; Tian, R.-C.; Fan, F.-D. Spatio-temporal evolution of rocky desertification and its driving forces in karst areas of northwestern Guangxi, China. Environ. Earth Sci. **2010**, 64, 383–393. [

14. Zhang, M.; Wang, K.; Liu, H.; Zhang, C.; Wang, J.; Yue, Y.; Qi, X. How ecological restoration alters ecosystem services: An analysis of vegetation carbon sequestration in the karst area of northwest Guangxi, China. Environ. Earth Sci. **2015**, 74, 5307–5317.

15. Feng, X.; Fu, B.; Lu, N.; Zeng, Y.; Wu, B. How ecological restoration alters ecosystem services: An analysis of carbon sequestration in China's Loess Plateau. Sci. Rep. **2013**, 3.

16. Piao, S.L.; Fang, J.Y.; Ciais, P.; Peylin, P.; Huang, Y.; Sitch, S.; Wang, T. The carbon balance of terrestrial ecosystems in China. Nature **2009**, 458, 1009–1013.

17. Brown, D.R.; Dettmann, P.; Rinaudo, T.; Tefera, H.; Tofu, A. Poverty alleviation and environmental restoration using the clean development mechanism: A case study from Humbo, Ethiopia. Environ. Manag. **2011**, 48, 322–333.

18. Harris, N.L.; Brown, S.; Hagen, S.C.; Saatchi, S.S.; Petrova, S.; Salas, W.; Hansen, M.C.; Potapov, P.V.; Lotsch, A. Baseline map of carbon

emissions from deforestation in tropical regions. Science **2012**, 336, 1573–1576.

19. Evans, J.; Geerken, R. Discrimination between climate and human-induced dryland degradation. J. Arid Environ. **2004**, 57, 535–554.

20. Xin, Z.B.; Xu, J.X.; Zheng, W. Spatiotemporal variations of vegetation cover on the Chinese Loess Plateau (1981–2006): Impacts of climate changes and human activities. Sci. China Ser. D **2008**, 51, 67–78. [

21. Field, C.B.; Randerson, J.T.; Malmstrom, C.M. Global net primary production: Combining ecology and remote sensing. Remote Sens. Environ. **1995**, 51, 74–88.

22. Zhu, W.Q.; Pan, Y.Z.; He, H.; Yu, D.Y.; Hu, H.B. Simulation of maximum light use efficiency for some typical vegetation types in China. Chin. Sci. Bull. **2006**, 51, 457–463.

23. Liu, Z.J.; Shao, Q.Q.; Liu, J.Y. The performances of MODIS-GPP and -ET products in China and their sensitivity to input data (FPAR/LAI). Remote Sens. **2015**, 7, 135–152.

24. Running, S.W.; Coughlan, J.C. A general-model of forest ecosystem processes for regional applications 1. Hydrologic balance, canopy gas-exchange and primary production processes. Ecol. Model. **1988**, 42, 125–154.

25. Thornthwaite, C.W. Atmospheric moisture in relation to ecological problems. Ecology **1940**, 21, 17–28.

26. Su, C.-H.; Fu, B.-J.; He, C.-S.; Lü, Y.-H. Variation of ecosystem services and human activities: A case study in the Yanhe watershed of China. Acta Oecol. **2012**, 44, 46–57.

27. Miyazaki, M.; Ouyang, C.; Zhou, X.Z.; Murdoch, J.B.; Fushimi, Y.; Okada, T.; Fujimoto, K.; Kido, A.; Arakawa, Y.; Miyamoto, S.; et al. Z-spectrum analysis provides proton environment data (ZAPPED): A new two-pool technique for human gray and white matter. PLoS ONE **2015**, 10.

28. Zhu, H.Y. Underlying motivation for land use change: A case study on the variation of agricultural factor productivity in Xinjiang, China. J. Geogr. Sci. **2013**, 23, 1041–1051.

29. Yan, J.H.; Wang, W.T.; Zhou, C.Y.; Li, K.; Wang, S.J. Responses of water yield and dissolved inorganic carbon export to forest

recovery in the Houzhai Karst Basin, Southwest China. Hydrol. Process **2014**, 28, 2082–2090.

30. Zhang, M.-Y.; Wang, K.-L.; Liu, H.-Y.; Zhang, C.-H.; Duan, Y.-F. Spatio-temporal variation of vegetation carbon storage and density in karst areas of northwest Guangxi based on remote sensing images. Chin. J. Eco-Agric. **2013**, 21, 1545–1553.

31. Yue, Y.; Wang, K.; Zhang, B.; Chen, Z.; Jiao, Q.; Liu, B.; Chen, H. Exploring the relationship between vegetation spectra and eco-geo-environmental conditions in karst region, southwest China. Environ. Monit. Assess. **2010**, 160, 157–168.

32. Yue, Y.; Zhang, B.; Wang, K.; Liu, B.; Li, R.; Jiao, Q.; Yang, Q.; Zhang, M. Spectral indices for estimating ecological indicators of karst rocky desertification. Int. J. Remote Sens. **2010**, 31, 2115–2122.

33. Ngugi, M.R.; Johnson, R.W.; McDonald, W.J.F. Restoration of ecosystems for biodiversity and carbon sequestration: Simulating growth dynamics of brigalow vegetation communities in Australia. Ecol. Model. **2011**, 222, 785–794.

34. Yu, Q.; Wang, S.; Mickler, R.; Huang, K.; Zhou, L.; Yan, H.; Chen, D.; Han, S. Narrowband bio-indicator monitoring of temperate forest carbon fluxes in northeastern China. Remote Sens. **2014**, 6, 8986–9013.

35. Cheng, J.; Ji, Y.; Liu, H. Segmentation-based PoLSAR image classification using visual features: RHLBP and color features. Remote Sens. **2015**, 7, 6079–6106.

5

CHAPTER

SINGLE EVENT TIME SERIES ANALYSIS IN A BINARY KARST CATCHMENT EVALUATED USING A GROUNDWATER MODEL (LURBACH SYSTEM, AUSTRIA)

C. Mayaud a,⇑, T. Wagner a, R. Benischke b, S. Birk

a Institute for Earth Sciences, University of Graz, Heinrichstraße 26, A-8010 Graz, Austria

b Department of Water Resources and Environmental Analytics, Institute for Water, Energy and Sustainability, Joanneum Research Forschungsgesellschaft mbH., Elisabethstraße 18/II, A-8010 Graz, Austria

SUMMARY

The Lurbach karst system (Styria, Austria) is drained by two major springs and replenished by both autogenic recharge from the karst massif itself and a sinking stream that originates in low permeable schists (allogenic recharge). Detailed data from two events recorded during a tracer experiment in 2008 demonstrate that an overflow from one of the sub-catchments to the other is activated if the discharge of the main spring exceeds a certain threshold. Time series analysis (autocorrelation and cross-correlation) was applied to examine to what extent the various available methods support the identification of the transient inter-catchment flow observed in this binary karst system. As inter-catchment flow is found to be intermittent, the evaluation was focused on single events. In order to support the interpretation of the results from the time series analysis a simplified groundwater flow model was built using MODFLOW. The groundwater model is based on the current conceptual understanding of the karst system and represents a synthetic karst aquifer for which the same methods were applied. Using the wetting capability package of MODFLOW, the model simulated an overflow similar to what has been observed during the tracer experiment. Various intensities of allogenic recharge were employed to generate synthetic discharge data for the time series analysis. In addition, geometric and hydraulic properties of the karst system were varied in several model scenarios. This approach helps to identify effects of allogenic recharge and aquifer properties in the results from the time series analysis. Comparing the results from the time series analysis of the observed data with those of the synthetic data a good agreement was found. For instance, the cross-correlograms show similar patterns with respect to time lags and maximum cross-correlation coefficients if appropriate hydraulic parameters are assigned to the groundwater model. The comparable behaviors of the real and the synthetic system allow to deduce that similar aquifer properties are relevant in both systems. In particular, the heterogeneity of aquifer parameters appears to be a controlling factor. Moreover, the location of the overflow connecting the sub-catchments of the two springs is found to be of primary importance, regarding the occurrence of inter-catchment flow. This further supports our current understanding of an overflow zone located in the upper part of the Lurbach karst aquifer. Thus, time series analysis of single events can potentially be used to characterize transient inter-catchment flow behavior of karst systems.

Keywords: Binary karst aquifer, Single event, Overflow behavior, Time series analysis, MODFLOW, Single-continuum model

1. INTRODUCTION

Karst aquifers are of primary importance for supplying drinking water to nearly 25% of the world's population (Ford and Williams, 2007), but their significant reserves are highly vulnerable to contamination and to industrial or intensive agricultural land use. Because karst is a highly heterogeneous environment comprising aperture diameters varying over more than five orders of magnitude (from fracture openings less than 1 mm in the limestone matrix to conduits of more than 10 m width in large caves) there is a need to develop and improve existing tools helping to better understand the processes governing the hydrodynamic behavior of karst systems. As not more than a few percent of a karst aquifer are generally mapped (or explored), it can be defined as grey/black-box system, where the input is routed through to the output without a direct observation of the water transfer. Thus, indirect methods of characterization have been developed to obtain a maximum of information from the karst systems. They are mostly focused on the comparison between the available input and output data and include hydrograph and chemograph analyses using discharge, specific electric conductivity, water temperature, chemical parameters, isotopes and tracer experiments (e.g., Bakalowicz, 2005; Geyer et al., 2013; Kresic and Stevanovic, 2010; Pinault et al., 2001; Rehrl and Birk, 2010).

Time series analysis are signal processing methods belonging to this category, and are mostly used to improve the understanding of the hydrological behavior of karst systems. Mangin (1984) was the first to apply them to the field of karst hydrology. He compared the autocorrelation and power spectral density functions of the discharge of three Pyrenean karst aquifers under the same climatic conditions and deduced that their different responses were due to different degrees of karstification and storage capacities. Larocque et al. (1998) extended the analysis to a broader dataset (piezometric level, water discharge at the inlet and outlet, precipitation, specific electrical conductivity and water temperature) combined with new methods such as cross-correlation, cross-spectral density, coherence function, gain and phase functions, and proved their usefulness for the purpose of water management. Panagopoulos and Lambrakis (2006) applied time series analysis to two Greek karst aquifers well-known for their different karstification and found that the different results were in agreement with the differences in karstification. More recently Bailly-Comte et al. (2008) applied these methods to a small Mediterranean karst system and highlighted the interactions between the karst aquifer and an overflow river. Kovačič (2010) applied autocorrelation, cross-correlation, power spectral density and coherence function to the complex Unica river catchment and improved the understanding of its hydrodynamic behavior, allowing the differentiation of flow paths using two datasets of different time scale. Time series analysis were also successfully used by Amraoui et al. (2003), Bailly-Comte et al. (2011), Bouchaou et al. (2002), Jemcov and Petrič, (2010), Genthon et al. (2005) and Massei et al. (2006) using a broader dataset (rainfall, specific electric conductivity, turbidity, water temperature) and different methods (e.g. spectral and wavelet analyses).

Until now, time series analysis were mostly applied to time-periods covering a period of 1 year or more (Eisenlohr et al., 1997; Kovačič, 2010;

Larocque et al., 1998; Mangin, 1984; Panagopoulos and Lambrakis, 2006). Only a few studies (e.g. Bailly-Comte et al., 2008; Budge and Sharp, 2009; Covington et al., 2009, 2012; Valdes et al., 2006) applied the methods at a very short -or single event time scale to provide information about the hydrodynamic behavior of a karst system for short periods. Indeed, as opposed to long-term time series analysis, which is recommended to provide information about the "average" aquifer behavior/properties (Kovačič, 2010; Panagopoulos and Lambrakis, 2006), single event analysis has the potential to show how the system reacts at the scale of a single event only (Bailly-Comte et al., 2008; Covington et al., 2009, 2012; Valdes et al., 2006). This is of primary importance, because karst aquifers are highly dynamic non-linear systems whose behavior may evolve or vary temporarily depending on the hydrological conditions within the system (Mayaud et al., 2013; Wagner et al., 2013).

As demonstrated by the examples cited above, time series analysis has been frequently applied to karst catchments. Although the results from these applications were found to be in qualitative agreement with field observations, it has only rarely been attempted to verify the interpretation more quantitatively by applying the methods to synthetic catchments represented by a numerical model, where aquifer properties and hydrological stresses are known in detail. This approach was followed by Eisenlohr et al. (1997) who evaluated the results from times series analysis using a numerical groundwater flow model. These authors concluded that the results were not only dependent on the system geometry but also on the frequency and type (allogenic vs. autogenic) of the recharge events and on their intensity. In addition, an inappropriate length of the analyzed time series may cause errors in the interpretation. Jeannin and Sauter (1998) concluded that these methods were inappropriate without knowledge of the investigated area to characterize the underground geometry of karst aquifers. Nevertheless, Larocque et al. (2000) applied successfully autocorrelation and cross-correlation analysis to numerical data of a groundwater model representing the Larochefoucault karst aquifer (France). Later, Budge and Sharp (2009) applied short term cross-correlation analysis to a simplified synthetic MODFLOW catchment in order to develop a conceptual understanding of the Barton springs–Edwards aquifer. Their results showed that the cross-correlation was dependent on the input data but also on the geometrical properties of the aquifer.

The purpose of this paper is to improve the interpretation of time series analysis at the scale of single events in karst catchments that are characterized by the existence of a localized recharge component from a sinking stream and by temporarily varying drainage pattern due to the overflow from one spring catchment to another. This involves the need to investigate how physical characteristics of the karst system are reflected in the results from the time series analysis. To this end, two methods, autocorrelation and cross-correlation, are applied to a well investigated field site, the Lurbach karst system (Austria), where an overflow from one sub-catchment to another one is reported by numerous tracer experiments, and to a synthetic karst catchment represented by a numerical groundwater flow model, which accounts for the most relevant features of the field site in a simplified manner.

2. APPROACH

The following subsection 2.1 provides a brief introduction into the two time series analysis methods that are examined in this paper. These methods are evaluated in parallel using both a field site and a synthetic karst catchment represented by a numerical groundwater flow model, which are both described in the subsequent subsections 2.2 and 2.3, respectively. If the time series analysis of the synthetic and the field case yield comparable results, similar aquifer properties and geometries of the overflow section present in the model can be deduced for the Lurbach system.

2.1. Methods

2.1.1. Autocorrelation

The autocorrelation function examines how a value depends on the preceding values over a period of time. This function is represented with a correlogram. The slope of the correlogram is determined by the response of the system to an event. If the event has only a short-term influence on the response of the karst system, the slope of the correlogram will decrease steeply and quickly. In contrast, if the system is influenced by an event for a long time, the slope of the correlogram will decrease slowly. Generally the length of the influence of an event is given by the "memory effect" which is according to Mangin (1984) the lag number when $r(k)$ reaches the value of 0.2. The formula for autocorrelation is (Larocque et al., 1998; Mangin, 1984):

$$r(k)\frac{C(k)}{C(0)} \qquad (1)$$

with

$$C(k) = \frac{1}{n}\sum_{t=1}^{n-k}(x_t - \bar{x})(x_{t+k} - \bar{x}) \tag{2}$$

where k is the time lag and varies from 0 to m. According to Mangin (1984) m has to be taken as 1/3 of the length of the whole dataset to avoid stability problems.

Applied at a single event or short time scale, autocorrelation has the potential to allow an estimation of the inertia of the system (Valdes et al., 2006). Then, the memory effect shows how the karst conduits react to the event, and cannot be compared to memory effects resulting from analysis of a long time series.

2.1.2. Cross-correlation

Cross-correlation is used to determine the relationship between two variables x and y. In the case of karst hydrology they are mostly input–output relationships as for discharge–discharge, rainfall–discharge or water level–discharge. The cross-correlation is represented by a cross-correlogram, which has a positive and a negative part. A peak in the positive part means that the input signal has an influence on the output signal. If the cross-correlogram is symmetrical then the two signals respond at the same time. The maximum amplitude and the lag value of the cross-correlogram provide information about the delay which indicates the time of the pressure pulse transfer into the aquifer. If the input signal is a random process, then the cross-correlation function has the form of the impulse response of the system. According to Larocque et al. (1998) the formula for cross-correlation is:

$$r_{xy}(k) = \frac{C_{xy}(k)}{\sigma_x \sigma_y} \tag{3}$$

with

$$C_{xy}(k) = \frac{1}{n}\sum_{t=1}^{n-k}(x_t - \bar{x})(y_{t+k} - \bar{y}) \tag{4}$$

where σ_x and σ_y are the standard deviations of the two time series.

Applied at a single event scale, the cross-correlation shows how the energy is transferred and modified from the input to the output during a flood (Bailly-Comte et al., 2008; Covington et al., 2009) and represents the impulse response of the system.

2.2. Field site

The area under investigation is a binary karst catchment of 23 km^2 named Lurbach system, located about 15 km north of Graz (Styria, Austria) and belongs to the Central Styrian Karst (Fig. 1). The upper part of the catchment comprises an area of about 15 km^2 essentially composed of Paleozoic schists, and is drained by the Lurbach stream in an E–W direction towards the lower part, which is an 8 km^2 highly karstified unit. After passing the contact schist-limestone, the stream infiltrates along the streambed at a length of some hundred meters and finally disappears into a major sinkhole located right after the entrance of a big cave, the Lurgrotte (entrance at 633 m a.s.l.). Then, the water flows through the conduits and fissures of the limestone massif and resurges at the Schmelzbach outlet and the Hammerbach spring, both located in the valley of the Mur River on the western side of the catchment. The altitude of the whole area ranges between 1.109 m a.s.l. on the top of the Fragnerberg mountain (Fig. 1) and approximately 400 m a.s.l. at the bottom of the Mur valley close to the location of the Hammerbach spring.

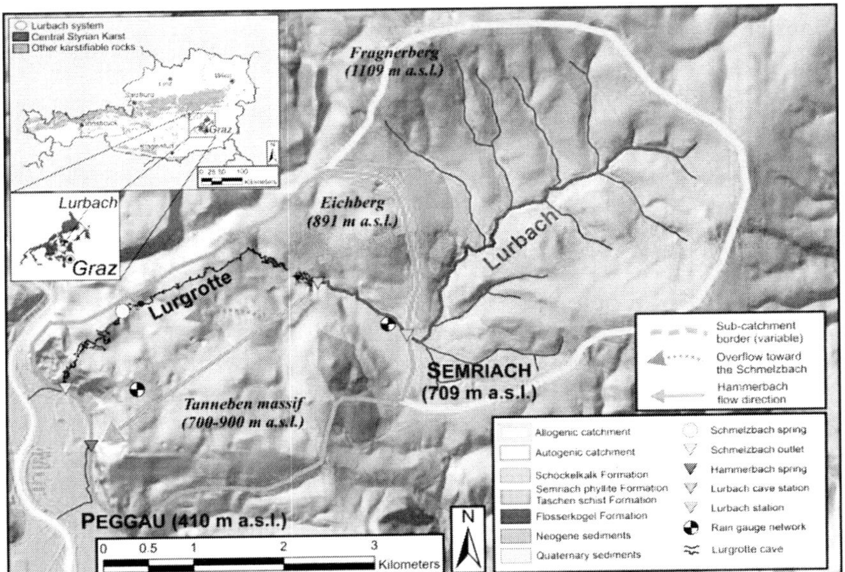

Figure. 1. Simplified geological map (modified after Geologische Bundesanstalt, 2005; sheet 164-Graz) of the Lurbach system including approximate flow directions and the monitoring network. The low permeable part (allogenic catchment) corresponds to the topographic ...

The Lurbach system is subject to a climate regime with low winter precipitation (approximately 50 days of snow cover per year) and frequent heavy thunderstorms during the summer (Harum and Stadler, 1992). The mean annual precipitation recorded from 1965 to 2010 at the station of Semriach (Fig. 1) is 880 mm. The maximum precipitation value was recorded at the same station during the summer 1975 with 93.5 mm rain in 1 day.

As it is regularly reported for karst aquifers, the behavior of the Lurbach system varies strongly according to its different hydrological conditions (Harum and Stadler, 1992):

At low and medium water conditions, the allogenic Lurbach waters (mean annual discharge of 141 l/s) supply only the Hammerbach spring (mean annual discharge of 193 l/s). The Schmelzbach outlet drains only autogenic recharge from the limestone massif and has a mean annual discharge of 79 l/s. Thus, both Hammerbach and Schmelzbach sub-systems are then totally separated.

When the Hammerbach discharge increases above a threshold discharge (about 200 l/s according to Behrens et al., 1992) an overflow from the Hammerbach sub-system to the Schmelzbach sub-system occurs.

At high water conditions the Lurbach stream can reach a maximum discharge of more than 10 m^3/s and the whole system is subject to catastrophic flood events. Then, the Lurbach flows directly through the Lurgrotte cave, toward the Schmelzbach, which becomes the main outlet of the karst aquifer and can reach peak discharges up to 10 m^3/s, whereas the Hammerbach spring shows a limited discharge capacity and cannot drain more than 2 m^3/s.

The Lurbach karst aquifer has undergone a complex speleological development: on the one hand the Lurgrotte (see Fig. 1) is a well-explored about 3 km long multi-level cave (Wagner et al., 2011). On the other hand, the Hammerbach conduit network is totally unexplored as all attempts to access it failed up to the present. The overflow between the Hammerbach and the Schmelzbach sub-systems is documented only by indirect observation of numerous tracer experiments (Behrens et al., 1992; Kübeck et al., 2013). Moreover, as no boreholes are available in the study area, there is a lack of information concerning the extent and local position of a phreatic zone within the karst massif. The only available information regarding phreatic conditions in the karst aquifer were derived from speleological observations in the cave Lurgrotte itself and by geomorphological observations of several dry caves located in the Tanneben massif (with more than 200 known smaller inactive caves).

The last field campaign in the Lurbach system was conducted from the 28th November 2008 to the 30th December 2008). At the beginning of this time period, a tracer experiment had been carried out (Oswald, 2009). One kilogram of uranine (Uranin AP; AppliChem GmbH, Germany) was continuously injected into the Lurbach stream at the contact between schist and limestone (marked as Lurbach station in Fig. 1) from the 28th November 2008 at 14:11 to the 29th November 2008 00:35. The water levels were recorded at the stations Lurbach cave, Hammerbach spring and Schmelzbach outlet with a frequency of 30 s, whereas the Lurbach station was recorded at a frequency of 5 min. Then, they were converted to discharge rates using control measurements. Precipitation and air temperature were recorded at a 5-min interval at the Ertlhube meteorological station (located at 763 m a.s.l.) on top of the Tanneben karst massif (Fig. 1). During the observation period, two hydrological events occurred. The results from this tracer experiment provide insight into the overflow behavior from the Hammerbach sub-catchment to the Schmelzbach sub-catchment, which will be taken into account when interpreting the results from the time series analysis.

2.3. Synthetic karst catchment

In order to evaluate the interpretation of results from single event time series analysis, the groundwater flow model MODFLOW-2005 (Harbaugh, 2005) is used to implement a simplified hypothetical karst setting similar to the Lurbach system. Since the model is applied to a karst setting, it is an obvious idea to employ the Conduit Flow Process (CFP; Shoemaker et al., 2008) for MODFLOW-2005 to account for turbulent flow conditions. However, neither CFP mode 1 (hybrid approach) nor CFP mode 2 (continuum approach) were found to be capable of simulating the rewetting and falling dry of cells representing the transient overflow from the Hammerbach to the Schmelzbach sub-system in the given model setting. Yet, it is important to note that in this work the model is intended to provide general insight into the dependency of the results from the time series analysis on the physical characteristics of a generic type of karst catchment. Thus, similar to the MODFLOW models presented by Ravbar et al. (2011) and Mayaud et al. (2013), the purpose is to represent the natural processes in a simplified manner rather than to provide a quantitative representation of the actual field site. For this purpose, using MODFLOW-2005 without CFP appears to be adequate, as it allows a robust, approximate representation of the overflow dynamics.

The general approach here is to apply single event auto- and cross-correlation to a synthetic catchment and to examine how far the results are similar to or different from those obtained for the Lurbach system. As the model parameters are all known, the differences found in the auto- and cross-correlation of the various scenarios can be clearly attributed to a controlling parameter. Thus, this approach helps to interpret the results from the real system. In addition to a general evaluation of the applied time series analysis methods, this will also improve the current understanding of the Lurbach system.

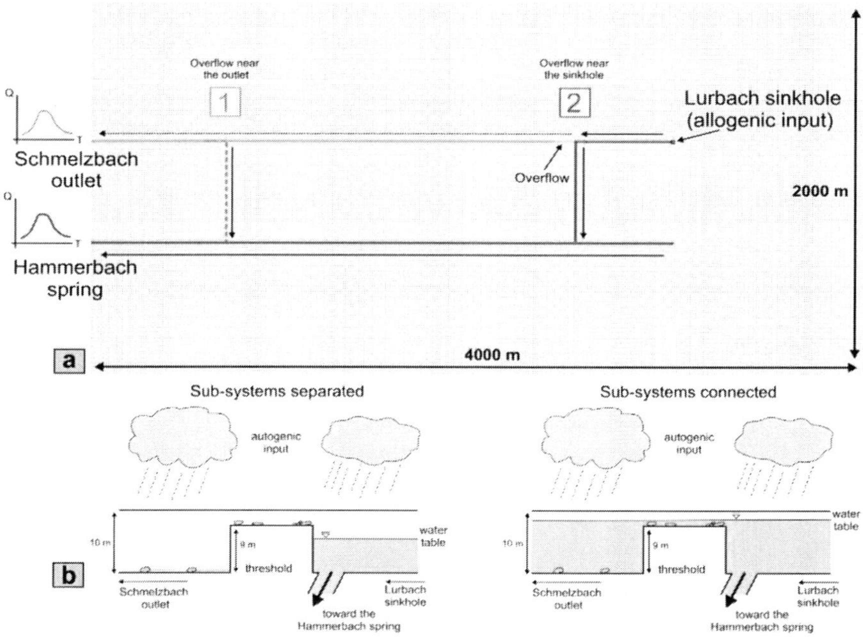

Figure. 2. Model setup. (a) Geometry of the MODFLOW model. The allogenic input is implemented using the Well package of MODFLOW in one cell representing the Lurbach sinkhole. The two different assumptions regarding the connection between the Hammerbach sub-catchment ...

The model setting is composed of a single unconfined layer with a dimension of 8 km² (4 km × 2 km), which provides a simplified, hypothetical representation of the autogenic part of the Lurbach system (Fig. 2a). The mesh size is set constant to 10 m × 10 m. The Lurbach allogenic input is introduced at a single point (Fig. 2a) using the Well package of MODFLOW, whereas autogenic recharge is given as a constant flux over the whole area. The model was built using the single-continuum approach (Sauter et al., 2006; Teutsch and Sauter, 1991) with two parallel cell lines of high hydraulic conductivity (representing the Schmelzbach and Hammerbach conduits) embedded in a low permeable matrix (see Fig. 2a). The conduit length was set to 3 km for each of the two conduits and is supposed to be close to the real length of the Schmelzbach and Hammerbach network sections (Behrens et al., 1992). The two outlets of the conduits were defined as constant head cells (to simulate the karstic springs) with a two meters difference of relative altitude (the Hammerbach spring being lower than the Schmelzbach outlet) and a linear distance of 550 m to each other. Similarly, the two conduits are connected to each other by a conduit of 550 m length with an overflow located after the first junction (Fig. 2b).

Two different assumptions with regard to the geographical location of the overflow are presented here (Fig. 2a). First, the overflow is located near the Schmelzbach outlet, whereas in the second case it is located close to the Lurbach sinkhole. These two extreme locations were chosen in order to obtain the maximum difference in the hydrological behavior of the springs. The overflow was simulated by elevating a hundred meter stretch of the bottom of the Schmelzbach conduit located just after the connection between the Hammerbach and Schmelzbach sub-systems (Fig. 2b). Within this part of the Schmelzbach conduit, the bottom of the cells was elevated by 9 m relative to the other model cells, such that flow through the conduit occurs only if a threshold water level is exceeded. The initial water table was defined below the overflow level to allow a separation of the two sub-systems at the beginning of the simulation. Then, the Wetting-Capability package of MODFLOW was used to allow rewetting of these cells and thus an activation of the overflow depending on the position of the water table (Fig. 2b). This implies that after a first stress period computed in steady-state the model simulation was transient with 384 stress periods of the same length (30 min), making a total simulation of 192 h. For both locations of the overflow between the two sub-systems, the values of constant head at the two outlets were adjusted such that the hydraulic gradient between the overflow and the Schmelzbach outlet remained approximately equal. First, the Schmelzbach and the Hammerbach catchment were assumed to consist of a homogeneous matrix and conduit system with one value of conductivity (1 m/s) in the conduits, one value in the matrix (10^{-5} m/s) and a constant specific yield for both matrix and conduit (0.01). Then, differences in the hydraulic conductivity and specific yield of the karstic conduits connecting to the two springs inferred from field observations were introduced, to account for heterogeneities of the two sub-systems.

3. RESULTS

3.1. Field site

During the one-month period shown in Fig. 3a, two precipitation events of different intensities occurred. The first was short and showed the stronger amplitude (maximum intensity of 5.75 mm per hour at the Ertlhube rain gauge) but the lower cumulative rainfall of 12.9 mm from the 30th November 2008 to the 2nd December 2008; the second had lower intensity (rainfall maximum of 2.6 mm per hour) but extended over a larger time span, resulting in a higher cumulative rainfall of 42.7 mm from the 10th December 2008 to 13th December 2008. Interestingly, no precipitation in form of snow was observed within this period. During these two events the air temperature recorded at the same station stayed mostly between 0 °C and 5 °C, although negatives temperatures were also reported for short periods.

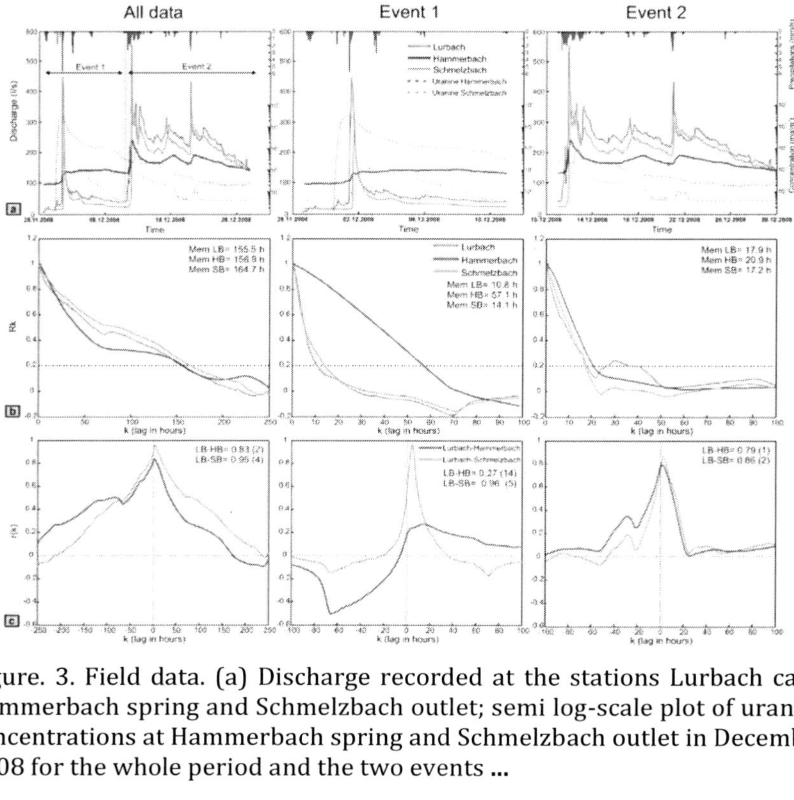

Figure. 3. Field data. (a) Discharge recorded at the stations Lurbach cave, Hammerbach spring and Schmelzbach outlet; semi log-scale plot of uranine concentrations at Hammerbach spring and Schmelzbach outlet in December 2008 for the whole period and the two events ...

The two precipitation events led to two hydrological events showing different spring responses (the first with a low baseflow, the second with a higher baseflow) recorded at the discharge gauging stations. The raw data of the gauging stations Lurbach cave, Hammerbach spring, Schmelzbach outlet and the precipitation data were transformed in hourly data in order to have an hourly scale for the time series analysis. A gap of approximately 30 h in the discharge dataset from the gauging stations Lurbach cave, Hammerbach spring and Schmelzbach outlet was filled with white noise based on the discharge recorded at Lurbach station which was complete.

The discharge and uranine concentrations show that a large part of the tracer was recorded at the Hammerbach spring a few hours before the first event was recorded at the station Lurbach cave (Fig. 3). Within this time period, no uranine was detected at the Schmelzbach outlet. This indicates that the Lurbach water was drained only toward the Hammerbach spring before the hydrological event. After the beginning of the first event, the tracer was still mainly recovered at the Hammerbach spring, and the breakthrough curve showed an undisturbed tailing. However, some small quantities of uranine were also recorded at the Schmelzbach outlet. This suggests that the overflow towards the Schmelzbach sub-system was activated due to the increasing water table during the event.

An interesting observation is that the overflow seemed to happen at lower discharge than the value of 200 l/s reported in Behrens et al. (1992), which agrees with the findings of Mayaud et al. (2013) and Wagner et al. (2013) that the Hammerbach spring exhibits a changed hydrological behavior since a major flood event happened in 2005. The shapes of the hydrographs further support the assumption of an activation of the overflow from the Hammerbach to the Schmelzbach sub-system because the hydrograph of the Schmelzbach appears to be similar to the hydrograph of the Lurbach, whereas the Hammerbach spring shows a damped response to this event and seems to drain mostly water from the aquifer storage.

Despite the lower maximum intensity of precipitation during the second event, the Hammerbach responds stronger to the resulting recharge pulse possibly due to the higher total amount of precipitation, which leads to an increase of the water table and the hydraulic gradient within the aquifer. The shape of the Schmelzbach hydrograph still appears to be almost identical to that of the Lurbach hydrograph. Uranine is still found in larger quantities at the Hammerbach spring but the concentration recorded at the Schmelzbach outlet responds faster than the Hammerbach to the recharge event. This suggests higher flow velocities and thus probably higher hydraulic conductivity of the Lurbach–Schmelzbach flow path and is in agreement with the assumption that the Schmelzbach still drained most of the Lurbach water during the second event when the overflow was already activated. The higher quantity of tracer recovered at the Hammerbach during the second event might be explained by a remobilization of uranine that was still stored within the Hammerbach sub-catchment down-gradient from the overflow.

Auto- and cross-correlations were computed for the three gauging stations Lurbach cave, Hammerbach spring and Schmelzbach outlet using the time series of the whole period (from the 28th November 2008 to the 30th December 2008 on the left hand side on Fig. 3), and for the two events separately (see Fig. 3 on the middle and right hand side). The two different events were defined taking into account the precipitation distribution and the baseflow discharge rather than the occurrence of discharge peaks. This allows the clear definition of two different periods: a first one with low baseflow (from the 28th November 2008 to the 11th December 2008) followed by a second one with a higher baseflow (from the 11th December 2008 to the 30th December 2008, respectively). As can be seen by comparing the results of auto- and cross-correlation of event 1 and event 2, the system behavior varied strongly during the period. During the first hydrological event, the Hammerbach autocorrelation indicates a damped behavior with a longer memory effect compared to the Lurbach and Schmelzbach autocorrelations. In contrast, the autocorrelation of the Schmelzbach is similar to that of the Lurbach. Correspondingly, the cross-correlation Lurbach cave–Schmelzbach outlet is higher in amplitude and shows a shorter lag time than the cross-correlation Lurbach cave–Hammerbach. This supports the idea that most of the Lurbach event-water was drained towards the Schmelzbach system after the overflow had been activated during the first event. The damped discharge of the Hammerbach is potentially explained by a larger storage capacity within this sub-catchment (Behrens et al., 1992).

When looking at the second event, the results from the time series analysis show a different trend: the Hammerbach autocorrelation function is very similar to those of the Lurbach and Schmelzbach. More specifically, the memory effect apparent in the autocorrelation of Lurbach and Schmelzbach (17.9 and 17.2 h in either case) is still similar to that of the first event (10.8 h and 14.1 h, respectively), whereas in the case of the Hammerbach it is clearly reduced from 57.1 h in the first event to 20.9 h in the second. This suggests that the aquifer storage of the Hammerbach sub-catchment is not able to attenuate the flood pulse of the second event to the same extent as in event 1, possibly because additional conduit pathways are activated at higher water levels (Kübeck et al., 2013). This further indicates that at this time, the Lurbach water pulse is transmitted toward the Hammerbach and Schmelzbach spring at a similar time-scale and that the two sub-catchments have more similar drainage behavior compared to event 1. This is also supported by the finding that the cross-correlation between Lurbach–Hammerbach and Lurbach–Schmelzbach at this time shows the same pattern (r_{xy} values of 0.79 and 0.86 with lags from 1 to 2 h respectively, opposed to r_{xy} values of 0.27 and 0.96 with lags of 14 and 5 h during the first event), which shows that the two sub-systems behave similar. Obviously, the different and varying behavior of the Hammerbach and Schmelzbach sub-catchments for these two particular events cannot be inferred from the autocorrelation and cross-correlation of the complete discharge dataset. In particular, the different behavior of Hammerbach and Schmelzbach during event 1 is not apparent if the events are not separated in the analysis (Fig. 3b and c, first column), as both springs then show similar memory effects (156 h for the Hammerbach, 164 h for the Schmelzbach) and similar cross-correlations with the Lurbach (0.83 amplitude and 2 h delay for the Hammerbach opposed to 0.95 in amplitude and 4 h delay for the Schmelzbach). Thus, single event analysis proves useful in this particular example, as it allows distinguishing the change of the global behavior of the Hammerbach sub-catchment during the two events.

3.2. Synthetic karst catchment

3.2.1. Homogeneous cases

Fig. 4 shows the modeling results of the synthetic karst catchment for the overflow located close to the Schmelzbach outlet (case 1 in Fig. 2a) and the overflow located near the Lurbach sinkhole (case 2 in Fig. 2a). The aquifer is unconfined and conduits and matrix are homogeneous, i.e. with the previously mentioned constant values of hydraulic conductivity and specific yield. In order to reproduce a hydrological situation similar to that occurring during the tracer experiment, two artificial allogenic recharge events were defined. The intensity of the allogenic input (Lurbach discharge) was increased from a first to a second event roughly by a factor of two (Fig. 4) in order to see the influence of event intensity on the overflow and discharge characteristics. Autogenic recharge remained constant during the whole simulation.

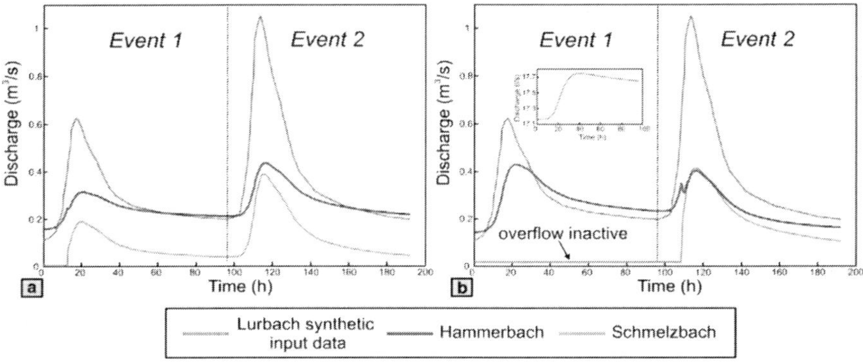

Figure 4. Results from the groundwater flow model: Simulated discharge response of Hammerbach and Schmelzbach due to two recharge events of different intensity. (a) Overflow located close to the Schmelzbach outlet (case 1 in Fig. 2a); (b) overflow located near ...

The comparison of the resulting responses of the synthetic Schmelzbach and Hammerbach hydrographs reveals an activation of the overflow similar to the observation at the field site: in the case of the overflow near the outlet (Fig. 4a) the Schmelzbach hydrograph stays at a constant value over a period of approximately 12.5 h during the first event when the Hammerbach has already started to respond to the Lurbach flood pulse by an increase in discharge. In the case of the overflow near the sinkhole (Fig. 4b), the Schmelzbach does not respond to the first Lurbach flood pulse, and responds later than the Hammerbach during the second event (after 108.5 h). Evidently, the intensity of the first event is too weak to activate the overflow in the case where it is located near the sinkhole (Fig. 4b). Yet, it should be noted that a slight increase in Schmelzbach discharge (see inset in Fig. 4b) is caused by the pressure propagation within the low-permeability matrix flow. It is further noteworthy that the peak discharge of the Schmelzbach slightly exceeds that of the Hammerbach for the second event when the overflow is located near the sinkhole (Fig. 4b), while it stays below in the case with the overflow near the Schmelzbach outlet (Fig. 4a). Likewise the baseflow of the Schmelzbach remains higher with the overflow near the sinkhole than with the overflow near the outlet. These observations are explained by the size of the autogenic sub-catchment of the Schmelzbach, which increases with increasing distance of the overflow from the outlet. Thus, the location of the overflow in the groundwater model is found to be of primary importance.

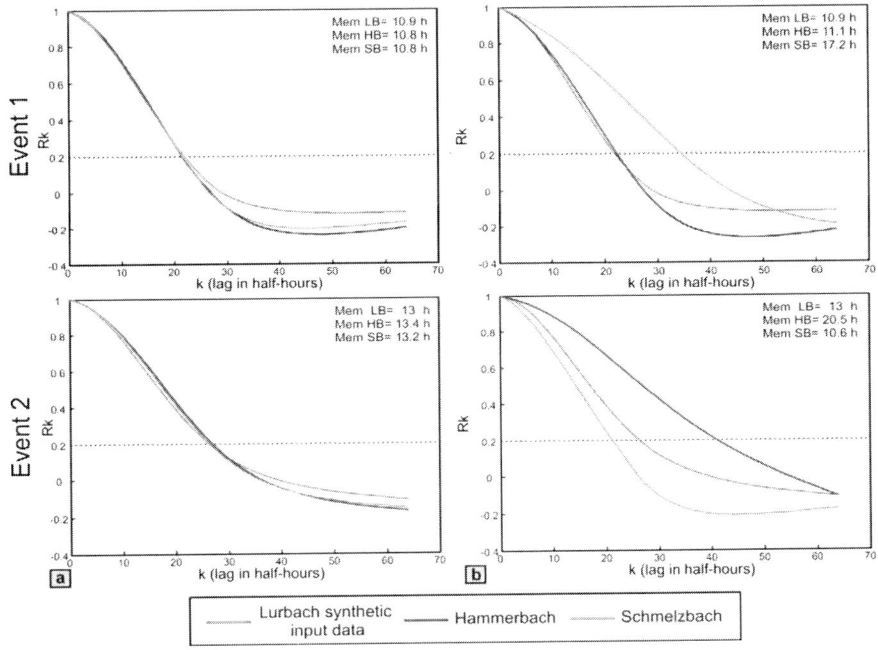

Figure 5. Autocorrelation functions of the simulated Lurbach (LB), Hammerbach (HB) and Schmelzbach (SB) discharges presented in Fig. 4 for the two separated events; (a) overflow near the outlet, (b) overflow near the sinkhole. Here the lags are given in half hours. ...

The autocorrelation of the hydrographs of the two models is presented in Fig. 5. With the overflow located near the outlet (Fig. 5a) the values of the memory effect for the Hammerbach and Schmelzbach are identical for both events and the values obtained for the first and the second event are similar (approximately 10.8 h during the first event; nearly 13 h during the second event). Contrary, the model with the overflow located near the sinkhole (Fig. 5b) shows different memory effects for Hammerbach and Schmelzbach and the memory effect is found to be different for the two events. The result from the first event can be explained by the inactivity of the overflow: almost all Lurbach allogenic water is drained towards the Hammerbach spring, whereas the Schmelzbach outlet is supplied only by a constant flux of autogenic water and a small amount of Lurbach water transferred through the matrix (see Inset in Fig. 4b). As a consequence, the memory effect for the Schmelzbach is higher than that for the Hammerbach. During the second event both Hammerbach and Schmelzbach are supplied by allogenic water from the Lurbach but still have different memory effects (11.1 h during the first event and 20.5 h during the second for the Hammerbach; 17.2 h during the first event and 10.6 h during the second for the Schmelzbach). As opposed to the first event the Schmelzbach responds faster (lower memory effect) than the Hammerbach. This is due to the sudden activation of the overflow forcing a high proportion of Lurbach water to flow towards the Schmelzbach system. Similarly, the memory effect of the real Schmelzbach was found to be lower than that of the real Hammerbach when the overflow was activated in the first event observed at the field site (see event 1 in Fig. 3b). The above findings thus demonstrate that the memory effect is influenced by the overflow. Yet, it should be noted that differences in the memory effects of different events can also be caused by different input signals, which suggests that the interpretation of the memory effect in a setting with strongly varying input signals is not straightforward and should be supported by additional evidence from other methods.

Single event cross-correlation results obtained with the two model set-ups are shown in Fig. 6. Corresponding to the results from the autocorrelation, the Lurbach–Hammerbach and Lurbach–Schmelzbach cross-correlation functions are similar in both events (more than 0.90 in amplitude for all cases) if the overflow is located near the outlet (Fig. 6a). In contrast, the model with the overflow near the sinkhole (Fig. 6b) shows a lower cross-correlation amplitude for the Lurbach–Schmelzbach during the first event when the overflow is not activated and a higher amplitude reaching almost the Lurbach–Hammerbach cross-correlation during the second event. Yet even during the second event the amplitude of the Lurbach–Hammerbach cross-correlation is higher than that of the Lurbach–Schmelzbach cross-correlation. In addition, the maximum of the lag time of the Lurbach–Schmelzbach cross-correlation is larger than that of the Lurbach–Hammerbach cross-correlation (14.5 h opposed to 3 h during the first event and 2.5 h opposed to 1 h during the second event). These findings are explained by the activation of the overflow during the second event, which causes a delayed response of the Schmelzbach relative to the Hammerbach. This is in striking contrast to the observation at the field site, where the real Lurbach–Schmelzbach cross-correlation exhibits a higher amplitude and lower lag time than the real Lurbach–Hammerbach cross-correlation when the overflow is activated (see event 1 in Fig. 3c). Thus, the autocorrelation of the observed hydrograph and that obtained with the model where the overflow is close to the sinkhole show some similarity (memory effect of Schmelzbach lower than that of Hammerbach when the overflow is activated), but the cross-correlation from the first event at the field site clearly suggests a more attenuated and damped response of the Hammerbach than that obtained with the model. In the above considered model scenarios the matrix and the conduit system were assumed to be homogeneous. As karst aquifers are highly heterogeneous an important question is how heterogeneities may influence the numerical spring response and the resulting shape of the auto- and cross-correlation functions. This is considered in the following sub-section.

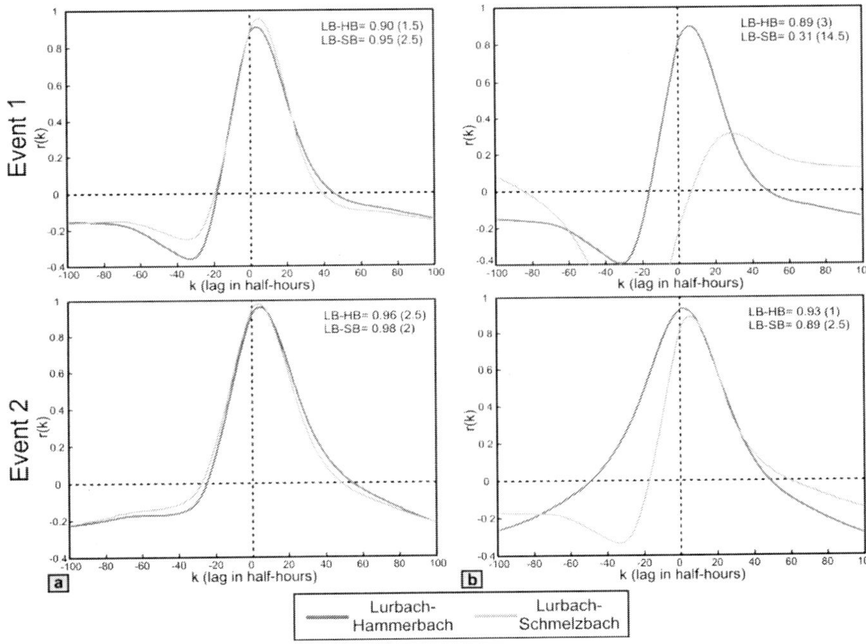

Figuer 6. Cross-correlation functions of the simulated Lurbach–Hammerbach (LB–HB) and Lurbach–Schmelzbach (LB–SB) discharges presented in Fig. 4 for the two separated events; (a) overflow near the outlet, (b) overflow near the sinkhole. ...

3.2.2. Heterogeneous cases

As the Schmelzbach system is assumed to be higher karstified than the Hammerbach system (Behrens et al., 1992), heterogeneities were introduced in the groundwater model by increasing the conduit hydraulic conductivity from 1 m/s to 1.2 m/s from the overflow location to the Schmelzbach outlet. Moreover, as the Hammerbach is reported to have the higher aquifer storage (Behrens et al., 1992), the value of specific yield of the conduit cells within this sub-catchment was consequently increased from 0.01 to 0.5.

Fig. 7 presents a comparison of hydrographs, auto- and cross-correlation functions between the heterogeneous case (dashed lines) and the homogeneous one (solid lines) for the overflow located near the outlet (Fig. 7a) and the overflow located near the sinkhole (Fig. 7b). Auto- and cross-correlation are shown only for the second event. For both geometries the Hammerbach hydrograph is evidently more damped in the heterogeneous than in the homogeneous case. In addition, the Hammerbach response in the heterogeneous model appears to be delayed compared to the homogeneous case if the overflow is located near the sinkhole (but not when the overflow is near the outlet). Another important result in the heterogeneous model with the overflow near the sinkhole is the activation of the overflow during the first event (after 14 h) whereas it remained inactive for the homogeneous case. This result is consistent with the lower value of 12.5 h found for the overflow located near the outlet.

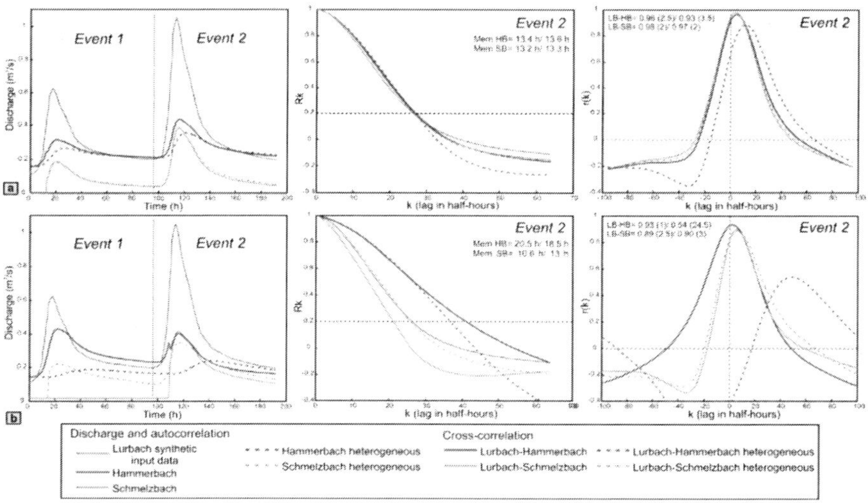

Figure 7. omparison of discharge, auto- and cross-correlation from the homogeneous model (solid lines) and the heterogeneous one (dashed lines); (a) overflow near the outlet, (b) overflow near the sinkhole. While matrix and conduit parameters of the Schmelzbach **...**

The autocorrelation for the model with the overflow located near the outlet shows almost no influence of the heterogeneities and yields similar memory effects for Hammerbach and Schmelzbach than those of the homogeneous case. Contrary, the model with an overflow located near the sinkhole shows different memory effects for the Hammerbach and the Schmelzbach and the values differ from those of the homogenous case (18.5 and 13 h as opposed to 20.5 and 10.6 h for the homogeneous case).

The results of the cross-correlation analysis correspond to those of the autocorrelation: the heterogeneous model with the overflow located near the outlet has cross-correlograms similar (Lurbach–Schmelzbach) or slightly damped and delayed (Lurbach–Hammerbach) compared to the homogeneous case. Evidently, the heterogeneity has little influence on the spring response if the overflow is close to the outlet. In this case the greater part of the aquifer is upstream of the overflow and thus shared by the two springs, while there is only a small part downstream of the overflow with two separate heterogeneous sub-catchments. In contrast, the model with the overflow located near the sinkhole is strongly influenced by the heterogeneities: the Lurbach–Hammerbach cross-correlation is clearly damped, whereas the Lurbach–Schmelzbach cross-correlation is only slightly changed. Thus, the Schmelzbach is the main outlet of the system and has a more flashy behavior than the Hammerbach, which corresponds well with the field observations and further supports a location of the overflow in the upper part of the aquifer.

Results from the aforementioned tracer experiment (Fig. 3a) also suggest an overflow location close to the sinkhole: since most of the tracer was recovered at the Hammerbach spring and even in the time period when the overflow was active only a small amount of tracer was detected in the Schmelzbach, the overwhelming majority of tracer must have passed the overflow location at the time when the overflow became active. In agreement with the results of the single event time series analysis these findings support an overflow location within the upper part of the Lurbach aquifer.

4. CONCLUSION

Single event time series analysis was combined with a groundwater model to examine how the inter-catchment flow in a karst aquifer varies during a period of 1 month and how this is reflected in the results from the time series analysis. Auto- and cross-correlation of the observed data differ for the two events considered here, thus showing the necessity to make a single event analysis rather than an analysis of the whole dataset. The numerical model implemented with MODFLOW was able to reproduce the general overflow behavior observed in the Lurbach system. Results of auto- and cross-correlation of the numerical model showed that aquifer heterogeneities and the overflow location were of primary importance. The model with an overflow located near the outlet was found to be relatively insensitive to a variation of hydraulic parameters. Contrary, the model with an overflow located near the sinkhole showed a high sensitivity to heterogeneities and was selected as the most probable option to better reproduce the observed behavior of the Lurbach system. Results from the model with the overflow near the sinkhole and heterogeneous aquifer parameters were found to be in good agreement with the field observations. Thus, in agreement with evidence from tracer tests an overflow location in the upper part of the aquifer rather than in the lower part is suggested. In summary, single event time series analysis was found to be useful for characterizing transient inter-catchment flow and aquifer properties (overflow location, aquifer heterogeneity) controlling the spring responses to recharge in this karst catchment. Yet, it is important to note that results from time series analysis need to be complemented by other aquifer characterization techniques to interpret them in terms of flow processes and aquifer properties. In this work, tracer testing and groundwater modeling proved useful for this purpose.

ACKNOWLEDGEMENTS

This project was supported by the Austrian Science Fund (FWF): L 576-N21 and the Austrian Academy of Sciences (ÖAW) Project: Global models of spring catchments. The authors would like to thank Gerfried Winkler and Stefan Oswald for the fruitful discussions and the owners of the Lurgrotte Peggau and Semriach for the access to the cave. The Hydrographische Dienst Steiermark is thanked for providing the meteorological data. Comments on the manuscript by three anonymous reviewers are gratefully acknowledged.

REFERENCES

1. Amraoui F., Razack M., Bouchaou L. Turbidity dynamics in karstic systems. Example of Ribaa and Bittit springs in the Middle Atlas (Morocco) Hydrol. Sci. J. 2003;48(6):971–984.

2. Bakalowicz M. Karst groundwater: a challenge for new resources. Hydrogeol. J. 2005;13:148–160.

3. Bailly-Comte V., Jourde H., Roesch A., Pistre S., Batiot-Guilhe C. Time series analyses for karst/river interactions assessment: case of the Coulazou River (southern France) J. Hydrol. 2008;349(1–2):98–114.

4. Bailly-Comte V., Martin J.B., Screaton E.J. Time variant cross correlation to assess residence time of water and implication for hydraulics of a sink-rise karst system. Water Resour. Res. 2011;47:W05547.

5. Behrens, H., Benischke, R., Bricelj, M., Harum, T., Käss, W., Kosi, G., Leditzky, H.P., Leibundgut, Ch., Małoszewski, P., Maurin, N., Rajner, V., Rank, D., Reichart, B., Stadler, H., Stichler, W., Trimborn, P., Zojer, H., Zupan, M., 1992. Investigations with Natural and Artificial Tracers in the Karst Aquifer of the Lurbach System (Peggau-Tanneben-Semriach, Austria); Steir Beitr z, Hydrogeologie 43, Graz.

6. Bouchaou L., Mangin A., Chauve P. Turbidity mechanism of water from a karstic spring: example of the Ain-Asserdoune spring (Beni Mellal Atlas, Morocco) J. Hydrol. 2002;265:34–42.

7. Budge T.J., Sharp J.M. Modeling the usefulness of spatial correlation analysis on karst systems. Ground Water. 2009;47(3):427–437.

8. Covington M.D., Wicks C.M., Saar M.O. A dimensionless number describing the effects of recharge and geometry on discharge from simple karstic aquifers. Water Resour. Res. 2009;45:W11410.

9. Covington M.D., Banwell A.F., Gulley J., Saar M.O., Willis I., Wicks C.M. Quantifying the effects of glacier conduit geometry and recharge on proglacial hydrograph form. J. Hydrol. 2012;414–415:59–71.

10. Eisenlohr L., Bouzelboudjen M., Király L., Rossier Y. Numerical versus statistical modelling of natural response of a karst hydrogeological system. J. Hydrol. 1997;202(1–4):244–262.

11. Ford D., Williams P. John Wiley & Sons Ltd.; 2007. Karst Hydrogeology and Geomorphology.

12. Geologische Bundesanstalt, 2005. Provisorische Geologische Karte der Republik Österreich, 164-Graz. – Geofast 1:50.000, Geologische Bundesanstalt, Wien.

13. Genthon P., Bataille A., Fromant A., D'Hulst D., Bourges F. Temperature as a marker for karstic waters hydrodynamics. Inferences from 1 year recording at La Peyrére cave (Ariège, France) J. Hydrol. 2005;311:157–171.

14. Geyer T., Birk S., Reimann T., Dörfliger N., Sauter M. Differentiated characterization of karst aquifers: some contributions. Carbonate Evaporite. 2013;28(1–2):41–46.

15. Harbaugh, A.W., 2005. MODFLOW-2005, The U.S. Geological Survey Modular Ground-Water Model – The Ground-Water Flow Process: U.S. Geological Survey Techniques and Methods 6-A16, variously p.

16. Harum, T., Stadler, H., 1992. Hydrologic and Climatologic Conditions, In: Behrens, H., Benischke, R., Bricelj, M., Harum, T., Käss, W., Kosi, G., Leditzky, H.P., Leibundgut, Ch., Małoszewski, P., Maurin, N., Rajner, V., Rank, D., Reichart, B., Stadler, H., Stichler, W., Trimborn, P., Zojer, H., Zupan, M., (Eds.), 1992. Investigations with Natural and Artificial Tracers in the Karst Aquifer of the Lurbach System (Peggau-Tanneben-Semriach, Austria); Steir Beitr z, Hydrogeologie 43, Graz.

17. Jeannin P.Y., Sauter M. Analysis of karst hydrodynamic behaviour using global approaches: a review. Bull. Hydrogéol. (Neuchâtel) 1998;16:31–48.

18. Jemcov I., Petrič M. Time serie analysis, modeling and assessment of optimal exploitation of the Nemanja spring, Serbia. Acta Carsologica. 2010;39(2):187–200.

19. Kovačič G. Hydrogeological study of the Malenščica karst spring (SW slovenia) by means of a time series analysis. Acta Carsologica. 2010;39(2):201–215.

20. Kresic N., Stevanovic Z. Elsevier; 2010. Groundwater Hydrology of Springs: Engineering, Theory, Management, and Sustainability.

21. Kübeck C., Małoszewski P., Benischke R. Determination of the conduit structure in a karst aquifer based on tracer data – Lurbach system, Austria. Hydrol. Process. 2013;27:225–235.

22. Larocque M., Mangin A., Razack M., Banton O. Contribution of correlation and spectral analyses to regional study of a large karst aquifer (Charentes, France) J. Hydrol. 1998;205(3–4):217–231.

23. Larocque M., Banton O., Razack M. Transient-state history matching of a karst aquifer groundwater flow model. Ground Water. 2000;38(6):939–946.

24. Mangin A. Pour une meilleure connaissance des systèmes hydrologiques à partir des analyses corrélatoire et spectrale. J. Hydrol. 1984;67:25–43.

25. Massei N., Dupont J.P., Mahler B.J., Laignel B., Fournier M., Valdes D., Ogier S. Investigating transport properties and turbidity dynamics of a karst aquifer using correlation, spectral, and wavelet analyses. J. Hydrol. 2006;329:244–257.

26. Mayaud C., Wagner T., Benischke R., Birk S. Understanding changes in the hydrological behaviour within a karst aquifer (Lurbach system, Austria) Carbonate Evaporite. 2013 doi: 10.1007/s13146-013-0172-3.

27. Oswald, S., 2009. Künstliche und natürliche Tracer in einem Karsteinzugsgebiet mit allochthoner Neubildung (Lurbach, Österreich). Karl-Franzens-University of Graz; unpubl. Bachelor thesis (in German).

28. Panagopoulos G., Lambrakis N. The contribution of time series analysis to the study of the hydrodynamic characteristics of the karst systems: application on two typical karst aquifers of Greece (Trifilia, Almyros Crete) J. Hydrol. 2006;329:368–376.

29. Pinault J.L., Plagnes V., Aquilina L., Bakalowicz M. Inverse modeling of the hydrological and the hydrochemical behavior of hydrosystems: characterization of karst system functioning. Water Resour. Res. 2001;37(8):2191–2204.

30. Ravbar N., Engelhardt I., Goldscheider N. Anomalous behaviour of specific electrical conductivity at a karst spring induced by variable catchment boundaries: the case of the Podstenjšek spring, Slovenia. Hydrol. Process. 2011;25:2130–2140.

31. Rehrl C., Birk S. Hydrogeological characterization and modelling of spring catchments in a changing environment. Aust. J. Earth Sci. 2010;103(2):106–117.

32. Sauter, M., Kovács, A., Geyer, T., Teutsch, G., 2006. Modellierung der Hydraulik von Karstgrundwasserleitern – Eine Übersicht. Grundwasser – Zeitschrift der Fachsektion, Hydrogeologie 3/2006. doi: 10.1007/s00767-006-0140-0.

33. Schubert, G., 2003. Hydrogeological Map of Austria 1: 500.000. Geologische Bundesanstalt, Wien.

34. Shoemaker, W.B., Kuniansky, E.L., Birk, S., Bauer, S., Swain, E.D., 2008. Documentation of a Conduit Flow Process (CFP) for

MODFLOW-2005. U.S. Geological Survey Techniques and Methods Book 6, Chapter A24. Reston, Virginia: USGS.

35. Teutsch, G., Sauter, M., 1991. Groundwater modelling in karst terranes: scale effects, data acquisition and field validation. In: Proc. 3rd Conference on Hydrogeology, Ecology, Monitoring and Management of Ground Water in Karst Terranes, Nashville, December 1991.

36. Valdes D., Dupont J.P., Massei N., Laignel B., Rodet J. Investigation of karst hydrodynamics and organization using autocorrelation and *T*-curves. J. Hydrol. 2006;329:432–443.

37. Wagner, T., Fritz, H., Stüwe, K., Nestroy, O., Rodnight, H., Hellstrom, J., Benischke, R., 2011. Correlations of cave levels, stream terraces and planation surfaces along the River Mur—Timing of landscape evolution along the eastern margin of the Alps. Geomorphology 134, 62–78.

6

CHAPTER

MEASURING AUTOGENIC RECHARGE OVER A KARST AQUIFER UTILIZING EDDY COVARIANCE EVAPOTRANSPIRATION

Nico M. Hauwert[1], John M. Sharp[2]

[1]*City of Austin Watershed Protection Department, Austin, USA*

[2]*Jackson School of Geosciences, University of Texas at Austin, Austin, USA*

ABSTRACT

Autogenic, or direct aquifer recharge can best be measured as the remainder of a water balance utilizing precise measurement of precipitation, evapotranspiration (ET) and runoff. ET is the largest component of a precipitation water balance and can be measured within 5% using an eddy covariance system with Bowen-ratio energy balance corrections. Water balance components of precipitation, evapotranspiration, internal runoff, soil moisture were measured using a eddy covariance system, tipping bucket and visual rain gauges, flumes, and soil-moisture sensors. The research site was located within a 0.19-km^2 (46-acre) internal drainage sinkhole basin where runoff never flows beyond the basin, but potentially reaches a cave serving as a drain to the sinkhole. Other than the cave drain, the basin slopes are indistinguishable from other slopes across the Barton Springs Segment of the Edwards Aquifer. Over a 505-day water balance interval where change in soil moisture was negligible and precipitation was 42% above average, ET was 68% of precipitation, discrete internal runoff was 6%, and remaining component of diffuse autogenic recharge was measured as the residual of total rainfall as 26% of rainfall. Over a longer period of average rainfall, internal runoff diminished to 3%, but was as high as 42% of precipitation during single storms when the soils were near saturation. These results closely match results from a fiveyear water balance over the Trinity Aquifer of Central Texas where ET was measured to be 65% of precipitation using a Bowen-ratio climate tower, runoff was measured to be 5% of precipitation, and recharge was calculated as the residual at 30% of rainfall. ET flux tower data from other sites across Central Texas indicate that under average precipitation conditions, autogenic recharge is about 28% and intervening recharge area runoff is about 3% of precipitation. During years of higher than average precipitation, authogenic recharge and intervening recharge area runoff combined increase within the range of 30% to 45% of precipitation.

Keywords: Evapotranspiration, Recharge, Edwards Aquifer, Karst, Water Balance

1. INTRODUCTION

In order to preserve increasingly scarce water supplies, understanding of recharge processes is vital. Recharge occurs below the surface where it is difficult or impossible to directly observe and measure. Recharge can be measured indirectly by accurately accounting for water balance components of precipitation or stream-flow loss.

Karst is a landscape underlain by soluble rock that typically contains large, focused springs and solution features such as caves, sinkholes, and sinking streams. Karst aquifers, such as the Edwards Aquifer, are some of the most productive water resources in the world. The rainfall to recharge cycle involves several steps including:

1) Evaporation of water on the surface, atmosphere, and plant surfaces (interception)2) Infiltration of rainfall and runoff into the soils, a portion of which is utilized by plants within their growing structure or is released to the atmosphere as evaporation and transpiration or interflow which repeats the runoff cycle3) Percolation of a portion of infiltrated water into

the bedrock and unsaturated zone, a portion of which may be temporarily stored over in perched pools or matrix4) Recharge of percolated water into the water table of the aquifer.

Components of natural surface recharge of a karst area include:

1) Allogenic recharge originating as flow from upstream source areas percolating within the major creek channels2) Autogenic (also authogenic and direct) recharge originating from direct precipitation over the recharge area of the karst aquifer.

a) Diffuse recharge that infiltrates through soil and small solution features and percolate into the bedrock of the intervening recharge area.

b) Discrete internal runoff (also internal drainage) recharge where runoff directly percolates into the bedrock of the intervening recharge area, typically through solution-enlarged sinkhole drains.

c) Major creek channel autogenic recharge is runoff originating as precipitation over the recharge area of a karst aquifer that percolates into swallets and other recharge features of the major creek channels.

3) Other natural leakage recharge sources potentially include leakage through overlying confining units or adjacent aquifers.

The amount of precipitation that recharges an aquifer, or recharge rate, is typically expressed as a percent or fraction of precipitation and is commonly measured using stream-flow loss, spring-flow hydrographs, isotopes, and chloride concentrations. The accuracy of recharge rate measurement from stream-flow loss depends on the accuracy of discharge measurements and precise delineation of groundwater basins contributing to spring discharge under varying hydrologic conditions. Recharge rate measurement using chloride assumes sufficiently precise characterization of average chloride concentrations in rainfall and groundwater and the assumption that significant chloride concentrations solely originate from rainfall—an assumption that is difficult to insure in modern marine settings and in rocks deposited in past marine settings [1] -[5] . In addition, the addition of a small percent of highly concentrated chemical or isotopic composition, such as a brine or leakage from an adjacent aquifer may be misinterpreted to indicate that a karst aquifer has much older age than a volumetric averaged age of sources would indicate. In general, an incomplete understanding of a karst aquifer tends to result in a bias of underestimating the recharge rate [6] .

Studies utilizing a variety of methods across the world typically demonstrate that a relatively large portion of precipitation recharges karst aquifers [7] -[9] . A study of the Carboniferous Limestone in Mendip Hills,

England found that no downstream runoff was generated and that 54% of precipitation recharged the aquifer [10] . Dublyanskii et al. [11] used stream flow loss to measure recharge amounting to 33% of precipitation in Crimea, Ukraine. In Guam, direct recharge amounted to 60% [12] to 67% [13] of precipitation. In Tunisia, Tixeront et al. [14] measured that 16% to 49% of annual precipitation recharged a Senonian monocline and two Jurassic limestone horsts over four years. Schoeller found that the recharge rate of a 5.8 km^2 Tunisian Eocene limestone syncline groundwater basin to Kef Spring varied annually over eight years from 30% to 90% depending on the amount of precipitation, intensity of rainfall and groundwater-level conditions [15] . Over 19 km^2 of the Djebel Bargou Aptial limestone syncline of Tunisia, Zebidi [16] measured 18% recharge rate over two years. Mero found that 53% of average precipitation recharged the 200 km^2 Cretaceous limestone aquifer discharging from Na'aman Spring in Israel over 30 years [17] . Goldschmidt found that 10% to 45% of rainfall recharged karst limestone of West Galilee of Israel [18] . Mandel and Shiftan indicated that 37% of precipitation recharged a Turonian/Cenomanian limestone aquifer of Israel [19] . In the 5123 km^2 Damascus basin of Syria, Burdon measured recharge as 41% of precipitation [20] . Voûte measured that 45% of precipitation recharged the Ghabbasin of Syria [21] . A recent study by Al-Charideh utilized the isotopic compositions of $\delta^{18}O$, $\delta^{2}H$, and chloride concentrations from 2001 to 2009 and found that 43% and 67% of precipitation recharged the Mountain karst aquifer and Figeh Spring of Damascus [22] . Hoetzl reports that in karst areas of Saudia Arabia, 47% of precipitation recharges [23] . In karst areas of Greece, Aronis et al. measured that an average of 52% of precipitation recharged a 95 km^2 groundwater basin to the five main Lialaia Springs over 20 years [24] . Burdon and Papakis measured that 51% of precipitation recharged the adjacent Parnassos-Ghiona carbonate aquifer of Greece [25] .

The Edwards Aquifer consists of several segments, including the San Antonio Segment, the Barton Springs Segment, the Northern Segment, and aerially extensive Edwards Plateau. The relatively compact size of Barton Springs Segment provides a suitable study area of the Edwards Aquifer. The Barton Springs Segment has 241 km^2 (94 mi^2) recharge area containing outcrop of the Edwards Aquifer, eastern drainage areas where overlying confining units drain runoff back to the outcrop area, as well as isolated topographically high or down faulted blocks of overlying confining units. The recharge area of the Barton Springs Segment that discharges at Barton Springs is a smaller subset of the Barton Springs Segment, 212 km^2 (82 mi^2) [26] . After eliminating areas covered with Del Rio Clay and other overlying confining units, the Edwards Aquifer outcrop area contributing to Barton Springs is only 176 km^2 (68 mi^2). A 30 km^2 (12 mi^2) recharge area of the Barton Springs Segment provides discharge to Cold Springs [26] —a group of springs partially submerged below the Colorado River that cannot be directly gauged [27] . The recharge area can be further distinguished into major creek channels and intervening (upland) area between the major creek channels. A 680 km^2 (264 mi^2) contributing area, underlain primarily by rocks of the Trinity Group, provides an upstream catchment of allogenic recharge for the major creek channels crossing the Barton Springs Segment recharge area.

Recharge to Central Texas karst aquifers has also been reported. Prior to the 1970's, because of the surface morphological characteristics and high transmissivity of the Edwards Aquifer, geologists assumed that a relatively high portion of rainfall recharged the aquifer [28] . Garza applied a flow loss water balance for the San Antonio Segment of the Edwards Aquifer that calculates potential recharge within a stream basin as the net difference in upstream and downstream gauging stations in addition to estimated runoff generated within the intervening area [29] . Slade et al. applied this method to estimate the amount of stream (85%) and intervening (15%) recharge to Barton Springs [30] , prior to delineation of groundwater basins using groundwater tracers that was later conducted. Using the Slade et al. (1986) results, Woodruff created a water balance to calculate that 85% of rainfall was lost through evapotranspiration, 5% of rainfall recharged in the major stream channels, and that 0.89% of rainfall recharged within the intervening upland areas [31] . The amount of rainfall attributed to direct recharge over the Edwards Aquifer (0.89%) was less than the 1% of rainfall that was estimated to percolate into the Eagle Ford Shale over a one-year interval in Central Texas [32] . Slade utilized the same presumed source areas and stream flow data from his earlier 1979 to 1983 study and determined that intervening areas contribute a maximum of 25% of total recharge and that this recharge constituted a maximum of 6.6% of precipitation using the same assumed groundwater basins [33] . Groundwater tracing conducted since 1996 has significantly refined groundwater basins contributing to Barton Springs [26] . Errors in groundwater basin delineation can account for the relatively low authogenic recharge to rainfall ratios calculated by the earlier recharge studies [26] .

Site specific water balances utilizing climate towers in Central Texas have more accurately measured evapotranspiration at site scale using eddy covariance or Bowen ratio systems along with weirs and flumes to measure surface runoff. Over a 5-year water balance of the Trinity Aquifer in Uvalde County, Dugas et al. used Bowenratio tower to calculate that 65% of rainfall evaporated or transpired, 5% of rainfall flowed as runoff into creeks, and 30% of rainfall recharged the underlying aquifer [34] . This study included two paired microbasins of Seco Creek, one of which was cleared of juniper, in order to examine the short-term effects of vegetation on the water budget. Based on Dugas et al. [34] results, Huang & Wilcox believed the recharge rate of the Edwards Aquifer must be greater than 10% of rainfall and that a stream-flow loss study in the same watershed significantly underestimated actual recharge [35] .

Slattery et al. reported rainfall and evapotranspiration on uplands adjacent to the Honey Creek tributary to the Guadalupe River using rain gauges and two Bowen-ratio climate towers [36] . The Honey Creek site is underlain by the Trinity Aquifer. Over 342 days, from March 19, 2002 to February 24, 2003, 1531 mm of precipitation and 632 mm of ET were measured, yielding an ET/precipitation ratio of 44%. Banda and Slattery later reported average values for 2002-2004 for two microbasins of ET as 78% and 81% of rainfall, runoff as 5% rainfall of rainfall, and a recharge rate as 14% and 17% of rainfall [37] . After clearing one microbasin of juniper, the untreated control averaged 85% ET, 2% runoff, and recharge rate of 13% of rainfall. Over the same six-year interval, the cleared microbasin had an ET rate of 74% of rainfall, runoff 2% of rainfall, and a recharge rate of 24% of rainfall [37] .

Heilman et al. reported five years of eddy covariance-measured evapotranspiration and rainfall over the Edwards Aquifer within the Blanco River watershed [38] . ET ranged from 57% to 112% of precipitation, averaging 76% of precipitation in the woodland station, and 79% of precipitation in the savanna station. ET values greater than 100% may be possible during years of low rainfall as moisture stored in drying soils and plants are released.

In this study, at site scale we measure the water balance components of rainfall, including runoff, soil moisture storage, and ET, to calculate recharge as the residual of precipitation. Each of the water balance components can be measured with high accuracy at site scale, consequently providing the most precise measurement of recharge. Site-specific ET and runoff data will then be compared with annualized data from other sites across Central Texas in order to provide recharge as a percent of rainfall under a range of precipitation conditions for the Edwards Aquifer.

2. METHODOLOGY

2.1. Study Area

Headquarters Flat Sinkhole Research station (HQ Flat) is a 0.19 km^2 (46-acre) internal-drainage sinkhole basin located on the "J-17" City of Austin

Water-Quality Protection Land within the Barton Springs Segment of the Edwards Aquifer (Figure 1).

Aside from the terminal runoff drain at HQ Flat Cave, the slopes of the basin are otherwise are indistinguishable from slopes outside the basin. About 10% of the recharge area for the Barton Springs Segment is contained within similar internal drainage microbasins, from which runoff rarely, if ever, flows to local creeks [26] .

2.2. Data Collection

Within a closed sinkhole basin of known area over a selected time interval, the amount of percolation into the bedrock can be calculated as:

$$I = P - ET + DR + S \quad (1)$$

where:

P = precipitation volume (m^3);

ET = evapotranspiration flux volume (m^3);

I = percolation from soil and lesser karst features into the underlying bedrock (diffuse recharge, m^3);

DR = internal runoff that enters the cave drain (discrete recharge, m^3);

S = change in water storage in soil during the reference period (m^3).

A 505-day period, from April 2, 2004, to August 20, 2005, was selected for determining the water balance where starting and ending volumetric soil moisture was measured to be 20% ± 2%. Because percolation occurs below the expected extinction depth beyond significant ET occurs, for the purposes of this study it is assumed that all percolated groundwater becomes recharge. The water table of the Edwards Aquifer is about 60 m (200 ft) deep at the research site.

Precipitation was measured using funnel cap plastic All-Weather rain gauges. These gauges have measurement precision of 0.0254 cm (0.01-inch) and are designed to reduce evaporative loss. In addition, three 0.0254 cm (0.01-inch) Rainwise tipping bucket rain gauges with HOBO data loggers continuously measured precipitation. Tipping-bucket rain gauges are prone to plugging from bird droppings and require careful maintenance. The Rainwise tipping buckets required a small calibration shift, based on verification with a graduated cylinder as well as correlations among other stations and visual rain gauges. Monthly totals of rainfall were compared at all gauges to find and correct potential data gaps.

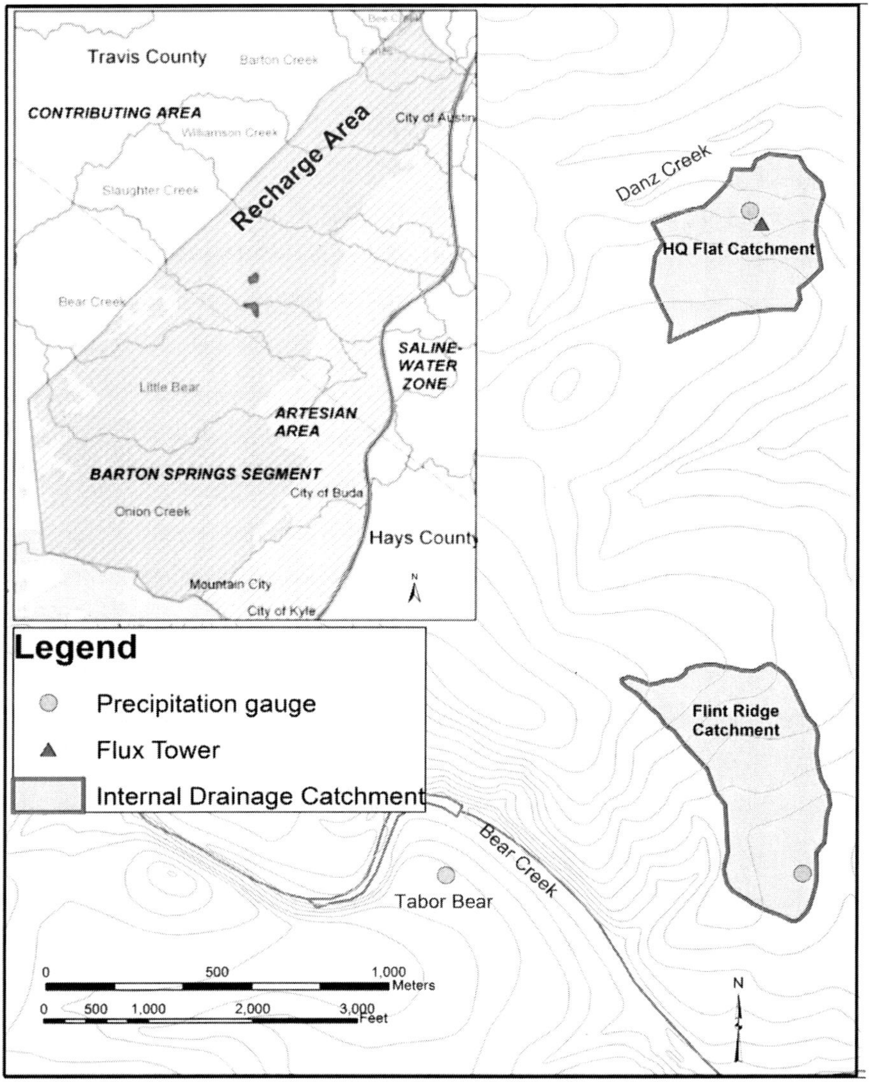

Figure 1. Location of study site (HQ Flat sinkhole microbasin is the primary water balance research site. Comparison rainfall and runoff was measured in Flint Ridge sinkhole catchment).

Flow is directed to HQ Flat Cave from 94% of its catchment area along a single channel which was outfitted with a 0.6 m (2 ft) aluminum H flume. An ISCO 3230 flow meter was used to continuously measure flow entering HQ Flat Cave. Outside of internal drainage basins this flow may be representative of runoff entering the major creeks. A second large internal drainage basin sinkhole, Flint Ridge (Figure 1), was also monitored for flow within its three terminal drainages in order to evaluate the representativeness of internal runoff measurements in other sinkhole microbasins.

Soil-moisture sensors measured soil moisture continuously near the ET flux tower. Continuous measurements of water content at 2.5 cm, 5 cm, and 9 cm depths were made using Campbell Scientific water content reflectometer probes (CS615) installed horizontally. Soil temperature and heat flux were measured continuously at depths of 2.5 and 10 cm using thermocouple probes and heat flux plates. Heat flux plates were installed at a depth of 5 cm, and storage heat flux in the zero to 5 cm layer was calculated from both soil temperature measurements at 2.5 cm depth and estimates of heat [39] . Gravimetric soil moisture samples were collected at various sites and depths near the ET flux tower over various surface moisture conditions, as described by Black [40] , in order to calibrate the soil-moisture sensors. The water balance interval was selected so that volumetric soil moisture was the same at the start and end, so that there is no significant change in soil moisture storage.

ET was measured nearly continuously using eddy covariance instruments placed 15-m high and above the tree canopy on a Rohn tower. Eddy covariance is the most direct method of measuring water vapor flux [41] . Measurements of wind speed and acoustic temperature were made with a three-dimensional sonic anemometer (CSAT-3, Campbell Scientific), and measurements of water vapor concentration were made with a Li-Cor LI-7500 open path infra-red gas analyzer at ten-hertz intervals. The Li-Cor LI-7500 was factory calibrated October 13, 2003 and field calibrated October 29, 2004.

The vertical component of airspeed and variations in absolute humidity were measured from mean humidity:

$$\textbf{Evapotranspiration} = \frac{w'\rho v'}{\rho w} \tag{2}$$

where:

w′ = vertical component of airspeed (time average in m/s);

ρv′ = variations in absolute humidity about the mean value (time average);

ρw = mean value of absolute humidity (converted from watts/m^2 to mmoles/m^2s to cm/30 min interval).

The readings represent average evapotranspiration for the fetch area, which varies with wind direction and sensor height. As a general guideline, the fetch ranges from 20 to 200 times the instrument height, with the farther fetch extending in the direction of prevailing winds [42] , although the actual fetch will vary with wind speed and other factors. The ET flux tower lies near the center of the Headquarters Flat sinkhole catchment, which has a 1-degree slope to the north, southeast, and southwest. The ground slope increases in elevation from the ET flux tower to the south at a maximum slope of 3 degrees. Because of the minimal slope in the direction of prevailing winds, the flux field was not rotated to account for slope.

Additional instruments used for meteorological and energy balance measurements include air temperature and humidity (HMP45C, Vaisala), net radiation (Radiation and Energy Balance Systems Q 7.1), incoming and upwelling solar radiation (LI-190 SB, Kipp and Zonen), and photosynthetic photon flux density (PPFD, LI-190SA, Li-Cor).

2.3. Data Analysis

The original 10-hertz datasets was pre-processed to calculate thirty-minute fluxes of latent heat and sensible heat for a 505-day test period following procedures outlined by Litvak et al. [43] . The ET data were filtered to remove poor-quality data, including readings taken when the wind approached the flux sensors from the north, where the tower distorts wind velocities, and during rain intervals. Initial data gaps were caused by power outages, lowering of the instrument for annual calibrations, and particulates (including raindrops) obscuring the glass lens of the gas analyzer. Of 24,285 thirty-minute intervals comprising the 505-day test period, 5585 data gaps were present after this quality filtering. The filtered ET values were corrected for Bowen ratio energy balance of sensible heat to latent heat, as described by Twine et al. [44] . To fill in the gaps remaining after filtering and energy balance, monthly correlations of post processed and filtered ET data were related to net radiation readings. On a monthly basis, the ET to net radiometer correlations had R^2 correlations greater than 0.7 and averaged 0.83. Where gaps existed in the J17 tower net radiation record, readings from an ET flux tower located 21 km south on the Texas State University Freeman Ranch were used. Measured nighttime ET values are relatively very small so that data gaps of nighttime intervals are assumed to be zero. Following net radiation gap filling, 813 daytime 30-

minute gaps remained, or 3% of the water-balance record. Where up to three consecutive 30-minute data gaps remained, the gaps were filled by averaging adjacent ET values. For 6 days, where larger data gaps remained, the daily total was calculated as 0.75 of the potential evapotranspiration values reported by Texas A & M University from a weather station at the City of Austin Davis Water Treatment Plant, based on an R^2 correlation of 0.56. The accuracy of measuring an average evapotranspiration across the fetch area using eddy covariance with energy balance corrections is about ±5% [45] .

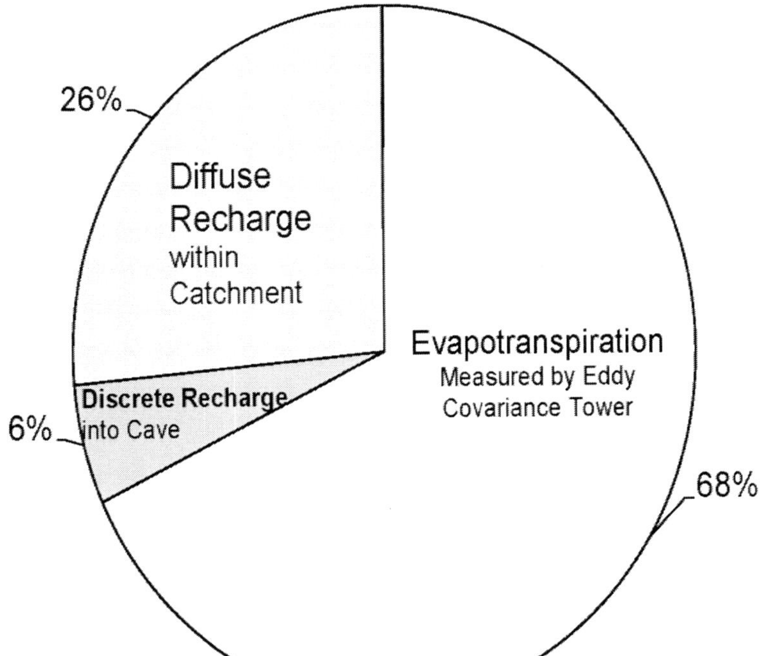

Figure 2. Water balance results for HQ Flat microbasin from April 2, 2004 to August 20, 2005 (505 days).

3. RESULTS

The water balance for HQ Flat sink shows recharge and evapotranspiration percentages similar to karst areas across the world. Over a 505-day period,

68% of the rainfall returned to the atmosphere as evapotranspiration. The remaining 32% of the rainfall falling on the HQ Flat basin recharged with 26% infiltrating through the soils and percolating into the underlying bedrock (diffuse recharge) and about 6% flowing as discrete internal runoff recharging into the cave drain at the lowest point of the sinkhole (Figure 2). The 160 cm of rainfall measured over the 1.4 year reference interval was 42% above the local annual average rainfall of 81 cm. Volumetric soil moisture was 20% ± 2% at the start and end of the selected water balance interval, so no significant change in soil moisture storage occurred.

4. DISCUSSION

The results from the 505-day water balance at HQ Flat Sink are strikingly similar to results a similar water balance study reported by Dugas et al. [34] where 65% of rainfall left as ET, 5% of precipitation became runoff, and 30% of precipitation recharged the Trinity Aquifer.

While measurements of ET were 68% of precipitation and discrete internal runoff into the cave entrance was 6% over an interval where precipitation was 42% higher than average, how much do these parameters, as well as diffuse recharge, vary at other sites and under climatic conditions? Over a single 2002 rain event, during which the ground was near saturation prior to precipitation, the amount of discrete internal runoff to HQ Flat Cave was 42% of the rainfall. Over a 1140-day interval under average precipitation conditions, discrete internal runoff to both HQ Flat Sink Cave and Flint Ridge Cave diminished to 3% of precipitation.

In order to examine variation in ET with precipitation conditions across Central Texas, ET/precipitation ratios reported from other climate towers were compiled and compared with annualized precipitation (Figure 3). These climate towers utilized either Bowen ratio or eddy covariance systems. Comparison of ET flux tower results indicate, as Dugas et al. [34] anticipated, that ET/precipitation ratios are inversely related to precipitation totals.

Distinct trends are indicated for above average and below average precipitation conditions. An equation that describes the relation between annual ET and precipitation during above average precipitation conditions is:

$$\mathrm{ET/Precipitation} = 4.62 \times \mathrm{PPT}_a^{-0.43} \tag{3}$$

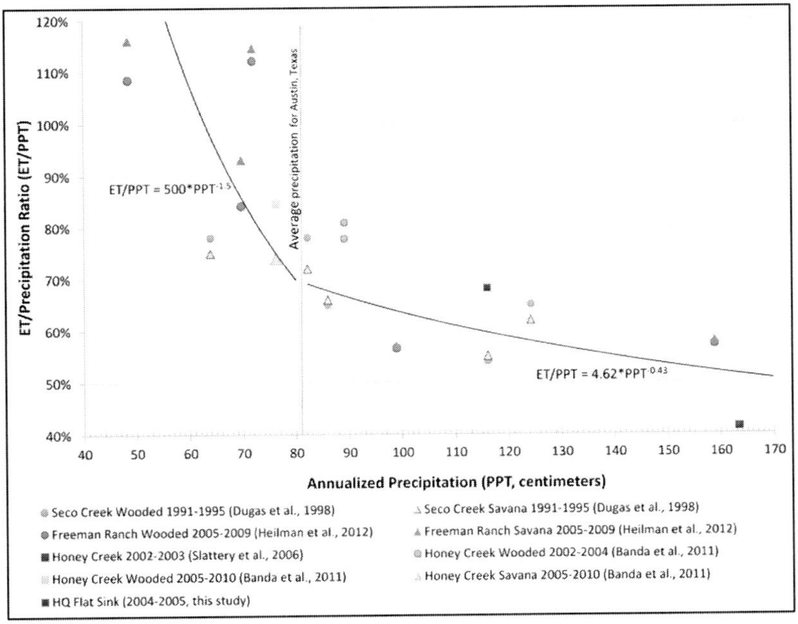

Figure 3. Central Texas ET flux tower data and annual precipitation.

where PPT_a = annualized precipitation in centimeters.

For below average precipitation, the relation between annual precipitation and ET may be steeper such as:

$$\text{ET/Precipitation} = 500 \times \text{PPT}_a^{-1.5} \quad (4)$$

While the towers are widely spaced both on the Edwards Aquifer and Trinity Aquifer, with variation in vegetation and soils, the trend has surprisingly limited variation. At average and above annual rainfall conditions, the data points for the ratio of ET/precipitation lie within 10% of the trend line. Under average Austin precipitation of 81 cm, an ET/precipitation ratio of 69% is indicated, although measured ET/precipitation ratio data points within 5 cm of 81 cm annual average precipitation actually range between 78% and 65%.

With 69% of rainfall lost to ET (ranging 65% to 78%) and 3% of rainfall allocated to runoff, the remaining 28% (ranging 19% to 32%) of precipitation would be expected to percolate into the underlying aquifer as autogenic recharge under average annual precipitation, based on the Equations (2) and (3) and actual data points under average conditions.

Aquifer recharge is focused during intervals of higher precipitation. In intervening areas between major creek channels, most autogenic recharge is expected to be focused for short periods around precipitation events. Shortterm ET measurements from the HQ Flat research station demonstrate that ET is minimal during the intervals where most autogenic recharge occurs (Figure 4). On an annual scale, most autogenic recharge occurs during years of higher than average precipitation when measured ET/precipitation ratios ranged from 70% to 55%, allowing 30% to 45% of precipitation to be distributed as the combination of recharge and runoff. Over sufficiently large intervals, the climate tower approach to measuring recharge accurately accounts at a site scale for the later release of water from storage in soil and plants as ET.

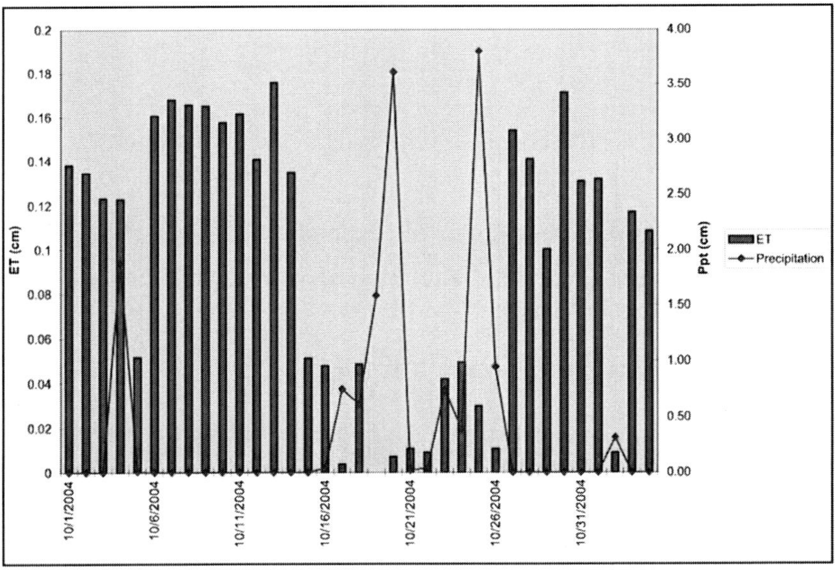

Figure 4. HQ Flat short-term response of ET to precipitation.

5. CONCLUSIONS

As the largest component of precipitation water balance, measurement of ET using eddy covariance climate tower provides one of the most precise measurements of recharge, the next largest component of the water balance. Precipitation, ET, runoff, and soil moisture were monitored over a 505-day water balance interval, during which time precipitation was 42% higher than average and volumetric soil moisture was 20% ± 2% at the beginning and end. ET measured 68% of precipitation and internal runoff into the cave drain was 5%, and diffuse recharge was 26%. While internal runoff recharge and diffuse recharge totaled 32% over the water balance interval within the HQ Flat Sink research site, since internal drainage basins constitute only about 10% of the recharge area of the Barton Springs Segment, the diffuse recharge component of precipitation may be more representative overall of autogenic recharge over the Barton Springs Segment.

Based on compilation of ET data from other flux towers in Central Texas under a wide variety of annual precipitation conditions, it can be estimated that under average precipitation conditions, 69% of rainfall leaves as ET; 28% of rainfall percolates as autogenic recharge into the Edwards Aquifer. Based on a longer 1140-day measurement of internal runoff at HQ Flat and Flint Ridge under average rainfall conditions, the remaining 3% of rainfall is representative for runoff in intervening in the recharge area. During years which precipitation is greater than average, ET decreases to between 55% and 70% of precipitation as recharge and runoff proportionally increases to the remaining 30% to 45% of precipitation.

ACKNOWLEDGEMENTS

The expertise and experience of Dr. Marcy Litvak, currently of University of New Mexico, was critical for constructing and operating the flux tower, including flux sensor calibration and data analysis. William Asquith of the US Geological Survey assisted in establishing the flux tower. The University of Texas provided use of Li Cor LI-7500 flux sensors. The City of Austin Watershed Protection Department provided a majority of purchased supplies and labor, including concrete foundation, aluminum flumes, rain gauges, and ISCO 3230 flow meters. The Hill Country Conservancy provided $10,000 in supplies for the tower, including data logger and laptop. The University of Texas Lady Bird Johnson Wildflower Center provided a net radiometer used in the study. The City of Austin Water Quality Protection Land program provided the research site, assisted in site security, and provided supplies including the Rohn tower. City of Austin Electric Department provided use of a crane and crew to lift the tower. David Johns of the City of Austin Watershed Protection Department provided valuable review comments for this paper.

REFERENCES

1. White, D.E. (1965) Saline Waters in Sedimentary Rocks. In: Young, P. and Galley, J.E., Eds., Fluids in the Subsurface Environment, American Association of Petroleum Geologists Memoirs, 4, 342-365.
2. Sukhija, B.S., Reddy, D.V., Nagabhushanam, P., Hussain, S., Giri, V.Y. and Patil, D.J. (1996) Environmental and Injected Tracers Methodology to Estimate Direct Precipitation Recharge to a Confined Aquifer. Journal of Hydrology, 177, 77-97.
3. Hendrickx, J.M. and Walker, G. (1997) Recharge from Precipitation, Chapter 2. In: Simmers, I., Ed., Recharge of Phreatic Aquifers in (Semi)-Arid Areas, IAH Contributions to Hydrogeology Series, Taylor and Francis, Balkema, Rotterdam.
4. Scanlon, B.R. (2000) Uncertainties in Estimating Water Fluxes and Residence Times Using Environmental Tracers in an Arid Unsaturated Zone. Water Resources Research, 36, 395-409.
5. Moller, P., Weise, S., Tesmer, M., Dulski, P., Pekdeger, A., Bayer, U. and Magri, F. (2008) Salinization of Groundwater in the North German Basin: Results from Conjoint Investigation of Major, Trace Element and Multi-Isotope Distribution. International Journal of Earth Science, 97, 1057-1073.
6. White, W.B. (1977) Conceptual Models for Carbonate Aquifers: Revisited. In: Diliamarter, R.R. and Csallany, S.C., Eds., Hydrologic Problems in Karst Regions, Western Kentucky University, Bowling Green, 176-187.
7. Zebidi, H. (1984) Chapter 4 Hydrology of Carbonate Areas, 4.3 Infiltration. In: La Moreaux, P., Wilson, B.M. and Memon, B.A., Eds., Guide to the Hydrology of Carbonate Rocks, Studies and Reports in Hydrology 41, UNESCO, Paris, 134.
8. Lerner, D.N., Issar, A.S. and Simmers, I. (1990) Groundwater Recharge: A Guide to Understanding and Estimating Natural Recharge. IAH International Contributions to Hydrogeology, 8, Taylor and Francis, Balkema, Rotterdam.
9. De Vries, J.J. and Simmers, I. (2002) Groundwater Recharge: An Overview of Processes and Challenges. Hydrogeology Journal, 10, 5-17.

10. Atkinson, T.C. (1977) Diffuse Flow and Conduit Flow in Limestone Terrain in the Mendip Hills, Somerset, UK. Journal of Hydrology, 35, 1-2, 93-110.

11. Dublyanskii, V.N., Pribluda, V.D. and Kodzhaspirov, A.A. (1984) Evaluation of the Karst-Water Balance of the Southwestern Upland Crimea. In: Castany, G., et al., Eds., Hydrogeology of Karstic Terrains, Case Histories, 1, 18-20.

12. Mink, J.F. and Vacher, H.L. (1997) Hydrogeology of Northern Guam. In: Vacher, H.L. and Quinn, T., Eds., Geology and Hydrogeology of Carbonate Islands, Elsevier Science, Amsterdam, 743-761.

13. Jocson, U., Jenson, J.W. and Contractor, D.N. (2002) Recharge and Aquifer Response: Northern Guam Lens Aquifer, Guam, Mariana Islands. Journal of Hydrology, 260, 231-254.

14. Tixeront, J., Berkaloff, E., Caine, A. and Mauduech, E. (1951) Biland'eau des massifs calcaires en Tunisie Gaz des captages de Tumis et de Bizerte. IAHS, Assemblee de Bruxelles.

15. Schoeller, H. (1948) Le regime hydrogeologique des calcaires, Eocene du synclinal du Dyr El Kef (Tunisie). Bulletin de la Societe Geologique de France, 5, 167-180.

16. Zebidi, H. (1963) Contribution a l'etude du Bilan Hydrogeologique du Djebel Bargou. These Publication, B.I.P.H., Tunisie.

17. Mero, F. (1958) Hydrological Investigations of the Na'aman Spring Region. Unpublished Water Planning for Israel Limited Report, Tel Aviv, Tahal.

18. Goldschmidt, M.J. and Jacobs, M. (1958) Precipitation over and Replenishment of the Yarqon and Nahal Hatteninim Underground Catchments. Hydrological Service, Ministry of Food Agriculture, Israel.

19. Mandel, S. and Shiftan, Z.L. (1981) Groundwater Resources: Investigation and Development. Academic Press, New York.

20. Burdon, D.J. (1961) Groundwater Development and Conservation in Syria. FAO/ETAP Report 1270.

21. Voute, C. (1961) A Comparison between Some Hydrological Observations Made in the Jurassic and Cenomian Limestone Mountains Situated to the West and to the East of the Ghab Graben. In: Eaux souterraines dans les zones arides: Colloque d'Ath è nes,

Association international d'hydrologie scientii que, Gentbrugge, 160-166.

22. Al-Charide, A. (2012) Recharge Rate Estimation in the Mountain Karst Aquifer System of Figeh Spring, Syria. Environmental Earth Sciences, 65, 1169-1178.

23. Hoetzl, H. (1995) Groundwater Recharge in an Arid Karst Area (Saudi Arabia). International Association for Housing Science, 195-207.

24. Aronis, G., Burdon, D.J. and Zeris, K. (1961) Development of a Karst Limestone Spring in Greece. Report from UNESCO, Ground Water Resources, Athens.

25. Burdon, D.J. and Papakis, N. (1963) Handbook of Karst Hydrogeology with Special Reference to the Carbonate Aquifers of the Mediterranean Region: Athens, Greece. United Nations Special Fund Karst Groundwater Investigations, Institute for Geology and Subsurface Research, 276.

26. Hauwert, N.M. (2009) Groundwater Flow and Recharge within the Barton Springs Segment of the Edwards Aquifer, Southern Travis County and Northern Hays County, Texas. PhD Dissertation, University of Texas at Austin, Austin, 328.

27. Hill, R.T. (1892) On the Occurrence of Artesian and Other Underground Waters in Texas, New Mexico, and Indian Territory: Together with the Geology and Geography of Those Regions. US Geological Survey Report, 166.

28. De Cook, K.J. (1957) Geology of San Marcos Quadrangle, Hays County, Texas. MA Thesis, University of Texas at Austin, Austin, 90.

29. Garza, S. (1962) Recharge, Discharge, and Changes in Ground-Water Storage in the Edwards and Associated Limestones, San Antonio Area, Texas—A Progress Report on Studies, 1955-59. Texas Board of Water Engineers Bulletin 6201.

30. Slade, R., Dorsey, M. and Stewart, S. (1986) Hydrology and Water Quality of the Edwards Aquifer Associated with Barton Springs in the Austin Area, Texas. US Geological Survey Water-Resources Investigations Report 86-4036, Austin.

31. Woodruff, C.M. (1984) Water Budget Analysis for the Area Contributing Recharge to the Edwards Aquifer, Barton Springs Segment. In: Woodruff, C. and Slade, R., Eds., Hydrogeology of the

Edwards Aquifer-Barton Springs Segment, Travis and Hays Counties, Texas, Austin Geological Society Guidebook 6, 36-42.

32. Harrison, A. (1996) Recharge Mechanisms of Swelling Clays and Shales, Central Texas. MS Thesis, Baylor University, Waco.

33. Slade, R. (2014) Documentation of a Recharge-Discharge Water Budget and Main Streambed Recharge Volumes, and Fundamental Evaluation of Groundwater Tracer Studies for the Barton Springs Segment of the Edwards Aquifer. Texas Water Resources Institute, Austin. Texas Water Resources Journal, 5, 12-23.

34. Dugas, W.A., Hicks, R.A. and Wright, P. (1998) Effect of Removal of Juniperus ashei on Evapotranspiration and Runoff in the Seco Creek Watershed. Water Resources Research, 34, 1499-1506.

35. Huang, Y. and Wilcox, B.P. (2005) How Karst Features Affect Recharge—Implications for Estimating Recharge to the Edwards Aquifer.10th Multidisciplinary Conference on Sinkholes and the Engineering and Environmental Impact of Karst, Geotechnical Special Publication No. 144, 201-206.

36. Slattery, R.N., Furlow, A.L. and Ockerman, D.J. (2006) Hydrologic and Water-Quality Data, Honey Creek State Natural Area, Comal County, Texas, August 2001-September 2003. US Geological Survey Data Series 200.

37. Banta, J.R. and Slattery, R.N. (2011) Effects of Brush Management on the Hydrologic Budget and Water Quality in and Adjacent to Honey Creek State Natural Area, Comal County, Texas, 2001-10. US Geological Survey SRI Report 2011-5226, Reston, 34.

38. Heilman, J.L., Litvak, M.E., McInnes, K.J., Kjelgaard, J.F., Kamps, R.H. and Schwinning, S. (2012) Water-Storage Capacity Controls Energy Partitioning and Water Use in Karst Ecosystems on the Edwards Plateau, Texas. Ecohydrology, 7, 127-138.

39. Kimball, B.A. and Jackson, R.D. (1979) Modification of the Aerial Environment of Crops: Soil Heat Flux. ASAE Monograph 2, American Society of Agricultural Engineers, 211-229.

40. Black, C.A. (1965) Methods of Soil Analysis: Part I Physical and Mineralogical Properties. American Society of Agronomy, Madison.

41. Burba, G. (2013) Eddy Covariance Method for Scientific, Industrial, Agricultural and Regulatory Applications: A Field Book on Measuring Ecosystem Gas Exchange and Areal Emission Rates. Li-Cor Biosciences, Lincoln.

42. Monteith, J.L. and Unsworth, M.H. (1990) Principles of Environmental Physics. 2nd Edition, Butterworth-Heinemann, Elsevier, Oxford.

43. Litvak, M.E., Miller, S., Wofsy, S.C. and Goulden, M. (2003) Effect of Stand Age on Whole Ecosystem CO2 Exchange in the Canadian Boreal Forest. Journal of Geophysical Research, 108.

44. Twine, T.E., Kustas, W.P., Norman, J.M., Cook, D.R., Houser, P.R., Meyers, T.P., Prueger, J.H., Starks, P.J. and Wesely, M.L. (2000) Correcting Eddy-Covariance Flux Underestimates over a Grassland. Agricultural and Forest Meteorology, 103, 279-300.

45. Ham, J.M. and Heilman, J.L. (2003) Experimental Test of Density and Energy Balance Corrections on Carbon Dioxide Flux as Measured Using Open-Path Eddy Covariance. Agronomy Journal, 95, 1393-1403.

7

CHAPTER

GROUNDWATER QUALITY CHANGES IN A KARST AQUIFER OF NORTHEASTERN WISCONSIN, USA: REDUCTION OF BROWN WATER INCIDENCE AND BACTERIAL CONTAMINATION RESULTING FROM IMPLEMENTATION OF REGIONAL TASK FORCE RECOMMENDATIONS

Kevin Erb [1,*], Eric Ronk [2], Vikram Koundinya [3] and John Luczaj [4]

[1] *University of Wisconsin-Extension Environmental Resources Center, 1150 Bellevue St, Green Bay, WI 54302, USA*
[2] *University of Wisconsin-Extension, 206 Court Street, Chilton WI 53014, USA*
[3] *University of Wisconsin-Extension Environmental Resources Center, 445 Henry Mall, Madison, WI 53706, USA*
[4] *Department of Natural & Applied Sciences, University of Wisconsin-Green Bay, Green Bay, WI 54311, USA*

ABSTRACT

In the Silurian Dolostone region of eastern Wisconsin, the combination of thin soils and waste application (animal manure, organic waste) has led to significant groundwater contamination, including Brown Water Incidents (BWIs—contamination resulting in a color or odor change in well water) and detections of pathogen indicator bacteria such as E. coli and others. In response, a Karst Task Force (KTF) was convened to identify risks and recommend solutions. This article looks at the impact eight years after the 2007 Karst Task Force report—both the actions taken by local resource managers and the changes to water quality. We present the first regional analysis of the 2007 Karst Task Force report and subsequent regulatory changes to determine if these regulations impacted the prevalence of wells contaminated with animal waste and the frequency of BWIs. While all of the counties in the KTF area promoted increased awareness, landowner/manager and waste applicator education alone did not result in a drop in BWIs or other water quality improvements. The two counties in the study that adopted winter manure spreading restrictions on frozen or snow-covered ground showed statistically significant reductions in the instances of BWIs and other well water quality problems.

Keywords: Silurian aquifer; karst task force; Wisconsin; karst; animal waste; manure; well contamination; regulations

1. INTRODUCTION

Karst regions are widespread across the world, and approximately 20%–25% of the global population depends on groundwater resources obtained directly from karst aquifers [1]. Karst aquifers are particularly vulnerable to microbial pathogens and other introduced substances resulting from surface land use activities due to a lack of filtration in the aquifer and short subsurface residence times [2]. Microbial pathogens include bacteria, viruses, and protozoan parasites. Conduits, such as sinkholes and swallow holes, provide direct access points that connect water in the surface environment to the karst aquifer below, often bringing with it contaminants that would not normally enter the aquifer by diffuse recharge, such as phosphate, pesticides, and pathogens [2]. Regions of karst bedrock with little or no soil or unconsolidated sediment cover are especially vulnerable.

Impacts to karst aquifers from pathogens and nitrates have been reported in karst aquifers from many parts of the world (e.g., [2,3,4,5,6]), and contamination from multiple sources, including animal waste, are well documented in parts of Wisconsin and the Midwestern United States where a regional Paleozoic karst aquifer is present (e.g., [7,8,9,10,11,12,13]), In

Wisconsin, contamination from pathogenic indicator bacteria is often associated with "brown water" incidents (BWIs), resulting in a color or odor change in well water. BWIs have occurred for decades throughout northeast Wisconsin, often in the spring during snowmelt after application of bovine (dairy) manure on agricultural fields. Between 2006 and mid-2014, sixty-four well replacements were subsidized throughout Wisconsin due to confirmed contamination by livestock manure [7]; three-quarters of those wells were located in areas rated as having a significant to extreme vulnerability to groundwater contamination related to karst-type landscape features (e.g., sinkholes, disappearing streams, surface carbonate rock outcrops, and fracture traces) [13].

Thin soils are a particular risk factor for microbial impact and nitrate contamination of karst aquifers (e.g., [13,14]), Risk for groundwater contamination in karst aquifers as a result of manure application is higher when manure is liquid (< ~12% solids); surface applied outside the normal growing season to wet, frozen, or bare frozen soils; applied to a wet snowpack ready to melt, or immediately prior to significant rainfall [14]. Ronk and Erb [15] noted that in Wisconsin, a majority of animal waste surface water contaminations due to runoff occur in the late winter runoff period. In particular, land application of manure on frozen soils in the Midwestern United States has seen increased attention because of concerns of the negative effects on surface water and groundwater quality [13,16].

This article focuses on a four-county region in northeastern Wisconsin, USA (Figure 1a). Luczaj [17] provides a more detailed description of the geology in the study area, but a basic description of the geology is included here as necessary context. Paleozoic age sedimentary rocks, including dolostone and shale, underlie the region and dip gently eastward toward the ancestral Michigan basin [17]. The landscape displays relatively moderate topography, except along the Niagara Escarpment, which locally has cliffs and steep slopes as high as 70 m. The escarpment occurs along and near the western edge of the Silurian dolostone in the region (Figure 1b) where the underlying Maquoketa shale has been preferentially removed by erosion. Bedrock across the entire study area is overlain by Pleistocene glacial till, glaciofluvial sediments and lacustrine sediments that range in thickness from <1 m to >100 m in buried bedrock valleys. The portions of this four-county area that are most heavily impacted by nitrates, pathogens and pathogenic indicators, and BWIs are located east of the Niagara Escarpment, typically in areas where the Silurian bedrock is mantled with thin soils or Pleistocene sedimentary cover [7,9]. West of the Niagara Escarpment, bedrock consists of Ordovician age shale and dolostone (Figure 1b), with generally thicker amounts of unconsolidated sedimentary cover [17]. Karst features are not typically observed west of the Niagara Escarpment in the four-county region identified in the study.

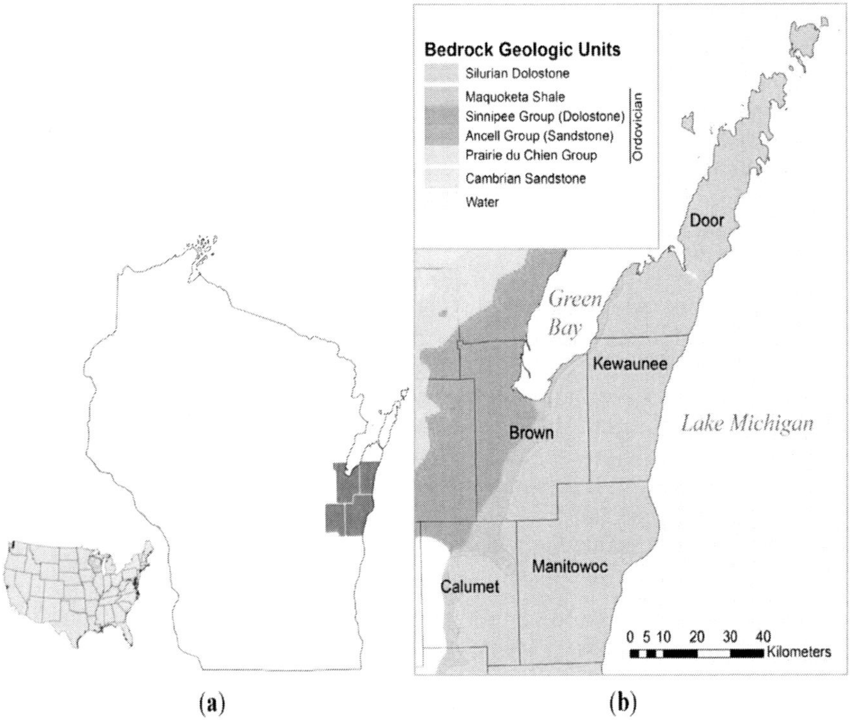

Figure 1. Maps showing the location of the four-county study area in Wisconsin, USA. (**a**) shows an inset map of the lower 48 contiguous United States along with an outline of the State of Wisconsin and counties (local units of government) for the study area highlighted; (**b**) is the bedrock geologic map showing geologic units for this portion of northeastern Wisconsin with names of counties in the region. Bedrock map after [18].

The Silurian bedrock east of the Niagara Escarpment displays significant karst development, albeit heavily modified by Pleistocene age glacial activity (Figure 1). Many karst features are exposed, such as sinkholes, solution-enlarged joints, and caves, which act as preferential recharge points to the aquifer. Most karst features are concealed beneath a variably thick mantle of sediment, but exposed areas of Silurian dolostone bedrock are common within about 15 km of the escarpment edge. Thin to moderate soils (defined for this region as <15.25 m of soil over the bedrock) are also common in this area (Figure 2).

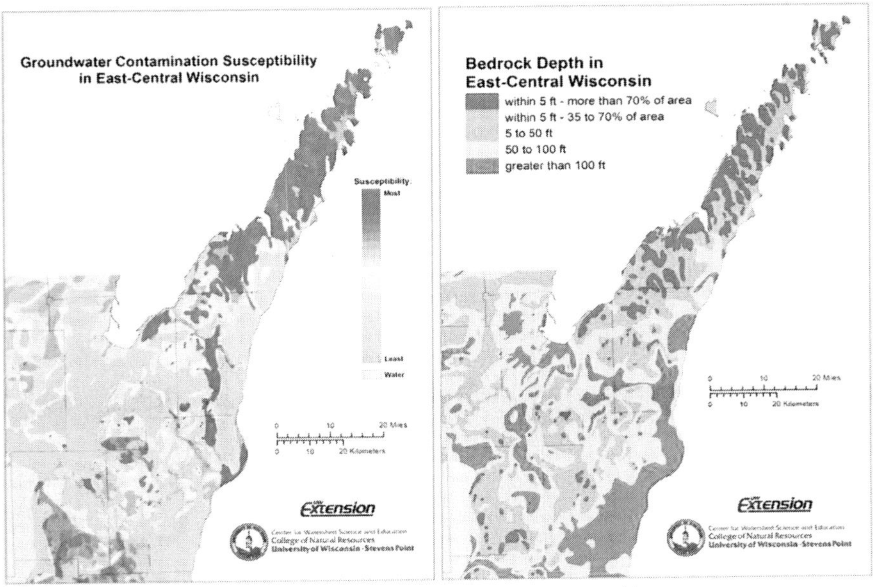

(a) (b)

Figure 2. Maps showing soil depth (or unconsolidated sediment depth) over bedrock (**a**) and groundwater contamination susceptibility (**b**), based on soil type and depth to bedrock [13]. Original depth to bedrock map was constructed with categories of <5 feet (1.5 m), 5 to 50 feet (1.5 to 15 m), 50 to 100 feet (15 to 30.5 m), and >100 feet (>30.5 m).

Climatologically, the four-county study area (Figure 1 and Figure 2) lies in a portion of the United States with a humid continental climate and cold winters. Mean annual precipitation (liquid equivalent) varies from about 75 to 84 cm (29.5 to 33 inches) [19], with about two-thirds of the precipitation falling during the growing season. Winter snowfall averages about 105 to 130 cm (41.3 to 51.4 inches) [19]. Average annual temperatures for the four-county study area range from about 6.6 to 7.4 °C (43.8° F to 45.3° F) [20]. The lowest mean monthly temperatures in the region occur during January, −9.4 to −7.2 °C (15 to 19° F), and the highest mean monthly temperatures occur during July, 20.0 to 21.7 °C (68–71° F) [20].

The presence or absence of seasonally frozen ground is an important climatological factor that can influence a number of groundwater quality indicators, including BWIs and variability in nitrates, chloride, alkalinity, and conductivity [9,14,16]. In the study area between 2006 and 2014, the average number of days of frozen ground per winter at Green Bay (Brown County) ranged from 78 to 134 days (average 109) while Manitowoc (Manitowoc County) ranged from 72 to 133 days (average 102) between 2003 and 2014. [21], In Kewaunee County, there were 120 days of frozen ground during the winter of 2013–2014 [9]. The average number of days with >2.5 cm (1 inch) of snow cover ranges from about 70 in the south along the Lake Michigan shoreline to about 90 in western Brown County [22].

The four-county study area includes Brown, Calumet, Kewaunee and Manitowoc counties (Figure 1). Land use in the study area is characterized by a mix of urban development, small rural communities and industries, including manufacturing, dairy (bovine) livestock production and agricultural field crops, but the majority of the study area is >75% cultivated land [23]. With the exception of areas near the Fox River and the Bay of Green Bay in Brown County, the region's municipal boundaries are largely based on the U.S. Public Land Survey System, in which Towns (93.2 km^2 (36 mi^2)) are the most basic unit of local government in rural areas. Several towns, villages, and cities occur within counties.

The residents and manufacturing industries in the region rely on both ground and surface water [24], while the livestock production industry relies entirely on groundwater. In the four-county study area, only the City of Manitowoc in Manitowoc County and major municipalities near Green Bay in Brown County utilize surface water for municipal supplies. Rural regions making up most of the four-county study area depend on either domestic wells or municipal wells to provide potable water resources [7]. The vast majority of agricultural field crops rely on annual precipitation as their sole source of moisture [25,26,27]

The dairy industry in the four counties has recorded a 30% increase in milking cows between 2002 and 2012 (from 131,500 to 172,500) [25,26,27] and a corresponding increase of 5.2 million kiloliters to 6.53 million kiloliters of animal waste annually (Figure 3). In the study area, 100% of the animal waste (manure) is land applied. More than half of the waste is stored and applied seasonally (usually in the fall and spring months), however, year to year variations in weather result in some stored manure being land applied to frozen ground. Less than 5% of the volume is treated with anaerobic digestion prior to land application. The increase in bovine numbers, and associated increase in manure volume applied to the land increases the risk of manure contamination occurring.

Statewide, restrictions on the land application of manure (in both karst and non-karst areas) are based on the USDA NRCS 590 Nutrient Management Standard [28]. This technical standard is referenced in several state laws that govern manure and fertilizer application including ATCP 50 and NR 151.

A statewide survey in 1994 of 538 wells [29] showed positive detections for coliform bacteria in 23.3% of the state's wells, with E. coli in 2.5% of the wells, and 6.5% of wells exceeding 10 ppm nitrate-nitrogen. In the region, sediment and pathogen contamination of the shallow aquifer has long been a common occurrence. Concerns regarding coliform bacteria, E. coli, and nitrate in the region's wells became greater during the late 1990s and early 2000s, as the number of contaminated wells and the severity of the contamination continued to rise [13,30]. WDNR policy denotes any well with any detectable coliform, any type of E. coli or any pathogen as contaminated by a pathogenic indicator [30]. Wells over 2.0 ppm nitrate-nitrogen are also considered impacted by human activity, while those over 10 ppm are considered unsafe (above the WNDR heath standard) for sensitive populations (young children, pregnant women) [13,30].

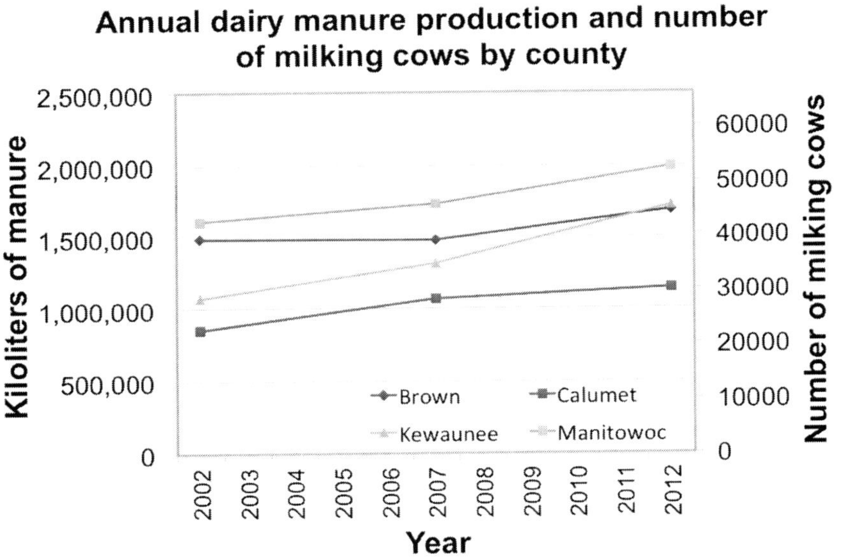

Figure 3. Estimates of annual dairy manure production by county (left axis) and the number of milking cows by county in the four-county study area (right axis). Values show increasing population trends and manure production volumes over time [25,26,27].

Voluntary well testing programs in two to six townships in Calumet, Brown, Kewaunee, and Manitowoc counties documented between 20% and 30% of wells exceed the nitrate-nitrogen standard of 10 ppm [8,13]. Targeted testing programs of more than 1000 wells between 2002 and 2005 in Calumet County revealed that 35% of the samples were positive for coliform bacteria, and 4.6% were positive for E. coli. The testing also showed that 53% of wells had elevated levels of nitrate-nitrogen (2–10 ppm), with 25% above the health standard of 10 ppm for nitrate-nitrogen. Altogether, 47% of the wells tested were considered unsafe for either bacteria or nitrate, with 12% unsafe for both bacteria and nitrate [13].

In Door County (adjacent to the study area with the same geologic, soil and climate features), a well sampling program (2000 samples) showed that on any given day, over one-third of the tests indicated the presence of coliform bacteria [2].

Each spring, snowmelt and rainfall are typical precursors to clusters of contaminated wells in the area, as documented by The Karst Task Force [13] and shown in Table 1. Weather events in February and March of 2006 resulted in both snowmelt and precipitation runoff carrying manure into shallow karst bedrock features and improperly abandoned wells, leading to coliform and E. coli detections in 78 of the 422 tested wells in the Town of Morrison [13]. Public pressure resulting from this event and previous clusters of contamination led to the formation of the Northeast Wisconsin Karst Task Force (KTF) [13].

Table 1. Example contamination cases [13,30].

Incident Location	Date	Impact
Town of Morrison, Brown County	February–March, 2006	78 wells tested unsafe for coliform bacteria and/or *E. coli*
Town of Franklin, Manitowoc County	2005	10 wells tested unsafe for coliform bacteria; 6 positive for *E. coli*
Town of Luxemburg, Kewaunee County	March, 2004	A single well contamination by manure that tested positive for *E. coli* resulted in severe illnesses and hospitalizations

The objective of this study was to evaluate groundwater quality changes that occurred in a four-county region of northeastern Wisconsin to determine whether or not implementation of recommendations from a regional task force had an impact on groundwater quality in the region. A brief description of the Karst Task Force results and regulatory/policy changes that resulted are presented below.

2. RESULTS OF THE 2007 KARST TASK FORCE (KTF)

2.1. Purpose and Justification

The KTF was charged with examining the available scientific data and with making recommendations on how to address the groundwater contamination problem. The Task Force consisted of researchers and experts from five University of Wisconsin institutions, resource managers from county Land and Water Conservation Departments, and the state environmental agencies (Wisconsin Department of Ag Trade and Consumer Protection and Department of Natural Resources (WDNR)), a crop consultant, farmer, a professional manure hauler, a USDA Natural Resources Conservation Service (NRCS) District Conservationist, and a well driller. Outside contributors included hydrogeologists from the Minnesota DNR, the University of Minnesota, and a private engineering consultant.

The Task Force focused on agricultural issues, as the primary land use in the karst area is livestock and cash crop agricultural production, and the trends noted in NASS [25] showed the number of livestock increasing and with a concurrent decrease in available cropland for manure application. Task Force members agreed that because of the aquifer type, overlying soils and land use practices, it would be impossible to prevent every instance of contamination, but that landowners can take significant steps to reduce the potential for animal and human waste, and other materials from entering the groundwater [13]. It also became clear that the physical environment cannot be characterized, understood, or protected by merely locating and managing karst features visible at the surface. Rather, the controlling factor is the underlying fractured carbonate bedrock and the pathways through which surface contaminants may enter the bedrock environment.

This task force created a set of recommendations, of which two counties chose to implement restrictions on the application of animal waste (manure) to reduce the risk of groundwater contamination in 2007. This article represents the first regional analysis of the KTF and subsequent regulatory changes to determine if these regulations impacted the prevalence of wells contaminated with animal waste and the frequency of BWIs. This paper is the first analysis of the impacts of the task force report and the effectiveness of actions taken by local resource managers that have been examined with regard to groundwater quality.

2.2. Regulatory and Policy Responses to the 2007 KTF Report

The two counties that implemented regulations varied in their approach. Manitowoc County was the stricter of the two, implementing both year-round and frozen soil restrictions on manure application [31,32,33] in 2007. Year round restrictions included (1) No application within 30.5 m of a known karst feature, (2) requiring incorporation (injecting or mixing into the soil) of animal manure within 48 h of application on any site (a) within 30.5–151.5 m of a known karst feature, and (b) any area that drained into a known karst feature. Seasonal (frozen soil) restrictions include no applications of liquid manure on slopes >6% when unable to incorporate within 48 h. Solid manure can be applied on slopes between 6% and 12% if at least 40% of the soil surface is covered with crop residue. Manitowoc County does not require pre-approval of application sites, but has created maps of the above restrictions and made them available to livestock farmers, agronomic consultants and the general public. Exemptions are not granted for incorporation or setbacks. Violators may be fined $500 (US) plus fees. The county ordinance was voted on by each town in the county, and passed on a popular vote.

Regulations implemented in Brown County in 2007 [34] required the land manger to gain pre-approval for seasonal land application by submitting a winter manure spreading plan (WSP). The focus of the WSP is to prioritize fields from lowest to highest risk, with maps showing karst features provided to farmers and agronomic consultants. There are no mandatory additional incorporation requirements or setbacks from karst features. Manure application rates are pre-approved by the SWCD. Non-mandatory recommendations include limiting manure application to 46,670 L/hectare and 44,834 kg/hectare and encouraging larger setbacks from wells and karst features. The Brown County ordinance addresses only manure and not other waste applications.

Three other counties (Calumet, Door, Kewaunee) used the 2007 KTF report to begin building the case for additional setbacks and regulations, but did not make changes to their county regulations impacting manure and waste application. State regulations continued to be enforced in all five counties.

Calumet County's LWCD created a focused educational approach—identifying the areas of highest risk of contamination and working one on one with landowners and land managers in these areas to identify and map karst features and encourage better management. This was done on a volunteer basis because there were no revisions made to the ordinances, limiting any enforcement and incentive for landowners to change management practices [35,36,37].

Door County has also documented BWIs. The county did not make any ordinance changes as a result of the 2007 KTF. County Conservationist William Schuster, however, states that the KTF, by providing a solid scientific consensus, has eliminated debate over many issues and reaffirmed that contamination in the karst aquifer is a regional problem, and not just a county problem [38]. Recent (2014) BWIs have brought the issue to the forefront, and changes to manure and waste application regulations are likely to occur. In 2015, Door County is increasing enforcement of existing county ordinances surrounding manure application, however, these ordinances incorporate state rules by reference and do not include the additional winter restrictions that were adopted by Manitowoc and Brown counties [39]

Kewaunee County has continued to experience a high number of BWIs. While no regulatory changes were made in the years following the 2007 KTF, the KTF set the stage for a regional approach. Since the release of the KTF report, Kewaunee's past and current resource managers have focused their efforts on identifying features and implementing a detailed well water testing program to delineate the areas of concern. The county board approved detailed aquifer protection ordinance in 2014 and it passed with an 83% yes vote when placed on every municipal ballot in the county in April 2015. The ordinance prohibits animal waste application on both frozen/snow covered fields and applications between Jan 1 and April 15 on cropland fields with less than 6.1 m of soil over bedrock and those that drain to these areas [40].

All five counties implemented some aspects of the education recommendations made by the KTF by working with key people, such as for-hire manure applicators and consultants, who cross county lines. The county SWCDs and The University of Wisconsin - Extension partnered to increase outreach and education to three key audiences: (1) Commercial Manure Applicators: Responsible for >60% of the manure applied in the target area, educational modules were added to their Level 1 and Level 2 certification program to provide basic and advanced training to manure applicators, (2) Nutrient Management Plan Writers: Task force members communicated recommendations directly to plan writers at their statewide conferences and at local meetings and (3) Farmers: Information and KTF recommendations were included in the farmer training sessions for those farms writing their own Nutrient Management Plans.

3. METHODS

3.1. Data Sources and Statistical Analysis

Three sources of water quality data for domestic water wells were compiled during this study. These included state regulatory agency (Wisconsin Department of Natural Resources—WDNR) records, County health department (CH) records, and various county Land and Water/Soil and Water Conservation District (SWCD) records. Data were compiled from each source for as many years as records existed prior to 2007 (starting in 2002), and for the 2008–2014 timeframe. Data acquired from the sources included the date, well location, whether the color change in the water (BWI) was observed by a trained individual (WNDR, CH or SWCD staff), and the presence of a strong animal waste odor that would deter bathing or laundry (ODOR). As well testing is often suggested when a neighboring well is contaminated, data requested included either BWI or ODOR in that well or a nearby well along with a positive pathogenic indicator test, and lastly, a lack of either BWI or ODOR, but a nearby well had documented pathogenic indicators, ODOR or BWI. Data were received as Excel spreadsheets, and the data were then imported into SPSS (version 23.0) for statistical analysis. A Chi-Square test was computed to test for any statistically significant association between the treatments (year samples were taken (pre-2007 and post-2007) and frozen ground (presence or absence of frozen ground)) and the variables mentioned above.

WDNR records included staff investigations of homeowner-reported well contamination. Depending on the situation and the time between the initial report and staff visit, these may include records of visual documentation of BWIs, nasal detection of odors, laboratory detections of bacteria or nitrate, or if a specialized bacterial testing was conducted. A well was determined to be impacted by animal waste if trained staff sampled the well, followed WDNR or written CH QA/QC protocols, and one of the following tests produced conclusive results: (1) presence of Rhodococcus corpophilus, (2) presence of E. coli at levels higher than what would be attributable to human waste, and/or (3) presence of bovine Bacteroides, as determined by a Microbial Source Tracking (MST) analysis (post 2006) [41]. Data were summarized by WDNR staff at the authors' request. Reliable records were available from WDNR for the time period 2002–2014 for all counties except Door County; hence, it was excluded from this study.

The SWCD records from each county included staff investigations of homeowner-reported well contamination. The SWCD staff collaborated with WDNR and/or CH staff to conduct well water quality sampling. SWCD records closely paralleled the records of WDNR. WDNR records were provided to SWCD staff, who then compared them to their records and information they had on file from CH to complete missing data (such as visual verification of BWIs) or provide well testing results. Thoroughness of data tracking of reported well contamination varied between counties, with some having very complete records and others having no formalized tracking system [42,43,44,45,46,47,48].

The CH records included visual observations, along with E. coli, and Rhodococcus coprophilus test results. Counties with complete records were included in the dataset, either by the SWCD or by providing a copy of the WDNR data to the CH. Data from CH where no third party verification of BWI or odor existed for an incident, (such as where the well owner collected their own sample, or where it is unknown whether WDNR or CH QA/QC protocols were followed), were excluded from the dataset.

A total of 124 data points were included in the dataset. Sixty-two data points were from the 2002–2006 timeframe and 63 were from the 2008–2014 timeframe. Data from 2007 were excluded because Brown and Manitowoc Counties implemented their regulations during 2007. Twenty-nine of the data points were from Brown County, 12 were from Calumet, 22 were from Kewaunee, and 61 were from Manitowoc County. Multiple instances of contamination were documented in some wells during the studied time period. Multiple events of contamination that occurred within a 14-day period were counted as one contamination. If contamination occurred and the water clarity returned to normal or all pathogenic indicator tests were negative within 14 days and then re-occurred after the initial 14 day period, it was treated as a second instance.

Frozen ground has been identified as a critical factor in BWIs and well water contamination [49,50]. When the ground is frozen, snowmelt and/or precipitation infiltrate more slowly, as some of the pores between soil particles that normally allow for infiltration are filled with frozen water. This decrease in infiltration results in an increase in the volume of water leaving via surface flow. In some years, "concrete frost" or "dense frost" has developed, where a very high percentage of soil pores are filled with ice. This results in a much lower infiltration rate, higher runoff volumes, and has been tied to spikes in well water contamination [49].

For this study, the presence or absence of frozen ground was determined by using the National Oceanic and Atmospheric Administration's (NOAA) Cooperative Observer Database (COD) hosted by UW Madison [21]. A central location was chosen in each county, and for each well contamination event where a month and date of contamination was available, the COD data for that date was accessed. For the purposes of the data analysis, a central location for each county was selected to determine the presence or absence of frozen ground. A central location was selected to avoid the climate mitigating affects of Lake Michigan on the data analysis.

3.2. Data Limitations

The data presented in this article represent the most complete data available in the study area. However, the authors recognize several limitations to the dataset. Because the sampling of wells is triggered by an event and well user observations, not all instances of contamination will be documented. Pathogens or other microbial contamination can be present in water with no visual or other indication of problems that would trigger sampling. Illnesses caused by pathogens in well water may be misinterpreted as seasonal gastrointestinal illnesses or food poisoning by the well owner, and not necessarily connected to a well contamination problem. Asymptomatic illness and locally acquired immunity are further limitations in the context of illness as an indicator of water quality. All of our data providers [30,37,40,44,48,51,52,53] noted that some well owners are hesitant to report problems to avoid disrupting neighbor relations, or because of the additional expense of replacing a well or adding a treatment system. By requiring a third party verification of odor or BWI issues, and only including well samples where trained individuals following QA/QC protocols and the same analysis methods were used, our data represent the most reliable instances of positive well contamination indicators. Cow numbers and, as seen in Figure 3, did not hold steady, but increased during the entire study period [25,26,27].

The majority of human waste in rural areas of Wisconsin is discharged to septic systems with drain fields designed to allow for infiltration of liquids on a continual basis. While some instances of human pathogens in groundwater have been documented in the region (see [7] for details), analysis of human waste is beyond the scope of this paper, and septic system design and installation practices are not likely to have changed significantly during the study period.

The well water quality data gathered in the region are limited by the fact that there is not a scheduled or randomized testing program in place to document BWIs, ODORS or pathogenic indicators in rural wells. While it is clear that such contamination events occur, the true frequency and severity of these events remain unknown.

In addition, an ongoing WDNR well replacement subsidy program [7] and decisions by individual homeowners not participating in the subsidy program have resulted in numerous wells to be replaced in the karst region of the state. While this is a potential confounding factor, it is unlikely to be significant for the observed water quality changes because many of these wells were replaced in the two counties that did not adopt regulatory changes restricting winter spreading. If well replacement was the first-order cause of water quality improvement, we expect that this response should be seen in all four counties analyzed.

The 2007 KTF report focused on reducing BWIs in 5 counties. A combination of factors led us to exclude Door County from the analysis. Most importantly, complete datasets for 2002–2014 were not available from WDNR and SWCD for Door County. The WDNR was also unable to provide a complete dataset for this county as a result of staffing issues [54]. The SWCD stated [38] that they have not historically tracked BWIs and contamination, as it is considered to be a common occurrence. They noted that they are actively working with the Door County Health Department and landowners to address the problem.

4. RESULTS

Our results revealed that no statistically significant changes occurred in counties that only had education and training for manure application in a karst setting. However, in all scenarios tested, at least one statistically significant association occurred with counties that implemented regulatory changes as a result of the 2007 KTF report (Brown County, Manitowoc County, or both). Implementing seasonal restrictions on waste application has had a positive impact on ground water quality in the four counties in our study by reducing bacterial contamination from 35 documented cases (pre-2007) to 15 (post-2007) in the four county area, a fifty-seven percent decrease. Twenty-eight cases were documented in the two counties (pre-2007) that implemented regulations before implementation and eight after regulations were implemented (post 2007). The same number of cases was documented in the two non-regulated counties (7 total) in the pre and post 2007 timeframes. For BWIs, there was a 38% increase (from 5 to 8) in the counties with regulations, but a 70% increase (from 10 to 17) in the counties without. Since almost half of the data points originated in Manitowoc County due to a more aggressive testing program implemented by the County Health Department, the increased numbers are not unexpected.

4.1. Restrictions on Frozen Ground Spreading

The statistical analysis showed that the implementation of restrictions on winter (frozen ground) spreading of animal waste has likely had an impact on the documented instances of well contamination. In almost every variable examined, at least one of the regulated counties (Brown or Manitowoc) showed a statistically significant difference when the pre-regulation period (2002–2006) was compared to the post regulation period (2008–2014).

For BWIs, the Chi-Square analysis determined significance at the 0.01 level. The analysis showed Manitowoc County was statistically significant ($p = 0.003$), but not significant for the other three counties (Brown ($p = 0.204$), Calumet ($p = 0.377$) and Kewaunee ($p = 1.00$)). For a confirmed ODOR, significance was at the 0.01 level. The data were significant for Manitowoc County ($p = 0.003$), with non-significance for the other three counties (Brown ($p = 0.204$), Calumet ($p = 0.190$) and Kewaunee ($p = 0.35$)).

Where BWI and/or ODOR was present, and a pathogenic indicator test produced conclusive results with the presence of Rhodococcus coprophilus, presence of E. coli at levels higher than what would be attributable to human waste, and/or presence of bovine Bacteroides, as determined by a Microbial Source Tracking (MST) analysis, significance was at the 0.05 level. The data were statistically significant for Brown County ($p = 0.017$), with non-significance for the other three counties (Manitowoc County ($p = 0.06$), Calumet County ($p = 0.855$) and Kewaunee County ($p = 0.448$)).

In cases where the well owner did not have a BWI or ODOR, but the well was sampled because a neighboring well did, significance was at the 0.01 level. Both Brown ($p = 0.000$) and Manitowoc counties ($p = 0.020$) were significant, but the other two counties were not significant (Calumet ($p = 0.345$) and Kewaunee (p = number not recorded as there was only one level of this variable for this county and is hence a constant)). In cases where the well had a positive pathogenic indicator test with or without BWI or ODOR, significance occurs at the 0.01 level. Significant for Brown County ($p = 0.000$), but not significant for the other three counties (Calumet ($p = 0.554$), Kewaunee ($p = 0.448$) and Manitowoc ($p = 0.945$)).

4.2. Presence or Absence of Frozen Soil

The statistical analysis shows that the presence or absence of incidents involving spreading of animal waste on frozen ground likely had an impact on the documented instances of well contamination. In almost every variable examined, at least one of the regulated counties (Brown, Manitowoc) showed a statistically significant difference compared to the non-regulated counties.

For BWIs, significance occurs at the 0.01 level. Results were significant for Manitowoc County (p = 0.001), but not significant for the other three counties (Brown (p = 0.842), Calumet (p = 0.490) and Kewaunee (p = 0.211)). For confirmed strong manure odor (ODOR), Significance occurred at the 0.01 level. The results were significant for Manitowoc County (p = 0.001), but not significant for the other three counties (Brown (p = 0.842), Calumet (p = 0.301) and Kewaunee (p = 0.061)).

In cases with a positive pathogenic indicator test and either BWI or ODOR in the well or a neighboring well, the Chi-Square test showed a statistically significant difference (p = <0.05). Values for each county were: Brown (p = 0.045), Calumet (p = 0.490), Kewaunee (p = 0.377) and Manitowoc (p = 0.13). In cases with a positive pathogenic indicator test and BWI or ODOR, Significance occurs at the 0.05 level. The results were significant for Manitowoc County (p = 0.015), but not significant for the other three counties (Brown (p = 0.152), Calumet (p = 0.325) and Kewaunee (p = 0.371)).

In cases where the well did not have BWI or ODOR but was sampled because of neighbor's well did, significance occurs at the 0.01 level. The results were significant for Brown County (p = 0.000), but not significant for the other three counties ((Calumet (p = 0.490), Kewaunee and Manitowoc counties (p = number not recorded as there was only one level of this variable for these counties and is hence a constant)). In cases where the well tested positive for a pathogenic indicators, regardless of other factors, significance occurred at the 0.01 level. The results were significant for Brown County (p = 0.000) and Manitowoc County (p = 0.005), but not significant for the other two counties (Calumet (p = 0.175), Kewaunee (p = 0.371)).

5. DISCUSSION

The objective of this study was to evaluate groundwater quality changes that occurred in a four-county region of northeastern Wisconsin to determine whether or not implementation of recommendations from a regional task force had an impact on groundwater quality in the region. By comparing the pre-regulation and post-regulation statistics, the counties that implemented WSPs had a statistically significant impact in one or both regulated counties on reducing the number of BWIs, ODORS, and pathogenic indicator bacteria in groundwater. This study also verifies that the presence of frozen ground and implementing the WPSs reduced BWIs, ODORS and pathogenic indicator bacteria in groundwater.

The data showed an overall decrease in the number of pathogenic indicator bacteria in the counties that implemented regulations, with 28 recorded incidents in the pre-2007 period and 8 in the post 2007 period. Pathogenic indicator bacteria remained constant (7) in the other two counties that did not implement regulations. BWIs showed a lower rate of increase (38%) in regulated counties (5 pre/8 post) versus 70% in non-regulated counties (10 pre/17 post). The increase in manure volume (Figure 3) applied in the target area may or may not be a factor, and was beyond the scope of this study.

The average snowfall totals, number of days with snow cover, and number of days with frozen ground were higher during the second half of the study due to decadal-scale climate anomalies. [20,55]; Manitowoc recorded an average of 93 days frozen ground in the first part of the study and 107 in the second. Only one year of frozen ground data prior to 2007 was available for Green Bay. [21]. However, it is unlikely that these variables had a strong influence on the observed trends for two principal reasons. First, increased snowfall would be expected to produce increased spring runoff and greater groundwater quality impacts during the 2008–2014 period, which was not observed. Second, and more importantly, these inter-annual climatological variations would not be expected to impact only Brown and Manitowoc counties, rather, if significant they should have impacted all four analyzed counties. What is more likely important for limiting spring infiltration is the percentage of ice-filled soil pores, which is a complex function of freeze-thaw cycles and winter rain precipitation events [49]. Therefore, we feel that these potential confounding factors were not likely of first-order importance in the observed changes in water quality.

6. FUTURE TRENDS

Within the past year, Kewaunee County passed a winter spreading ordinance that is more restrictive than either the Brown or Manitowoc ordinances [40]. Starting in 2015, this regulation prohibits the land application of animal waste during the frozen ground months and extends into the thawed period (15 April). Future study will be needed to determine the impact of this regulation on well water.

On a statewide level, the proposed revisions to the USDA NRCS 590 Nutrient Management Standard [56,57,58] includes the designation of "Silurian Dolomite" soils as "areas where Silurian dolomite bedrock is present within 1.52 m (60 inches) of the surface." It is clear that the 2007 KTF report is reflected in the current draft of the WI NRCS 590 practice standard including: the immediate (subsurface) incorporation of manure within 24 h in areas known to deliver surface water runoff to direct conduits to groundwater and no winter application of liquid manure in February and March when soils are frozen or snow covered [57].

7. CONCLUSIONS

The combination of the geology of northeastern Wisconsin and the seasonal spreading of animal waste (manure) on frozen soil are contributing factors to the detection of pathogen indicator bacteria in the aquifer and Brown Water Incidents (BWIs). The implementation of seasonal winter manure spreading restrictions on these sensitive areas did not eliminate the contamination, but did significantly reduce the risk of both pathogen contamination and BWIs, resulting in improved aquifer water quality.

ACKNOWLEDGMENTS

We are very grateful to the many individuals cited in this article that provided water quality data and other information, without which this study could not have been completed. Steve Buan (NOAA) and Rick Wayne (UW Madison) helped with finding some of the climatological data used in the study. Dave Mechenich (UW Stevens Point Center for Watershed Science and Education) and Sarah Congdon (UW Extension Environmental Resources Center) provided drafting and graphics assistance. Jenna Klink helped review some of the statistical analysis results. We are also grateful to three anonymous reviewers, whose comments and suggestions greatly enhanced this paper.

AUTHOR CONTRIBUTIONS

Kevin Erb was responsible for the data collection, and worked with WDNR, CH and SWCD staff in four counties to gather the data and coordinating the writing of the paper. Eric Ronk's contribution includes early data gathering and summarization of contamination issues in the region (prior to his current position) researching and meeting with Calumet County LWCD staff to draft and update part of the results, as well as the data on the trends in the dairy industry in the region. Vikram Koundinya performed the statistical analyses. John Luczaj provided information on geological and climatological data, and assisted in the development of the project, as well as edited the figures and tables.

CONFLICTS OF INTEREST

The authors declare no conflict of interest.

REFERENCES

1. Ford, D.; Williams, P. Karst Hydrogeology and Geomorphology, 2nd ed.; John Wiley & Sons Ltd.: West Sussex, UK, 2007; pp. 1–562.
2. Coxon, C. Chapter 5, Agriculture and Karst. In Karst Management, 1st ed.; van Beynen, P.E., Ed.; Springer: New York, NY, USA, 2011; pp. 103–138.
3. Guo, F.; Yuan, D.; Qin, Z. Groundwater Contamination in Karst Areas of Southwestern China and Recommended Countermeasures. Acta Carsologica **2015**, 32, 389–399.
4. Worthington, S.R.H.; Smart, C.C.; Ruland, W.W. Assessment of Groundwater Velocities to the Municipal Wells at Walkerton. In Ground and Water: Theory to Practice; Proceedings of the 55th Canadian Geotechnical and 3rd Joint IAH-CNC and CGS Groundwater Specialty Conferences, Niagara Falls, ON, USA, 20–23 October 2002; Stolle, D., Piggott, A.R., Crowder, J.J., Eds.; pp. 1081–1086.
5. Celico, F.; Musilli, I.; Naclerio, G. The impacts of pasture- and manure-spreading on microbial groundwater quality in carbonate aquifers. Environ. Geol. **2004**, 46, 233–236.
6. Hrudey, S.E.; Payment, P.; Huck, P.M.; Gillham, R.W.; Hruday, E.J. A fatal waterborne disease epidemic in Walkerton, Ontario: Comparison with other waterborne outbreaks in the developed world. Water Sci. Technol. **2003**, 47, 7–14.

7. Luczaj, J.A.; Masarik, K. Groundwater Quantity and Quality Issues in a Water-Rich Region: Examples from Wisconsin, USA. Resources **2015**, 4, 323–357.

8. Center for Watershed Science and Education (CWSE) WI Well Water Viewer. University of Wisconsin-Stevens Point. Available online: http://www.uwsp.edu/cnr-ap/watershed/Pages/ Well Water Viewer.aspx (accessed on 29 June 2015).

9. Bonness, D.; Masarik, K. Investigating Intra-annual Variability of Well Water Quality in Lincoln Township; Final Report; June; 2014; pp. 1–42. Available online: http://www.uwsp.edu/cnr-ap/ watershed /Documents/Lincoln_FinalReport.pdf (accessed on 2 July 2015).

10. Bauer, A.C.; Wingert, S.; Fermanich, K.J.; Zorn, M.E. Well water in karst regions of northeastern Wisconsin contains estrogenic factors, nitrate, and bacteria. Water Environ. Res. **2013**, 85, 318–326.

11. Borchardt, M.A.; Bradbury, K.R.; Alexander, E.C., Jr.; Kolberg, R.J.; Alexander, S.C.; Archer, J.R.; Braatz, L.A.; Forest, B.M.; Green, J.A.; Spencer, S.K. Norovirus outbreak caused by a new septic system in a dolomite aquifer. Ground Water **2011**, 49, 85–97.

12. Borchardt, M.A.; Bertz, P.D.; Spencer, S.K.; Battigelli, D.A. Incidence of enteric viruses in groundwater from household wells in Wisconsin. Appl. Environ. Microbiol. **2003**, 69, 1172–1180.

13. Erb, K.R.; Stieglitz, R. Final Report of the Northeast Wisconsin Karst Task Force; University of Wisconsin—Extension: Madison, WI, USA, 2007; pp. 1–46.

14. Czymmek, K.; Geohring, L.; Lendrum, J.; Wright, P.; Albrecht, G.; Brower, B.; Ketterings, Q. Manure Management Guidelines for Limestone Bedrock/Karst Areas of Genesee County, New York: Practices for Risk Reduction; Animal Science Publication Series No. 240; Cornell University: Ithaca, NY, USA, 2011; pp. 1–9. Available online: http://nmsp.cals.cornell.edu/publications/files/ Karst_2_15_2011.pdf (accessed on 8 August 2015).

15. Ronk, E.; Erb, K. A Preliminary Analysis of 300 Manure Incidents in Wisconsin. In Proceedings of the 2010 Wisconsin Crop Management Conference, Madison, WI, USA, 12–14 January 2010.

16. Gupta, S.; Munyankusi, E.; Moncrief, J.; Zvomuya, F.; Hanewall, M. Tillage and Manure Application Effects on Mineral Nitrogen

Leaching from Seasonally Frozen Soils. J. Environ. Qual. **2004**, 33, 1238–1246.

17. Luczaj, J.A. Geology of the Niagara escarpment in Wisconsin. Geosci. Wis. **2013**, 22, 1–34. Available online: http://wgnhs.uwex.edu/pubs/gs22a01/ (accessed on 2 July 2015).
18. Mudrey, M.G., Jr.; Brown, B.A.; Greenberg, J.K. Bedrock Geologic Map of Wisconsin; University of Wisconsin-Extension, Geological and Natural History Survey: Madison, WI, USA, 1982; scale= 1:1,000,000.
19. NOAA—National Weather Service, Index of Climate Images for Wisconsin. Available online: http://www.crh.noaa.gov /images /mkx/climate/ (accessed on 2 July 2015).
20. Wisconsin State Climatology Office (WSCO). Statewide Wisconsin Climate Data. Available online: http://www.aos.wisc.edu /~sco/clim-history/state/ (accessed on 2 July 2015).
21. University of Wisconsin-Extension, Wisconsin Cooperative Observer Database, Wisconsin State Climatology Office (WSCO). Available online: http://agwx.soils.wisc.edu/uwex_agwx/ weather/hyd (accessed on 7 July 2015).
22. Wisconsin State Climatology Office (WSCO). Wisconsin Winter Climate. Available online: http://www.aos.wisc.edu/~sco/ seasons/winter.html#Snow (accessed on 2 July 2015).
23. Department of Agriculture, Trade and Consumer Protection (DATCP). Agricultural Chemicals in Wisconsin Groundwater (April 2008). Available online: http://datcp.wi.gov/uploads/Environment /pdf/ARMPub180.pdf (accessed on 20 August 2015).
24. Krohelski, J.T. Hydrogeology and Ground-Water Use and Quality, Brown County, Wisconsin; Wisconsin Geological and Natural History Survey: Madison, WI, USA, 1986; Volume 57, pp. 1–42.
25. United States Department of Agriculture, National Ag Statistics Service (UDSA, NASS). 2002 Census of Agriculture. Available online: http://www.agcensus.usda.gov/Publications/2002/ (accessed on 8 July 2015).
26. USDA, NASS. 2007 Census of Agriculture. Available online: http://www. agcensus.usda.gov/Publications/2007/ (accessed on 8 July 2015).

27. USDA, NASS. 2012 Census of Agriculture. Available online: http://www.agcensus.usda.gov/Publications/2012/ (accessed on 8 July 2015).

28. United States Department of Agriculture, Natural Resources Conservation Service. March 2005 Wisconsin NRCS 590 Standard. 4 July 2015. Available online: http://efotg.sc.egov.usda.gov /references/public/WI/590.pdf (accessed on 8 July 2015).

29. Warzecha, C.; Gerhardt, R.; Kluender, S. Wisconsin Private Well Water Quality Survey; NGWIC Call Number GB1025.W6.W45; Wisconsin Department of Natural Resources: Madison, WI, USA, 2015.

30. Heinen, L.; Drinking Water and Ground Water Specialist, Wisconsin Department of Natural Resources. Personal Communication, 7 April 2015.

31. Smith, T.; Resource Conservationist, Manitowoc Soil and Water Conservation Department, Manitowoc, WI, USA. Personal Communication, 7 April 2015.

32. Riesterer, B.; Resource Conservationist, Manitowoc Soil and Water Conservation Department, Manitowoc, WI, USA. Personal Communication, 7 April 2015.

33. Manitowoc County, Wisconsin. Ordinances, Chapter 19, Animal Waste Management. Available online: http://www.manitowoc county.com/Upload/8/Chapter%2019%20-%202011-0118H.pdf (accessed on 5 May 2015).

34. Brown County, Wisconsin. Ordinances, Chapter 26, Animal Waste Management. Available online: http://www.co.brown.wi.us /i_brown/d/county_clerk/2013_web/chap026-updated_6-24-13.pdf?t=1373902597 (accessed on 5 May 2015).

35. Kleiber, A.; Land Resource Specialist, Calumet County Land and Water Conservation Department, Chilton, WI, USA. Well water incidents. Personal Communication, 4 April 2015.

36. Reali, A.; County Conservationist, Calumet County Land and Water Conservation Department, Chilton, WI, USA. Well water incidents. Personal Communication, 8 April 2015.

37. Santry, D.; Water Resource Specialist, Calumet County Land and Water Conservation Department, Chilton, WI, USA. Well water incidents. Personal Communication, 8 April 2015.

38. Schuster, W.; County Conservationist, Door County Soil and Water Conservation Department, Sturgeon Bay, WI, USA. Well water incidents. Personal Communication, 7 April 2015.

39. Schuster, W.; County Conservationist, Door County Soil and Water Conservation Department, Sturgeon Bay, WI, USA. Brown water incidents. Personal Communication, 27 June 2015.

40. Bonness, D.; County Conservationist, Kewaunee County, Wisconsin Land and Water Conservation Department, Luxemburg, WI, USA. Personal Communication, 7 April 2015.

41. Layton, A.; McKay, L.; Williams, D.; Garrett, V.; Gentry, R.; Sayler, G. Development of Bacteroides 16S rRNA Gene TaqMan-Based Real-Time PCR Assays for Estimation of Total, Human, and Bovine Fecal Pollution in Water. Appl. Environ. Microbiol. **2006**, 72, 4214–4224.

42. Wergin, A.; RN BSN, Health Officer, Manitowoc County Health Department, Manitowoc, WI, USA. Karst Data. Personal Communication, 17 June 2015.

43. Kinnard, C.; Registered Nurse, Public Health Director, Kewaunee County Health Department, Kewaunee, WI, USA. Karst Paper. Personal Communication, 18 May 2015.

44. Halverson, J.; County Conservationist, Manitowoc County, Wisconsin Soil and Water Conservation Department, Manitowoc, WI, USA. Research Paper. Personal Communication, 5 May 2015.

45. Bonness, D.; County Conservationist, Kewaunee County, Wisconsin Land and Water Conservation Department, Luxemburg, WI, USA. Personal Communication, 4 April 2015.

46. Jolly, J.; County Conservationist, Brown County Land and Water Conservation Department, Green Bay, WI, USA. Personal Communication, 3 April 2015.

47. Santry, D.; Water Resource Specialist, Calumet County Land and Water Conservation Department, Chilton, WI, USA. Karst Data Request. Personal Communication, 14 April 2015.

48. Gollman, R.; Environmental Lab Manager, Brown County (WI) Health Department, Green Bay, WI, USA. Personal Communication, 13 May 2015.

49. Komiskey, M.J.; Stuntebeck, T.D.; Frame, D.R.; Madison, F.W. Nutrients and sediment in frozen-ground runoff from no-till fields

receiving liquid-dairy and solid-beef manures. J. Soil Water Conserv. **2011**, 66, 303–312.

50. State of Wisconsin. Final Report of the Manure Management Task Force; Wisconsin Department of Ag, Trade and Consumer Protection: Madison, WI, USA, 10 March 2006.

51. Wergin, A.; RN BSN, Health Officer, Manitowoc County Health Department, Manitowoc, WI, USA. Karst Data. Personal Communication, 17 June 2015.

52. Jolly, J.; County Conservationist, Brown County Land and Water Conservation Department, Green Bay, WI, USA. Personal Communication, 3 April 2015.

53. Kinnard, C.; Registered Nurse, Public Health Director, Kewaunee County Health Department, Kewaunee, WI, USA. Maps for paper. Personal Communication, 18 May 2015.

54. Merry, J.; Water Supply Specialist, Hydrogeologist, WDNR, Green Bay, WI, USA. Karst Paper. Personal Communication, 8 June 2015.

55. Wisconsin State Climatology Office (WSCO). Green Bay Climate—Number of Days of 1 Inch or More Now Cover Per Snow Season: Green Bay Airport (1949/1950–2013/2014 season). Available online: http://www.aos.wisc.edu/~sco/clim-history/stations/grb/GRB-AP-D-snwcv.gif (accessed on 2 July 2015).

56. USDA, NRCS. Draft March 2015 Wisconsin NRCS 590 Standard. Available online: http://socwisconsin.org/wp-content/uploads/2015/03/Draft590Standard_BroadReview.pdf (accessed on 15 April 2015).

57. USDA, NRCS. Draft March 2015 Wisconsin NRCS 590 Tech Note 1. Available online: http://socwisconsin.org/wp-content/uploads/2015/03/590TechNote_BroadReview1.pdf (accessed on 15 April 2015).

58. Murphy, P.; State Resource Conservationist, United States Department of Agriculture Natural Resources Conservation Service, Madison, WI, USA. Personal Communication, 13 April 2015.

8
CHAPTER

ASSESSMENT OF HYDROGEOCHEMISTRY AND ENVIRONMENTAL ISOTOPES IN KARST SPRINGS OF MAKOOK ANTICLINE, KURDISTAN REGION, IRAQ

Omed Mustafa [1,*], Broder Merkel [1] and Stephan M. Weise [2]

[1] *Institute of Geology, Technische Universität Bergakademie Freiberg, Gustav Zeuner Str.12, 09599 Freiberg, Germany*

[2] *Department Catchment Hydrology, Helmholtz Centre for Environmental Research-UFZ, Theodor-Lieser-Str. 4, 06120 Halle (Saale), Germany*

ABSTRACT

Karst springs of the Makook anticline were investigated to get more insight into the hydrogeologic, hydraulic, and hydrodynamic behavior of this complex karst system. Eight springs were studied in terms of hydrogeochemical constituents as well as $\delta^{18}O$ and $\delta^{2}H$ during September 2011 to November 2012. For the first time, the local meteoric water line was plotted for the area based on precipitation data from the period November 2011 to April 2012. The regional meteoric line plots between the global and Mediterranean meteoric water lines. The majority of the spring samples plot between the local and Mediterranean meteoric water lines implying influences by Mediterranean air masses, rapid infiltration of rainfall through the karst system and a short residence time (shallow karst aquifers). The correlation of d-excess and saturation index of halite was used to evaluate the evaporation process in karst waters of the area. The temporal variation in isotopic composition of karst springs was observed making it possible to distinguish between different origins for springs and indicate a possible connection between the aquifers and depending on this, the karst system can be classified into three aquifers: Behkme aquifer, Kometan aquifer, and Shiranish aquifer.

Keywords: karst springs; Bekhme; d-excess; environmental isotopes; local meteoric water line; Kurdistan Region

1. INTRODUCTION

Karst springs often discharge substantial amounts of water, which draws the interest of water resources researchers especially in arid and semi-arid areas facing water scarcity. Different methods and techniques were used by researchers to understand the response of the karst system to recharge events. Thermal patterns of karst springs were used by [1] as a key base for identifying two types of systems: thermally ineffective springs corresponding to rapid flow of event waters and effective thermal systems characterized by small fractures and slower water movements leading to equilibration of recharge and aquifer water's temperature. Electrical conductivity of karst waters may represent a good indicator to outline the flow system, whether it has a conduit or diffuse flow (retain piston flow). A piston flow phenomenon is the displacement of more mineralized water by intense rainfall recharge water toward the spring [2,3]. The hydrodynamic and hydrogeochemical characteristics together can outline the behavior of karst springs [2].

Karst springs of the Makook anticline represent the main source of water for drinking, agriculture, and other uses in the areas of Balisan and Akoyan, Kurdistan Region, Iraq. The population of villages and towns in the part of Ranya district in which the studied area is located comprised more than 20,000 inhabitants. Water demand in the area has increased sharply in the last 20 years because of the increase in population and their needs. The mountainous area with poorly developed infrastructure creates certain limitations for studying the area. Few works have been done in the field of isotopes in the area in recent years. Only two local meteoric water lines (LMWL) were determined in nearby areas. The LMWL for Haji Omaran area was extrapolated by Mawlood [4] from isotopic data of Halab station in Syria and Adana station in Turkey, and the meteoric water line of Basara area was published by Hamamin and Ali [5]. During 2000–2003 geologic and hydrologic investigations were carried out by a team from the Food and Agriculture Organization (FAO), including groundwater monitoring of Sarwchawa spring and wells in the area. Al Manmi [6] did an environmental isotopic study of surface and groundwater in Ranya area, which is close to the studied area and includes some of the springs considered in this work. According to [6], water of the Ranya area is of good quality with low salinity and thus suitable for drinking and irrigation purposes. Some possible recharge sources were determined depending on isotopic fingerprints. Karst aquifers in Makook system are mentioned as one thick Bekhme aquifer, which, according to [6,7], may contain Bekhme, Qamchuqa, and Kometan Formations. Bekhme aquifer consists of highly fissured, well-karstified carbonate rocks containing many caves and channels.

The objective of this study is to explore the applications of environmental isotopes of $\delta^{18}O$ and $\delta^{2}H$ in the field of karst springs. This work includes an attempt to build a local water line for the area, which had not done before. The local meteoric water line was correlated to the regional lines to investigate the source and determine the residence time of this water till it reaches the springs. This study also addresses the relation between hydrodynamic and hydrogeochemical characteristics of the karst springs.

2. STUDY AREA

2.1. Geography

The area of interest is located in the Kurdistan region in the northeast of Iraq, mainly in Sulaimani governorate, Ranya district (Figure 1). The area is situated within latitudes 36°10′–36°35′ north and longitudes 44°30′–44°50′ east in the elevation range of 500 to >2000 meter (m) above sea level (a.s.l). Makook Anticline Mountain lies in a northwest–southeast direction and is surrounded by Shawre valley in the northeast and Balisan valley in the southwest. The area of interest comprises about 400 km^2 and is part of the Dokan lake catchment.

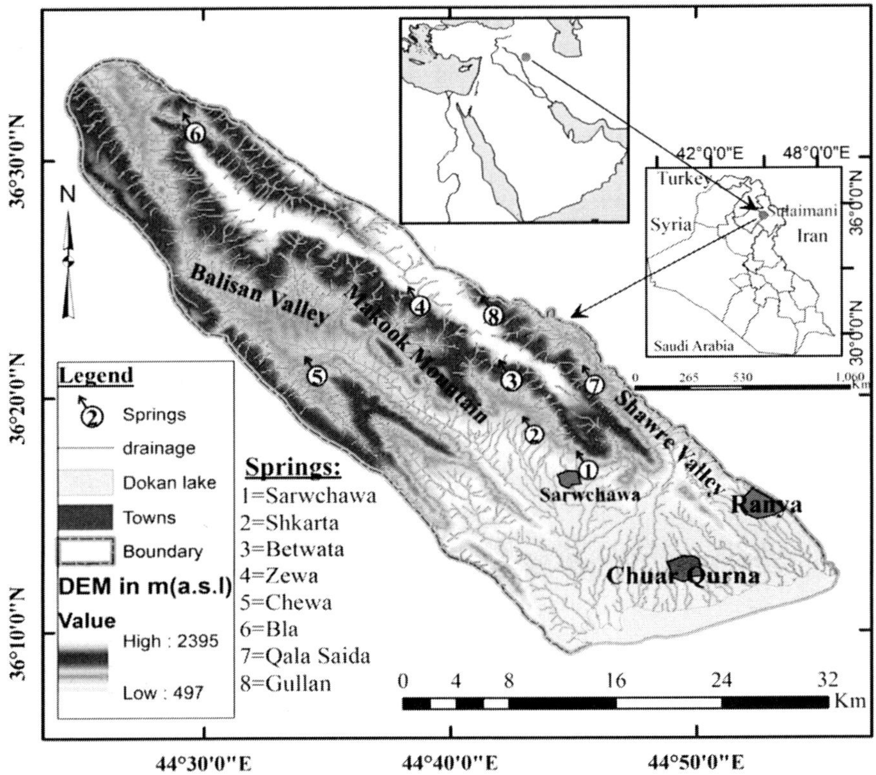

Figure 1. Digital elevation model (DEM) showing the area of interest in Kurdistan Region, Iraq. Numbers 1 to 8 refer to the springs where sampling was conducted.

The studied area is characterized by humid to moist climate. Rainy season starts from October and mainly ends in May (with minor showers in June, July, and September). The mean annual climatic parameters are presented in Figure 2 for the period 1980 to 2012. The highest precipitation rate, highest relative humidity, and lowest temperature are observed in February. The mean annual rainfall for the mentioned period is 570.5 mm and the average annual temperature is 20 °C. The mean annual evaporation (pan) is about 2000 mm and the mean relative humidity is about 47% [8].

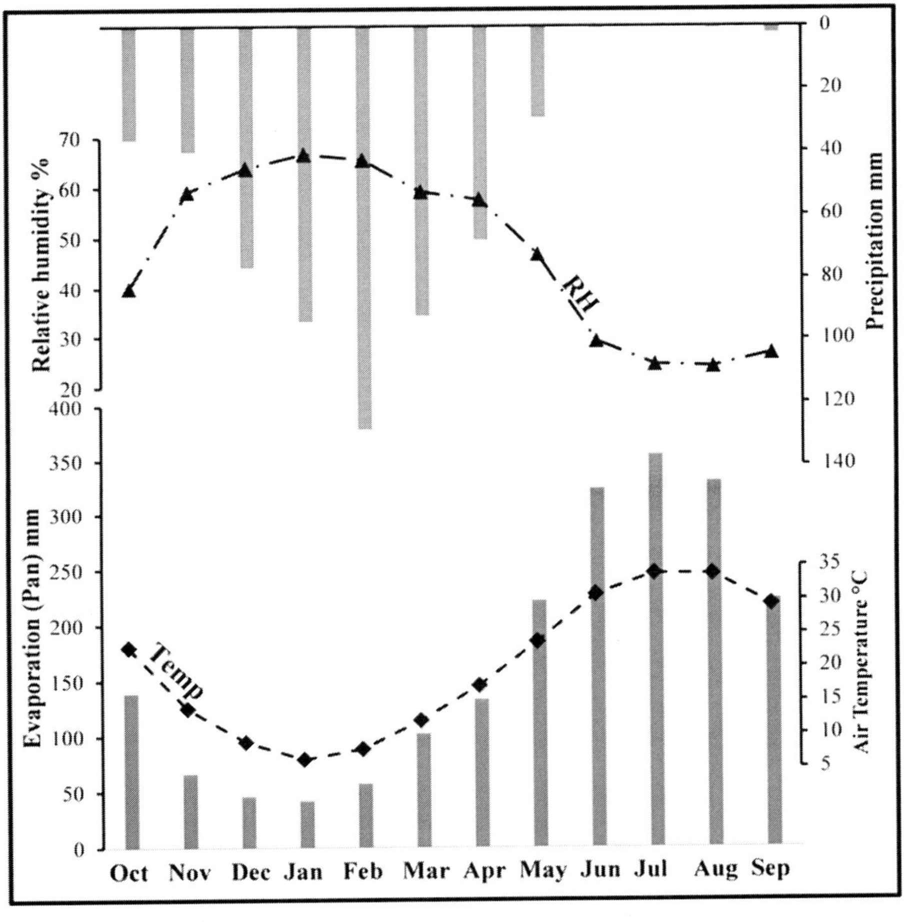

Figure 2. Climatological data of the Dokan Meteorological Station (690 m a.s.l) located in the investigation area. Averages for the period 1980–2012 are shown [8].

2.2. Geologic Setting

Rocks of Mesozoic and Cenozoic age can be found in the region of interest (Figure 3). Two units of Jurassic rocks are cropping out, representing older rocks: (I) Sarki and Sehkaniyan formation, which are composed mainly of limestone, dark dolomite, and shale [9]; and (II) Sargelu formation (Bajocian–Bathonian), Naokelekan formation (Late Oxfordian–Early Kimmeridgian), Barsarin formation (Kimmeridgian–Early Tithonian), and

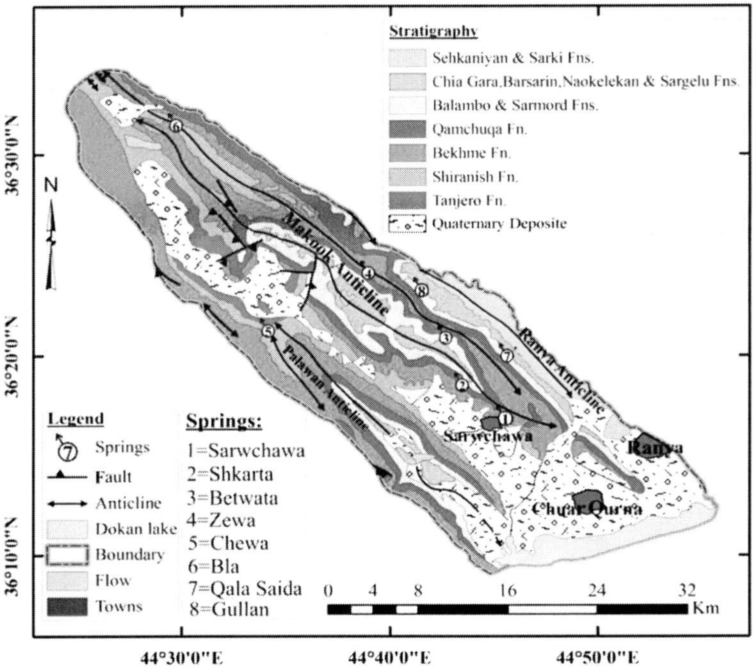

Figure 3. Geological map of the area (modified from [12]). Numbers 1 to 8 refer to the springs where sampling was conducted.

Chia Gara formation (Late Tithonoan), which include bedded to massive dolomite, limestone, bituminous limestone, and marl. The Cretaceous unit comprises Balambo and Sarmord formation (marly and bedded limestone, bedded dolomite, and marl), Qamchuqa formation (mainly massive limestone and dolomite), Bekhme formation and locally Kometan formation (well bedded to massive limestone), Shiranish formation (well bedded limestone and marl), and Tanjero formation (sandstone, claystone, and conglomerate) [10]. Kometan formation does not appear in the geological map (Figure 3), while it is more regional map and the existence of Kometan formation was in time of map creation still suspicious. Near Sarwchawa town, Kometan formation changed laterally to Bekhme formation through a transitional zone [11], and both formations are possibly overlain by Shiranish formation. Jurassic and Cretaceous rocks are occasionally overlain by Quaternary fan deposits (boulders, gravel, and fine clastics) and a thin layer of soil (mainly in the valleys, because of soil erosion in that mountainous area).

Makook Anticline represents a double plunging anticline (NW–SE) within in the parallel trend of Zagros folded structures between Ranya and Palawan anticlines and Sahwre and Balisan synclines (in between) in the NE and SW, respectively (Figure 3). The area is characterized by tectonic distortion, especially in the northwest part of Makook Anticline.

2.3. Hydrogeology

Generally, all springs are within the Makook Anticline Mountain, except Chewa spring, which is located in the Balisan valley. Two main Karst aquifers are recognized in the region of interest. First, Bekhme karst aquifer supplies major springs along Makook anticline, including Zewa, Chewa, Bla, and Gullan springs (Figure 3). Bekhme aquifer is distinguished as a well-karstified, thick, semi-confined aquifer underlain by Sarmord formation (aquiclude). The second one is Kometan formation, which represents the source of Sarwchawa and Betwata springs. Kometan formation is a well-karstified, highly fissured aquifer with confined to semi-confined conditions and overlain by Shiranish formation [7]. The aquifer is composed of carbonate rocks and contains a large amount of groundwater, varying in space and time. Strong karstification cycles took place in the aquifer system during the Paleocene–Miocene period [13]. It is assumed that karstification happens recently. One of the springs (Shkarta) drains poorly developed karst (with small fissures). This spring is located in the Shiranish formation (marl and marly limestone), which was inconsistently recognized as a fissured aquitard (locally in the marly limestone part) by [6]. The existence of Qamchuqa formation aquifer in studied area is possible, as mentioned by [6], but no field evidence has been observed.

3. METHODOLOGY

Eight karst springs (Sarwchawa, Shkarta, Betwata, Zewa, Chewa, Bla, Qala Saida, and Gullan) were monitored and sampled during six periods (September 2011, December 2011, April 2012, June 2012, September 2012 and November 2012). Precipitation samples (two snow and eight rain samples) were collected between November 2011 and April 2012 from three locations near Ranya town, Qala Saida and Gullan springs. Precipitation samples were collected by rain collector non-systematically depending on precipitation events. Field parameters (pH, conductivity and water temperature) were measured on-site (springs outlet). Redox potential (Eh) was measured via a WinLab redox meter (Windaus

Labortechnik, Germany) and WTW SenTix-ORP electrode (WTW: a Xylem Group brand, Germany) and checked by standard redox buffer (pH = 7 and 220 millivolts at 25 °C). The parameters pH, electrical conductivity (EC), dissolved oxygen (DO) and water temperature (T) were measured on-site using a multi-parameter WTW model 3430, WTW pH electrode SenTix-940, WTW TetraCon-925 conductivity electrode and WTW FDO-925 optical dissolved oxygen sensor. Three-point calibration was carried out for the pH electrode via technical WTW buffers (pH = 4.01, pH = 7 and pH = 10). The EC electrode was checked with WTW standard control solution; settings were chosen so that the output of the electrode was converted to a water temperature of 25 °C. The optical sensor for dissolved oxygen measurement was checked regularly using a water vapor saturated air vessel. A certified mercury-thermometer was employed for air temperature measurement. Alkalinity was determined immediately after sampling by means titration and converted to hydrogen-carbonate (HCO−3) according to [14]. Because no continuous discharge measurements of the springs were available, different methods (velocity–area using current meter and volumetric readings) were employed to record the discharge of the springs during the monitoring period from September 2011 to November 2012. Due to difficulties in the field, flow measurements and field parameters were measured only six times (September 2011, December 2011, April 2012, June 2012, September 2012 and November 2012) accompanying the sampling processes (Table S1).

All samples were filtered by 0.2 μm filters and stored at 4 °C until analysis in the laboratory. Samples analyzed by ICP-MS were acidified with an ultra-pure 1:1 nitric acid (30% HNO_3) mixture after filtration in the field. Batches of 48 samples were analyzed in the laboratories of the Chair of Hydrogeology, Technische Universität Bergakademie Freiberg, Germany. Lab work included determination of Si and Sr ions with an ICP-MS XSeries-2 model (Thermo Scientific, Bremen, NS, Germany). Major cations Ca^{2+}, Mg^{2+}, Na^+, K^+ were determined by ion chromatography (IC) using an 850 Professional IC Metrohm (Switzerland) and Metrosep C4-150 column with 2 mM dipicolinic acid eluent. Major and minor anions F^-, Cl^-, PO3−4, SO2−4, Br^-, and NO−3 were determined by a Metrohm Compact IC Pro 881 (Switzerland) and Metrosep A sup 15–150 column with 3 mM $NaHCO_3$ and 3.5 mM Na_2CO_3 eluent. The reproducibility of IC and ICP-MS determination were less than 2% and 5%, respectively. Stable isotopes of oxygen ($\delta^{18}O$) and deuterium (δ^2H) were determined by means of an LGR liquid-water isotope analyzer-DLT 100 (ABB Company, Mountain View, CA, USA) with a precision of <0.1‰ for $\delta^{18}O$ and <0.3‰ for δ^2H.

The ion balance, saturation index of calcite ($SI_{Calcite}$), dolomite ($SI_{Dolomite}$), gypsum (SI_{Gypsum}), fluorite ($SI_{Flourite}$), halite (SI_{Halite}), celestine ($SI_{Celestine}$) and partial pressure of CO_2 (P_{CO2}) in the spring samples were calculated by means of PHREEQC [15] with the WATEQ4F database.

Non-parametric two-tailed correlation analysis (Kendall and Spearman correlations) was performed for the hydrogeochemical and field parameters in addition to basic statistical treatment using SPSS software package.

4. RESULTS AND DISCUSSION

Hydrologic and hydrochemical data of the Makook Karst springs are shown in the electronic Tables S1 and S2. The hydrochemical data were assessed by PHREEQC [15]. The relative analytical error (E%) of the most samples was within an acceptable range of ±2%, except Shkarta and Zewa springs, which have a higher deviation in April 2012 and September 2012, respectively, but generally the accuracy is acceptable (Table S2).

4.1. Discharge and Recharge Estimations

From the discharge of springs, mean annual precipitation, catchment area, and average chloride concentration in precipitation and groundwater, different hydrologic parameters were estimated. Maximum discharge of karst springs has an important impact on hydrological modeling [16]. The maximum estimated discharge (D_{me}) for an area is defined as:

$$D_{me} = P \times A_c \tag{1}$$

where P is mean annual precipitation in mm and A_c is catchment area in square meter (m^2). The mean annual precipitation is 570.5 mm (1.81×10^{-5} $L/m^2/s$) and the studied area is 4×10^8 m^2; therefore D_{me} of the studied karst area is 7240 L/s.

The recharge in the area of interest was estimated using two methods: chloride mass-balance (CMB) and empirical Kessler method. The CMB method has often been used for recharge estimation [17,18] and is applicable in semi-arid regions when additional sources of chloride are absence [19,20]. The precipitation in the area of interest was sampled 10 times during hydrological year (2011–2012). The temporal variation of chloride seems to be rather low (Table 1). Thus it is assumed that this average can be used for recharge calculation. Most of the chloride in the karst water is from precipitation (about 0.08 meq/L). Only 0.7% (0.02

meq/L Cl) of the chloride in karst waters stems from carbonate rocks, except in Shkarta spring where the chloride concentration is about 0.2 meq/L by the interaction of water with clay minerals of Shiranish aquifer (Table 1 and Table S2). Most of the recharge areas of the springs are not arable land (except some areas related to Sarwchawa and Shkarta springs). Therefore fertilizers as additional source of chloride were excluded. Surface runoff in the recharge area of the springs is not significant and can thus be omitted from the recharge estimation. Recharge (R) in the area of interest was estimated by the CMB method:

$$R = P \times (Cl_p/Cl_{sp}) \tag{2}$$

where Cl_p is average reference chloride concentration in precipitation and Cl_{sp} is average chloride concentration in spring water (both in mg/L). The re-calculation of Cl in precipitation to reference Cl is to consider the impact of evapotranspiration on the mass-balance estimation. The average reference Cl of precipitation is 2.52 mg/L and the average Cl in studied springs is equal to 2.9 mg/L, thus the estimated recharge is 496 mm/year. The recharge estimation by Kessler's method [21] was originally adapted for karst aquifers in semi-arid regions, and is based on the pluviometric data of last four months of the year. The estimated recharge in the area of interest by Kessler method is equal to 525 mm/year (the methodology of recharge calculation is described in [21]).

Table 1. Isotopes, altitude, and hydrochemistry of precipitation.

Sample Code	Date of Sampling	$\delta^{18}O$ ‰	$\delta^{18}O$ *StD*	δD ‰	δD *StD*	*d*-Excess ‰	Altitude m.a.s.l	T_{Air} °C	*Evap.* mm	*Precip.* mm	pH	Cl (mg/L)	Na (mg/L)
R-1	25.11.11	−5.18	0.05	−23.07	0.34	18.4	882	11	1.4	0.2	6.3	1.9	0.41
R-2	31.12.11	−9.18	0.08	−52.8	0.16	20.6	882	8.5	1.4	12.5	6.5	3.2	0.2
R-3	18.01.12	−3.35	0.08	−1.95	0.29	24.8	882	7.5	0.6	9	6.2	2.4	0.1
R-4	31.01.12	−12.29	0.07	−82.31	0.18	16	882	7.5	1	45	6.3	2.8	0.17
R-5	07.02.12	−7.91	0.11	−44.95	0.52	18.3	882	7	2	20.2	6.1	1.9	0.06
R-6	18.02.12	−9.5	0.12	−57.6	0.57	18.4	882	8.5	2.1	9	6.4	2.5	0.43
R-7 *	18.02.12	−8.27	0.08	−39.17	0.27	27	1460	8.5	2.1	9	6.3	2.5	0.7
R-8	02.04.12	−9.27	0.08	−54.57	0.38	19.6	882	7.5	1.1	15	6.5	2.4	0.76
R-9 *	04.04.12	−10.66	0.03	−55.77	0.25	29.5	845	3.5	1.3	25	6.5	2.3	0.22
R-10	15.04.12	−5.38	0.04	−28.85	0.24	14.1	845	12.5	4.3	29	6.5	4.2	0.43

StD = Standard deviation; * = snow samples; T_{Air} = Air temperature; ***Precip.*** = Precipitation; **Evap.** = Evaporation; Meteorological parameters represents mean daily values of the sampling date.

The recharge rate according to Kessler's estimation is 92% and of CMB is 87%. The difference between the two estimations is normal (being different approaches were used), but the latter is more reasonable depending on the field observation, and conclusions of [22] in karst systems of other semi-arid region. These estimations are convenient with the calculated effective recharge (R_E) that defined as:

$$R_E = (D_{sp}/D_{me}) \times 100 \quad (3)$$

where R_E is the effective recharge (the amount of water that reach the groundwater), D_{sp} is the total average discharge of the springs (4037 L/s), and D_{me} is maximum estimated discharge. The effective recharge in the studied area is 55.8%, and depending on these estimations it can be assumed that minority of the precipitation will loss by evaporation and transpiration.

4.2. Springs Hydro-Chemograph Analysis

Relating to the hydro-chemographs, different interpretations for the studied springs can be concluded (Figure 4). Flow rates in the springs (discharge) of the study area vary according to the capacity of the aquifer and degree of karstification in the spring aquifer (Figure 4a and Table S1). The Sarwchawa spring shows higher flow rates (2370 and 4630 L/s in dry and rainy seasons, respectively) and represents the most developed karst spring in the Makook karst system. The Shkarta spring (Shiranish formation) represents the less developed karst or non-karst spring with a low flow rate of 0.75 L/s, originating from fissured marly limestone [6]. Sarwchawa, Shkarta and Qala Saida springs show a direct response of increasing discharge with respect to October 2011/April 2012 precipitation events (Figure 4a,b). In Shkarta spring, the response of discharge to the recharge events is less compared to the majority of the springs. As results of low flow rate in marly rocks, the residence time then will be high [23]. Therefore, low flow rates in micro-fissures of Shiranish marly limestone can be the reason of high residence time and gradually varied discharge confirms this hypothesis. Betwata, Zewa, Chewa, Bla and Gullan springs show a nearly two months delay (maximum peak discharge appear in June 2012). The hydrograph of Betwata spring shows two peaks in response to the October–December 2011 and April–June 2012 rainfall events. A similar pattern can be noticed in Chewa spring's pattern but with lesser response to the October/December 2011 raining event. A gentle rise in Zewa spring's hydrograph was recorded and a steep decrease of discharge was recorded, which refers to a low storage capacity in the karst aquifer that feeds Zewa spring. Decrease in Bla's hydrograph in the December 2011 rainfall event was observed, which makes this spring different from all other studied springs.

The pH of the springs ranges between 7.1 and 8.4 (average 7.5) showing neutral to slightly alkaline water (Figure 4b and Table S1). The variability in pH of the springs is caused by changes in flow paths, water discharge, and residence times within the aquifer [24]. Two of the springs (Zewa and Gullan) show decrease in pH during the December 2011 precipitation event. The similarity in pH patterns refers to the similarity in origin of water (same aquifer). The other springs show increasing of pH within the same recharge event (December 2011 precipitation). The addition of the newly recharged precipitation (December 2011–April 2012) caused an obvious decrease in pH of all the springs, and this was repeated in the November 2012 precipitation event. The pH value of precipitation collected during the current study indicates slightly acid conditions (Table 1). According to the nature of the karst aquifer, different impacts of this acidic recharged water are expected. Therefore, temporal variation in pH value can be used as a tracer in karst waters.

The redox potential of the springs showed partially oxidizing waters (323–478 mV, with an average of 383 mV). Chemistry of the karst waters in the unsaturated zone is modified by local flow variations and changes in redox condition [25]. Analysis of the redox graphs reveals a falling pattern of Eh for all springs (Figure 4c). Variation of Eh graphs is controlled by changes in karst water temperature, and this relation is more significant in Shkarta, Chewa, Bla and Gullan springs. The increasing of Eh in Shkarta, Betwata, Chewa and Bla springs can be interpreted as a result of recharge with water containing more oxygen [26] in a shallower depth compared with the first group of springs.

Electrical conductivity ranges between 265 and 1117 μS/cm outlining differences in recharge rate and seasonal effects between wet and dry periods (Figure 4d). Generally, conductivity decreases in wet seasons, reflecting dilution by precipitation water. But this situation became completely the opposite (except Zewa and Gullan springs) during the November 2012 precipitation event when EC increases drastically during dilution by recharge. The dilution processes followed by a piston flow effect assumed to be the reason of the EC increase during the recharge period [3].

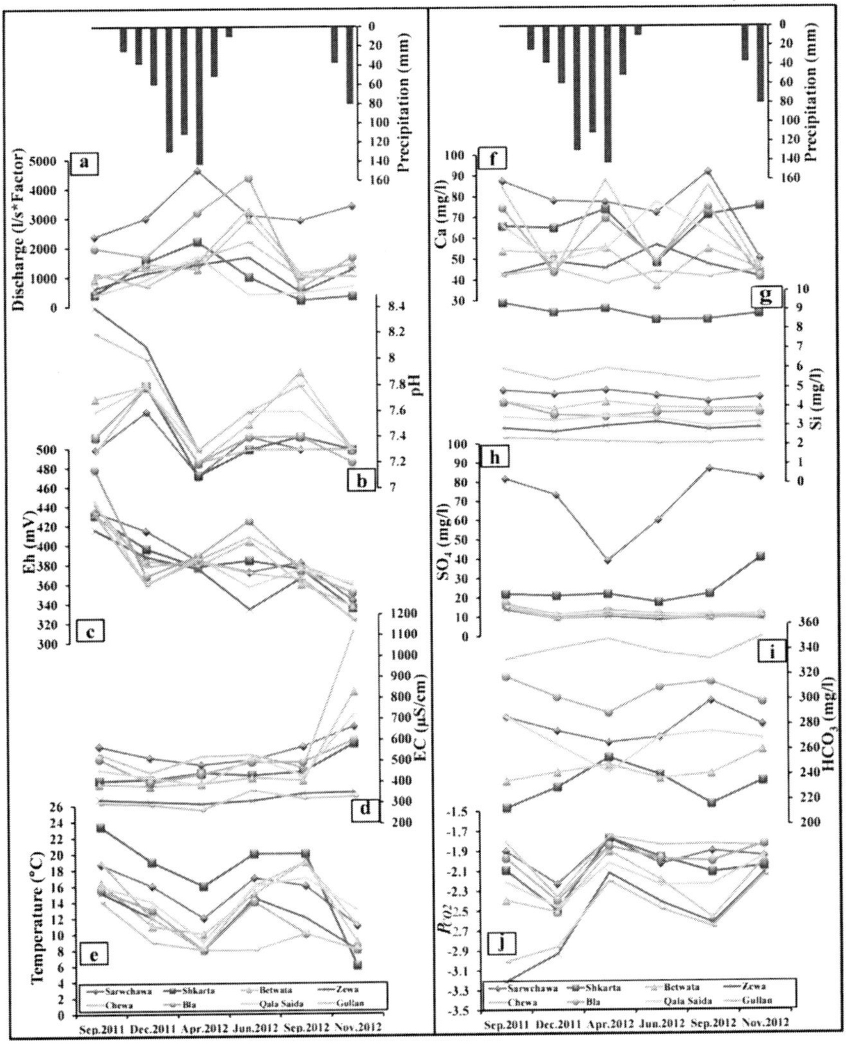

Figure 4. Hydro-chemograph of the karst springs. The recorded precipitation is from September 2011 to November 2012 period. (**a**) Hydrograph; discharge values were multiplied by a factor because of the differences in the magnitude and to better magnify their response to precipitation as follow: Sarwchawa × 1, Shkarta × 2000, Betwata × 50, Zewa and Chewa × 10, Bla × 5, Qala Saida × 250 and Gullan × 25; (**b**) pH; (**c**) Eh; (**d**) EC; (**e**) water temperature; (**f**) Ca; (**g**) Si; (**h**) SO_4; (**i**) HCO_3; and (**j**) P_{CO2}.

The water temperature of the springs ranges between 6 °C and 23.4 °C (Table S1). The water temperature pattern in Shkarta spring was like that of Sarwchawa spring, but with higher amplitude in Shkarta (water temperature variation: 17.7 °C), which was recorded as the maximum among the springs studied (Table S1). A lateral groundwater influx from one aquifer to another with presence of clay and marl layers in the vicinity of springs leads to mixing and causes a decrease in water temperature [27]. This scenario is the reason for the drastic decrease of spring water temperature (as obvious in the thermal pattern of Shkarta spring) in the rainfall period of October–November 2012 (Figure 4e). According to the stratigraphic superposition, a lateral influx of water from Kometan aquifer to Shiranish aquifer assumed to be the reason of the October–November 2012 temperature decrease in the Shkarta spring. The water temperature pattern in Betwata is the same as in the two former springs, but with nearly two-month time lag until recovery. The time lag in temperature recovery in karst springs results from low transit times, low development in karstification, and diffuse flow pattern [3]. The variability of temperature in karst springs is considered a tracer for recharge and geometry of the aquifer [1]. Slow movement of recharge water through Sarwchawa and Gullan spring aquifer results in thermal equilibrium between recharge and discharge water and indicates long residence time and rather deep karst. The enrichment of Sarwchawa spring water with Ca, Mg and SO_4 indicates higher residence time, more intense water-rock interaction, and slower movement of recharged water through the aquifer. Gullan spring shows smoother thermal fluctuation at lower temperature (8 °C–14.1 °C) compared to Sarwchawa (11 °C–18.7 °C), which may be caused by differences in altitude of the springs outlet, elevation of recharge area, flow type, and depth of the aquifer (lower altitude, deeper aquifer and conduit flow for Sarwchawa spring). There is some evidence that convective heat transport of infiltrating water (recharge) is responsible for the observed temperature patterns [28]. Unfortunately the data given are not sufficient to calculate the recharge by inverse modeling of the temperature distribution in the aquifer over time.

Temporal variation of major cations (Ca^{2+} and Mg^{2+}) and anions (SO2−4 and HCO−3) between wet and dry periods correlates to corresponding precipitation events (September 2011 to September 2012) and EC as illustrated in Figure 4. The calcium chemographs of spring water mainly show decreasing calcium concentration during recharge periods, except Zewa spring (Figure 4f), which shows the reverse (increasing Ca concentration with increasing recharge). Similar decreasing of Ca through dilution by low mineralized recharge water was reported by Vesper and White [29] for karst springs in Kentucky, USA. A zigzag-like pattern in Ca chemographs of Chewa and Bla springs was observed and indicates the similarity in recharges and flow systems for both springs. A time lag was enhanced by the most noticeable changes in hydrogeochemistry after the recharge events. This time lag suggests the presence of piston phenomena [29].

Chemograph of Si shows low variation, implying that silica in karst waters is conservative (Figure 4g). Shkarta spring shows higher Si concentration compared to the other springs, indicating presence of silicate mineral in this aquifer (Shiranish marly limestone).

Sulfate concentrations show the same variation as Ca corresponding to the dilution with recharge water, except in Zewa and Gullan springs, which showed the opposite patterns (Figure 4h). The sulfate concentration is higher in Sarwchawa spring. Sulfate-rich waters may form by oxidation of pyrite in carbonate rocks [30]. Oxidation of pyrite concretions [9,31,32] and gypsum impurities along the flow path of Kometan aquifer leads to formation of this sulfate rich water. The direct response of sulfate to recharge is interpreted as presence of conduit type of flow and is reported from different karst springs worldwide [29,33,34,35].

A general trend of increasing P_{CO2} with increasing discharge was observed in the karst springs. This reflects the decrease of diffuse atmospheric CO_2 and increase of CO_2 in soil [36]. The pattern of P_{CO2} in karst springs is consistence with sulfate patterns rather than the Ca ones (Figure 4j). Increasing of P_{CO2} is accompanied by increasing of Ca, sulfate, EC, T and Eh.

4.3. Water–Rock Interaction and Geochemical Processes

The variation in hydrochemistry of water can be used as an indicator to understand water–rock interactions and geochemical processes during flow in the aquifers [37]. In order to identify the hydrochemical type of the spring waters different techniques were used. Karst waters were classified depending on the relative abundance of Ca^{2+}, Mg^{2+}, Na^{+}, and K^{+} as major cations and HCO_3^-, SO_4^{2-}, Cl^-, and NO_3^- as major anions (in meq/L). According to the abundance, all springs in all sampling periods were classified as Ca-HCO_3 water except Sarwchawa spring, which was classified as Ca-HCO_3-SO_4 water. The excess of Ca and SO_4 in Sarwchawa spring relates to the longer residence time. A Chadha diagram classification [38] was also used for classifying the spring water, and all springs show Ca-Mg-dominant-HCO_3 type (5th-sub field), which indicates that alkaline earths and weak anions exceed both alkali metals and strong anions.

Different hydrogeochemical relations indicate different sources of geologic material for the karst waters of the area (Figure 5). The significant relations of Na–Cl and Na–(NO_3 + PO_4) refers to different aquifers and sources of geochemical materials in addition to meteoric water (Figure 5a,b). Up to 0.017 meq/L of sodium stems from rainwater, while 0.093 of Na stems from limestone and dolomite of Kometan and Bekhme formation. Excess Na in karst water stems from clay minerals and feldspars [39]. Therefore clay minerals of Shiranish marly limestone seem to be the source of excess Na in water of Shkarta spring. According to this variety of sources of geochemical materials, three different aquifers were recognized in the karst system of Makook anticline. (1) Springs feed by Shiranish aquifer (marly limestone) are characterized by high Na, Cl, NO_3 and PO_4; (2) springs feed by Kometan aquifer (limestone) shows medium content of Na, Cl, NO_3 and PO_4; and (3) springs feeds by Bekhme aquifer (dolomite) characterized by lower Na, Cl, NO_3 and PO_4 compared to the other aquifers (Figure 5a,b).

Calculated saturation indices (Table S3 and Figure 5c) indicate that Zewa, Chewa and Qala Saida springs are over-saturated ($SI \geq 0.05$) with calcite while Sarwchawa, Shkarta, Betwata and Bla springs are in equilibrium ($-0.05 \leq SI \leq 0.05$). De-dolomitization processes are in progress in waters in equilibrium with calcite and under-saturated with dolomite [40]. Only Gullan spring was under-saturated ($SI \leq -0.05$) with respect to calcite, which suggests shorter flow path and residence time in the aquifer compared to the other springs (Figure 5c). Regarding the $SI_{Dolomite}$, only Qala saida spring was in equilibrium, and the rest were under-saturated with respect to dolomite. All springs were under-saturated with respect to gypsum, fluorite, and halite. Dissolution is assumed to be incongruent if a mineral dissolves while another one is oversaturated and thus precipitate. Obviously, incongruent dissolution occurs in waters of Sarwchawa, Shkarta, Betwata and Bla springs. SI_{Gypsum} is significantly correlated with $SI_{Celestine}$ (Figure 5d) indicating presence of gypsum and celestine ($SrSO_4$) impurities in karst rock, especially in springs drained from limestone. The sources of Sr are controlled by the stratigraphy [41] as well as the lithology. Thus, limestone of Kometan formation provides the karst water with more Sr than the two other aquifers (Figure 5d).

A nonparametric Kruskal-Wallis-H test was performed to test the significance of the above grouping. The data were taken randomly and independently of each other. The results confirm that the three groups of spring water are different by means of p-value ≤ 0.05 for Ca, Mg, Na, K, Cl, SO_4, HCO_3, F, NO_3, Si and Sr, except PO_4, which shows higher p-values.

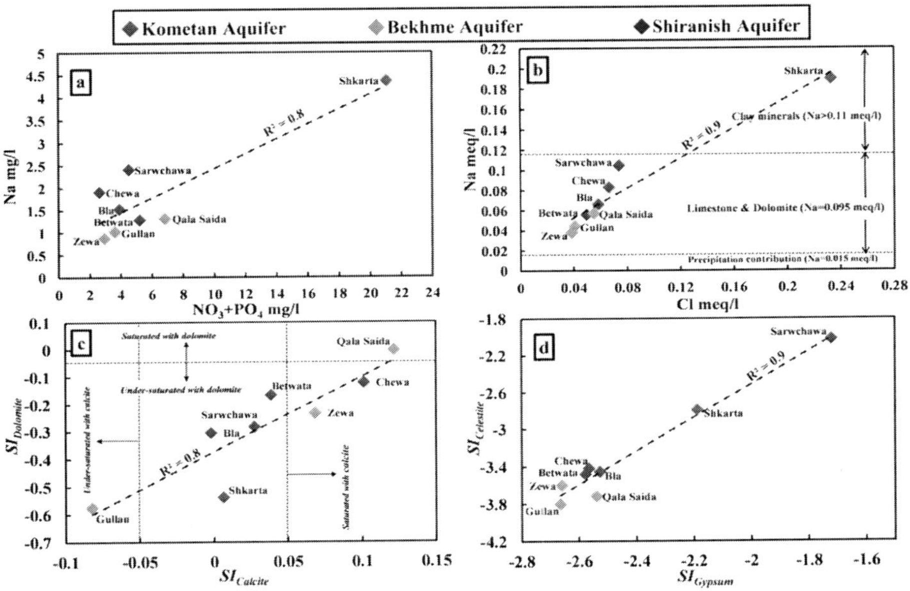

Figure 5. Hydrogeochemical relations among karst waters. (**a**) Na–(NO_3 + PO_4) plot, (**b**) Na–Cl plot, (**c**) $SI_{Calcite}$ vs. $SI_{Dolomite}$ and (**d**) SI_{Gypsum} vs. $SI_{Celestine}$; concentrations of Na and Cl in rainwaters are mean values from 10 precipitation samples, which are listed in Table 1.

4.4. Isotopes in Precipitation and Karst Springs

4.4.1. Development of a Local Meteoric Line

Oxygen ($\delta^{18}O$) and hydrogen (δ^2H) isotope compositions for the precipitation and spring samples are given in Table 1 and Table 2, respectively. The isotopic composition of precipitation in the studied area ranged between −12.28 to −3.35‰ for $\delta^{18}O$ and −82.3 to −1.95‰ for δ^2H. The regression equation of the LMWL is δ^2H‰ = 20‰ + 7.9 × $\delta^{18}O$ (R^2 = 0.95, $p \leq 0.001$, and n = 10). In relation to the Global Meteoric Water Line (GMWL), given by equation δ^2H = 10‰ + 8 × $\delta^{18}O$ [42], and the Mediterranean Meteoric Water Line (MMWL), given by equation δ^2H‰ = 22‰ + 8 × $\delta^{18}O$ [43], the LMWL is located between these two (Figure 6), and it is closer to MMWL. The LMWL of the studied area is very close to that of Haji Omaran area (50 km northeast of the studied area, 36°40′00″N, 45°03′00″E and with higher altitude) δ^2H‰ = 20‰ + 8 × $\delta^{18}O$ [4], but far away from that of Basara area (100 km southeast of the studied area, 35°20′30″N, 44°57′40″E and with lower altitude) δ^2H‰ =14.4‰ + 7.7 ×

$\delta^{18}O$ [5]. The d-excess higher than 20‰ in precipitation results from arid vapor sources [44]. The vapor sources of some precipitation samples (R-2, R-3, R-7, and R-9) are arid (low humidity), because the d-excess is higher than 20‰. The evaporation of precipitation can reduce the d-excess [44]. Therefore, the d-excess in samples R-1, R-4, R-5, R-6, R-8 and R-10 was reduced as a result of water loss by evaporation (Table 1). A continental effect is a possible reason for the high d-excess in some precipitation samples (R-3, R-7 and R-9), because they were higher than that of the MMWL. Another possible reason for this high d-excess in R-7 and R-9 is the origin and nascence condition [45], because they represent snow samples.

4.4.2. Factors Affecting Stable Isotopes in Precipitation

The air temperature is correlated with d-excess (d-excess ‰ = 31.9‰ − 1.4‰/°C × T_{Air} (°C), R^2 = 0.67, p = 0.03, n = 10). This correlation may emphasize the seasonal effects and thermodynamic conditions in the atmosphere [45,46] (Figure 7a) and may also be underlined by the increase of evaporation with temperature (Evaporation (mm) = 0.26 mm/°C × T_{Air} (°C) − 0.44 mm, R^2 = 0.6, p = 0.05, n = 10). The amount effect (Figure 8 and Figure 9b) shows a negative correlation (higher rain amount shows a lower isotopic signature), which was reported worldwide except in polar areas [47].

Table 2. Isotopic composition of spring water samples.

Sample Code	$\delta^{18}O$ ‰							$\delta^{2}H$ ‰							*d*-Excess (Mean) ‰
	2011		2012					2011		2012					
	15-Sep.	6-Dec.	26-Apr.	4-Jun.	5-Sep.	11-Nov.	Mean ‰	15-Sep.	6-Dec.	26-Mar.	4-Jun.	5-Sep.	11-Nov.	Mean ‰	
Sarwchawa (1)	−7.42	−7.63	−7.47	−7.41	−7.51	−7.59	−7.42	−39.91	−40.02	−39.18	−39.2	−39.81	−43.39	−40.25	19.8
Shkarta (2)	−6.27	−6.42	−6.26	−6.14	−6.26	−6.18	−6.25	−32.33	−32.35	−32.7	−31.9	−32.43	−35.57	−32.88	17.1
Betwata (3)	−7.8	−7.85	−7.68	−7.74	−7.73	−7.79	−7.77	−40.95	−40.42	−40.39	−40.92	−40.56	−43.87	−41.18	20.9
Zewa (4)	−8.29	−7.98	−8.07	−8.1	−8.19	−8.08	−8.12	−43.57	−42.79	−43.14	−44.03	−43.59	−45.93	−43.84	21.1
Chewa (5)	−7.67	−7.4	−7.91	−7.51	−7.69	−7.66	−7.64	−40.26	−39.57	−41.84	−40.38	−40.7	−43.75	−41.08	20
Bla (6)	−7.98	−7.68	−7.44	−7.86	−7.88	−7.78	−7.77	−41.57	−41.42	−40.2	−41.79	−42.08	−44.71	−41.96	20.2
Qala Saida (7)	−7.86	−7.51	−7.42	−7.66	−7.7	−7.36	−7.59	−40.64	−39.55	−37.9	−39.28	−38.95	−41.62	−39.65	21
Gullan (8)	−8.16	−7.55	−7.87	−8.1	−8.15	−7.84	−7.95	−42.64	−41.32	−41.1	−41.79	−41.81	−45.02	−42.28	21.3
StD	0.05	0.05	0.06	0.05	0.08	0.11	0.07	0.31	0.33	0.41	0.33	0.63	1.16	0.53	1.6

StD = Analytical standard deviation; NA = Not available; Spring codes can be used to trace the locations of the springs in Figure 1.

4.4.3. Isotopic Pattern of the Karst Springs

The spring water samples have isotopic compositions between −8.29‰ and −6.14‰ for $\delta^{18}O$ and −45.02‰ to −31.89‰ for δ^2H (Table 2). Two regression lines for the karst springs were developed (Figure 6), one for all spring samples, excluding those of November 2012 period ($\delta^2H‰ = 2.3‰ + 5.56 \times \delta^{18}O$, $R^2 = 0.95$, $p < 0.001$ and $n = 40$), and the other for November 2012 samples ($\delta^2H‰ = 5.56 \times \delta^{18}O - 1.36‰$, $R^2 = 0.99$, $p < 0.001$ and $n = 8$). The majority of the samples plot between LMWL and MMWL. From this it can be concluded that the precipitation that recharges the karst aquifers originates from Mediterranean vapor masses, reflecting rapid infiltration of rainfall through the karst system and a short transit time (the karst aquifer is shallow, therefore, short residence time does not lead to reliable change in isotopic composition of the recharged water). Shkarta spring samples are isotopically enriched compared to the other springs. In argillaceous carbonates, evaporation leads to enrichment of isotopes in the pore water [48]. The isotopic enrichment of Shkarta spring relates to slow infiltration because of low permeability in the marly limestone of Shiranish aquifer and thus increased evaporation. The first regression line of the springs (green dashed line in Figure 6) is assumed to be a local evaporation line (LEL), which may occur due to the influence of Shkarta spring. The relation between δ^2H and chloride concentrations of the springs ($\delta^2H‰ = 21.83‰ + 0.473 \times Cl$, $R^2 = 0.9$, and $p < 0.001$) supports the evaporation trend in karst springs of the area (Figure 8a). Minerals like halite (NaCl) react fast with water attaining equilibrium within a short time compared to the residence time of groundwater [49]. Saturation of halite during evaporation of aqueous solution is accompanied by $\delta^{18}O$ enrichment [50]. A significant correlation between the d-excess and SI_{Halite} is notice (Figure 8b). This plot is interesting and supports the idea of evaporation, because it shows the same evaporation trend of δ^2H-Cl. According to this correlation (Figure 8a), increasing of SI_{Halite} corresponds to decrease in d-excess. The correlation between SI_{Halite} and d-excess is regarded to the effect of Mediterranean origin of the precipitation in the area of interest. Therefore, the saturation index of halite could represent a tool to evaluate the evaporation process in environmental isotopes of karst waters. The trend of evaporation is increasing from Zewa spring water, which is less enriched with deuterium and chloride, in comparison to Shkarta, which is the most enriched spring (Figure 8).

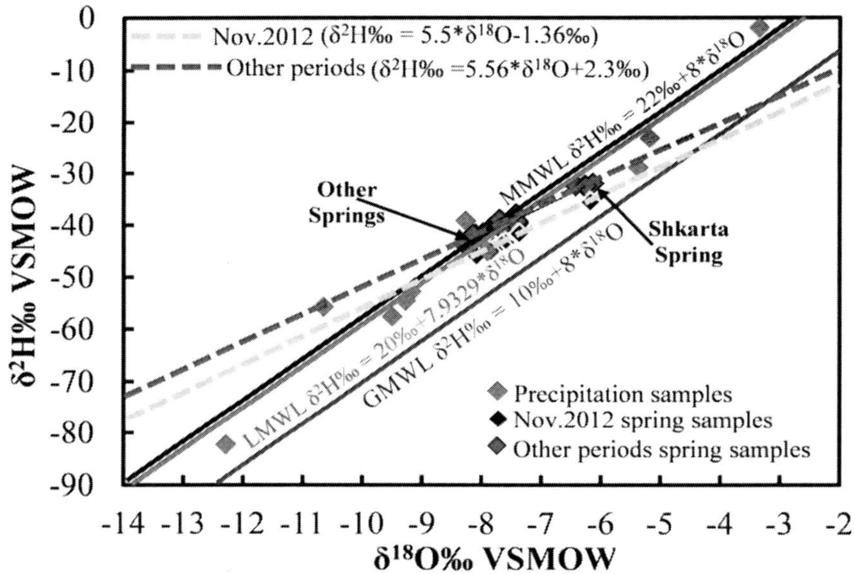

Figure 6. Meteoric water line (LMWL) and springs isotopic line for the studied area. Blue line is the global meteoric water line (GMWL) from [42]; black line is the Mediterranean meteoric water line (MMWL) derived by [43] and the red line is the local meteoric water line (LMWL) from the present study.

4.4.4. Altitude Effect

The isotopic composition of the groundwater in relation to altitude represents an indicator for locating the groundwater recharge area [51,52]. The plots of $\delta^{18}O$ and δ^2H in the spring water of the area versus the corresponding altitude of the springs are presented in Figure 9. The altitude effect on groundwater of the area is expressed by the linear correlation of $\delta^{18}O$ and δ^2H with altitude:

$$\delta^{18}O\ (‰) = -6.12 - 0.0017 \times A \quad (4)$$

$$\delta^{2}H\ (‰) = -33.24 - 0.008 \times A \quad (5)$$

where A is altitude in meter, R^2 is equal to 0.5 and 0.4, p-value is equal to 0.04 and 0.08 and n is 48 for $\delta^{18}O$ and δ^2H, respectively.

The local isotopic gradient ($LIG_{O\&D}$) shows depletion of stable isotopes by about −0.17‰ and −0.8‰ per 100 m elevation for $\delta^{18}O$ and δ^2H, respectively (Figure 9). The most depleted springs are Zewa, Bla, and Gullan, while Shkarta is the least depleted one (lower elevation).

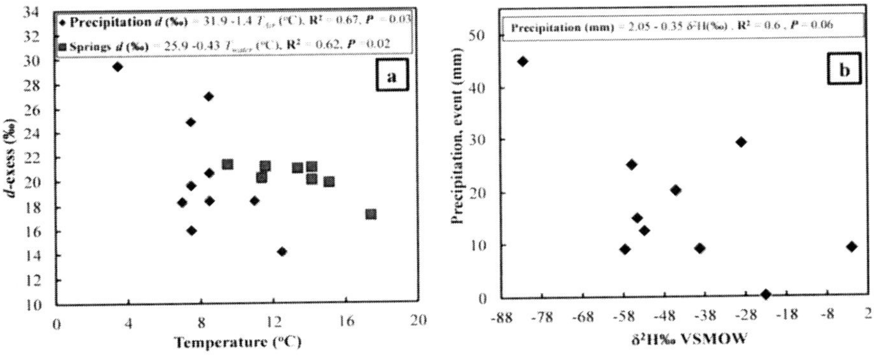

Figure 7. Relation between meteorological and isotopic parameters. (**a**) Plot of d-excess and temperature in precipitation and springs water, data of springs are represents mean d-excess and water temperature of six sampling period for each spring; (**b**) Plot of event precipitation amount and δ^2H in precipitation.

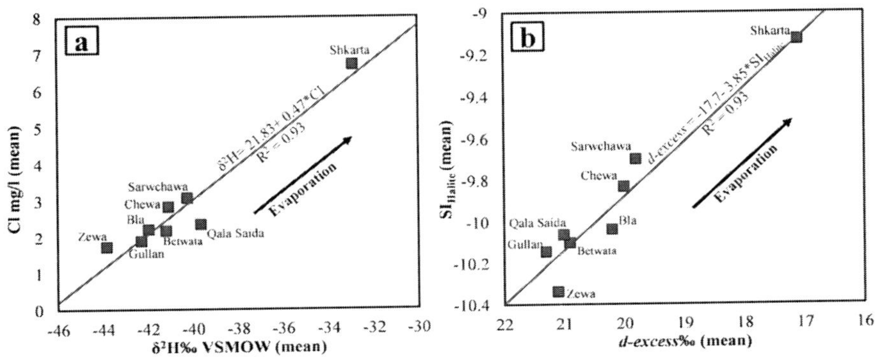

Figure 8. Illustration of evaporation trend from springs of Bekhme aquifer to Shiranish aquifer spring (Shkarta spring). (**a**) δ^2H—chloride plot; (**b**) d-excess and SI_{Halite} plot; values of δ^2H, d-excess, SI_{Halite} and chloride represent mean values of six samples for each spring: September 2011to September 2012.

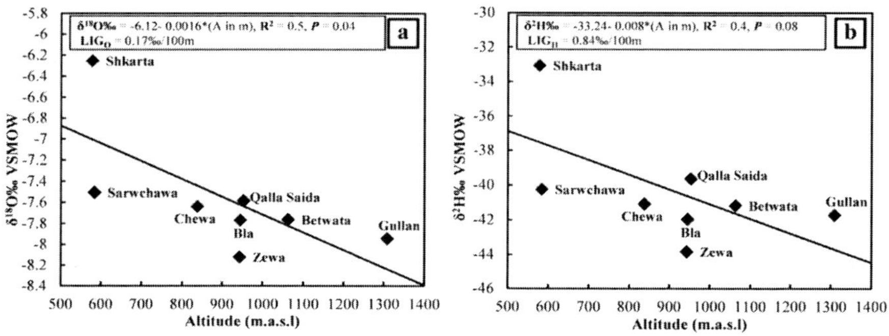

Figure 9. Relation between altitude and $\delta^{18}O$ and $\delta^{2}H$ in the studied area. (**a**) Plot of $\delta^{18}O$ in spring water and altitude of the spring outlet and (**b**) for $\delta^{2}H$. LIG_O is local isotopic gradient of $\delta^{18}O$ and LIG_H is for $\delta^{2}H$.

4.4.5. Temporal Variation

The temporal variation in the isotopic composition of the springs is obvious in all springs (Figure 10). This fluctuation shows more or less the different fingerprints of dry and wet periods. The variation in isotopic composition of karst springs indicates different origins of recharged water and possible connection between the aquifers [53]. The gap between Shkarta and other springs (Figure 10) indicates a connection between Bekhme and Kometan aquifers. Different isotopic signatures for the springs were recognized, separating them into three aquifers: Shiranish aquifer, represented by Shkarta spring (the most enriched isotopes), Kometan aquifer (Sarwchawa and Betwata), mainly represents intermediate isotopic signatures and Behkme aquifer (Zewa, Chewa, Bla, and Gullan springs), generally represented by depleted isotopic signatures if compared with the above two aquifers. The $\delta^{2}H$ is more sensitive to the temporal variation than the oxygen isotope one, because the mean difference ranges from 0.17 to 0.61‰ for $\delta^{18}O$ (about eight times the analytical standard deviation of $\delta^{18}O$, 0.07‰) and from 3.15 to 4.5‰ for $\delta^{2}H$ (about 11 times, 0.4‰). Depletion in $\delta^{2}H$ isotopic signature of the springs was noticed in samples of November 2012 (Figure 7) and this may be due to the nearly lowest recorded temperature of the spring waters during this sampling event and the amount of winter precipitation. This depletion–temperature dependency was also indicated in Figure 7 (lower water temperature characterized by lower $\delta^{18}O$). Depletion was also recorded in Chewa spring for both $\delta^{18}O$ and $\delta^{2}H$ April 2012, which probably occurred by the effect of precipitation amount of rainy season (January to April).

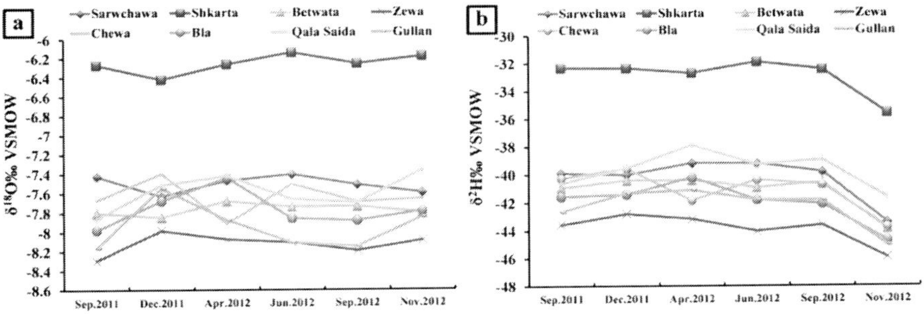

Figure 10. Temporal variation in isotopic composition of the karst springs. (**a**) Variation of δ18O during the period September 2011 and November 2012; (**b**) Variation of δ2H during the period September 2011 and November 2012.

4.4.6. Factors Affecting Stable Isotopes in Spring Waters

The water temperature has a significant effect on d-excess (d-excess‰ = $0.43 \times T_{water}$ (°C) − 25.9, $R^2 = 0.6$, $p = 0.02$, $n = 48$), which shows the same direction as in precipitation and suggests seasonal effects as well (Figure 8a), while air temperature is not significant (d-excess‰ = $0.06 \times T_{Air}$ (°C) − 21.4, $R^2 = 0.008$, $p = 0.5$, $n = 48$).

5. CONCLUSIONS

The main objective of this study was to understand the isotopic and hydrogeochemical relations in a complex karst system of Makook anticline. EC, pH, Eh, and water temperature temporal variations can be used as tracer for residence time, aquifer storage capacity and depth of the karst aquifers. Hydrogeochemical classification and isotopic composition can be used as indicators for the mean residence time in karst aquifers. The sources of water constituents and driving forces of hydrogeochemical processes are magnified by interrelations of karst water constituents. The sources of the geochemical properties were also indicated by these relations.

The local meteoric water line was derived for the area and is plotting between the global and Mediterranean meteoric water line (however, much closer to the Mediterranean MWL). From the d-excess in precipitation, the vapor source was found to be an arid, low humidity, and evaporation-dominant Mediterranean type. Temperature and amount effects are obvious from the isotopic behavior of the precipitation of the area. From the isotopic plot of springs it can be concluded that the recharged precipitation is of Mediterranean origin and that it infiltrates rapidly. Enrichment in stable isotopes gives rise to the hydraulic and hydrodynamic properties of karst aquifers. The local evaporation line is obvious from the oxygen–deuterium plot from Zewa to Shkarta spring and this trend was confirmed by the chloride–deuterium and d-excess–SI_{Halite} relations. It can be concluded from the temporal variation of stable isotopes that connections between the aquifers exist.

ACKNOWLEDGMENTS

First author is grateful to Diary Ali for his academic guidance. He is thankful to Soran Osman, Kardo Sardar and Ali Wahab for their great support during sampling. First author is also thankful to Lab staff in Chair of Hydrogeology/TU Bergakademie Freiberg for their support during Lab work. The paper has been significantly improved by means of three anonymous reviewers.

AUTHOR CONTRIBUTIONS

Omed Mustafa developed the project, collected and analyzed the water samples, interpreted the results and wrote the initial manuscript. Broder Merkel and Stephan Weise improved the manuscript and added certain sentences.

CONFLICT OF INTEREST

The authors declare no conflict of interest.

REFERENCES

1. Luhmann, A.; Covington, M.; Peters, A.; Alexander, S.; Anger, C.; Green, J.; Runkel, A.C.; Alexander, E. Classification of thermal patterns at karst springs and cave streams. Ground Water 2011, 49, 324–335.

2. López-Chicano, M.; Bouamama, M.; Vallejos, A.; Pulido-Bosch, A. Factors which determine the hydrogeochemical behaviour of karstic springs. A case study from the Betic Cordilleras, Spain. Appl. Geochem. 2001, 16, 1179–1192.

3. Liñán Baena, C.; Andreo, B.; Mudry, J.; Carrasco, C.F. Groundwater temperature and electrical conductivity as tools to characterize flow patterns in carbonate aquifers: The Sierra de las Nieves karst aquifer, southern Spain. Hydrogeol. J. 2009, 17, 843–853.

4. Mawlood, D. Application of Isotope Hydrology Studies Considering the Specific Climate, Hydrogeological and Geological Conditions in Order to Research Underground Water Resources in a Specific Area in the Near East. Ph.D. Thesis, Vienna University of Technology, Vienna, Austria, 2003. (In German).

5. Hamamin, D.; Ali, S. Hydrodynamic study of karstic and intergranular aquifers using isotope geochemistry in Basara basin, Sulaimani, North-Eastern Iraq. Arab. J. Geosci. 2013, 6, 2933–2940.

6. Al Manmi, D. Water Resources Management in Rania Area, Sulaimaniyah NE-Iraq. Ph.D. Thesis, University of Baghdad, Baghdad, Iraq, 2008.

7. Stevanovic, Z.; Marcovic, M. Hydrogeology of northern Iraq. In General Hydrogeology and Aquifer System, 1st ed.; FAO: Rome, Italy, 2004; Volume 2, pp. 42–47.

8. Ministry of Transportation and Communication (MTC). Directorate of Meteorology and Seismology; Kurdistan Regional Government: Sulaimani, KR, Iraq, 2012.

9. Jassim, S.Z.; Goff, J.C. Geology of Iraq, 1st ed.; Dolin, Prague and Moravian Museum: Prague, Czech Republic, 2006.

10. Bellen, R.C.; Dunnington, H.V.; Wetzel, R.; Morton, D.M.; Dubertret, L. Stratigraphic Lexicon of Iraq; Gulf PetroLink: Manama, Bahrain, 2005.

11. Karim, K.H.; Al-Hamadani, R.K.; Ahmad, S.H. Relations between deep and shallow stratigraphic units of Northern Iraq during Cretaceous. Iran. J. Earth Sci. 2012, 4, 95–103.

12. Sisakian, V. The Geology of Erbil and Mahabad Quadrangle Sheet NJ-38-14 and NJ-38-15, Scale 1:250 000; Iraq Geological Survey: Baghdad, Iraq, 1998.

13. Stevanovic, Z.; Iurkiewicz, A.; Maran, A. New insights into karst and caves of Northwestern Zagros (Northern Iraq). Acta Carsol. 2009, 38, 83–96.

14. APHA. Standard Methods for the Examination of Water and Wastewater, 20th ed.; American Public Health Association: Washington, DC, USA, 1998.

15. Parkhurst, D.; Appelo, C. Description of Input and Examples for PHREEQC (Version 3)-A Computer Program for Speciation, Batch-Reaction, One-Dimensional Transport, and Inverse Geochemical Calculations. Available online: http://pubs.usgs.gov/tm/06/a43/pdf/tm6-A43.pdf (accessed on 12 March 2015).

16. Bonacci, O. Analysis of the maximum discharge of karst springs. Hydrogeol. J. 2001, 9, 328–338.

17. Wood, W. Use and misuse of the chloride-mass balance method in estimating ground water recharge. Ground Water 1999, 37, 2–3.

18. Scanlon, R.; Healy, W.; Cook, G. Choosing appropriate techniques for quantifying groundwater recharge. Hydrogeol. J. 2002, 10, 18–39.

19. Bazuhair, S.; Wood, W. Chloride mass-balance method for estimating groundwater recharge in arid areas: Examples from western Saudi Arabia. J. Hydrol. 1996, 186, 153–159.

20. Marei, A.; Khayat, S.; Weise, S.; Ghannam, S.; Sbaih, M.; Geyer, S. Estimating groundwater recharge using the chloride mass-balance method in the West Bank, Palestine. Hydrol. Sci. J. 2010, 55, 780–791.

21. Kessler, H. Water Balance Investigations in the Karst Regions of Hungary; Act Coll Dubrovnik, AIHS-UNESCO: Paris, France, 1967.

22. Andreo, B.; Vías, J.; Durán, J.; Jiménez, P.; López-Geta, J.; Carrasco, F. Methodology for groundwater recharge assessment in carbonate aquifers: Application to pilot sites in southern Spain. Hydrogeol. J. 2008, 16, 911–925.

23. Pearson, F.; Scholtis, A. Controls on the chemistry of pore water in a marl of very low permeability. In Proceedings of the 8th International Symposium on Water-Rock Interaction (WRI-8), Vladivostok, Russia, 15–19 August 1995.

24. Parisi, S.; Paternoster, M.; Kohfahl, C.; Pekdeger, A.; Meyer, H.; Hubberten, H.W.; Spilotro, G.; Mongelli, G. Groundwater recharge areas of a volcanic aquifer system inferred from hydraulic, hydrogeochemical and stable isotope data: Mount Vulture, southern Italy. Hydrogeol. J. 2011, 19, 133–153.

25. Emblanch, C.; Zuppi, G.; Mudry, J.; Blavoux, B.; Batiot, C. Carbon 13 of TIDC to quantify the role of the unsaturated zone: The example of the Vaucluse karst system (southeastern France). J. Hydrol. 2003, 279, 262–274.

26. Guo, H.; Zhang, B.; Wang, G.; Shen, Z. Geochemical controls on arsenic and rare earth elements approximately along a groundwater flow path in the shallow aquifer of the Hetao Basin, Inner Mongolia. Chem. Geol. 2010, 270, 117–125.

27. Bundschuh, J. Modeling annual variations of spring and groundwater temperatures associated with shallow aquifer systems. J. Hydrol. 1993, 142, 427–444.

28. Anderson, M. Heat as a ground water tracer. Groundwater 2005, 43, 951–968.

29. Vesper, D.; White, W. Storm pulse chemographs of saturation index and carbon dioxide pressure: implications for shifting recharge sources during storm events in the karst aquifer at Fort Campbell, Kentucky/Tennessee, USA. Hydrogeol. J. 2004, 12, 135–143.

30. Eberts, S.; George, L. Regional Ground-Water Flow and Geochemistry in the Midwestern Basins and Arches Aquifer System

in Parts of Indiana, Ohio, Michigan, and Illinois (Vol. 1423); US Geological Survey: Denver, CO, USA, 2000.

31. Al Manmi, D. Chemical and Environmental Study of Groundwater in Sulaimaniyah City and Its Outskirts. Master's Thesis, University of Baghdad, Baghdad, Iraq, 2002.

32. Mustafa, O. Impact of Sewage Wastewater on the Environment of Tanjero River and Its Basin within Sulaimaniyah City, NE Iraq. Master's Thesis, University of Baghdad, Baghdad, Iraq, 2007.

33. Shuster, E.; White, W. Seasonal fluctuations in the chemistry of lime-stone springs: A possible means for characterizing carbonate aquifers. J. Hydrol. 1971, 14, 93–128.

34. Jacobson, R.; Langmuir, D. Controls on the quality variations of some carbonate spring waters. J. Hydrol. 1974, 23, 247–265.

35. Raeisi, E.; Karami, G. Hydrochemographs of Berghan karst spring as indicators of aquifer characteristics. J. Cave Karst Stud. 1997, 59, 112–118.

36. Delbart, C.; Barbecot, F.; Valdes, D.; Tognelli, A.; Fourre, E.; Purtschert; Couchoux, R.; Jean-Baptiste, P. Investigation of young water inflow in karst aquifers using SF6-CFC-3H/He-85Kr-39Ar and stable isotope components. Appl. Geochem. 2014, 50, 164–176.

37. Drever, J. The geochemistry of natural waters. In Surface and Groundwater Environments, 3rd ed.; Prentice Hall: Upper Saddle River, NJ, USA, 1997; p. 436.

38. Chadha, D. A proposed new diagram for geochemical classification of natural waters and interpretation of chemical data. Hydrogeol. J. 1999, 7, 431–439.

39. Han, D.M.; Liang, X.; Jin, M.G.; Currell, M.J.; Song, X.F.; Liu, C.M. Evaluation of groundwater hydrochemical characteristics and mixing behavior in the Daying and Qicun geothermal systems, Xinzhou Basin. J. Volcanol. Geotherm. Res. 2010, 189, 92–104.

40. Pavlovskiy, I.; Selle, B. Integrating hydrogeochemical, hydrogeological, and environmental tracer data to understand groundwater flow for a karstified aquifer system. Groundwater 2014.

41. Caetano Bicalho, C.; Batiot-Guilhe, C.; Seidel, L.; van Exter, S.; Jourde, H. Geochemical evidence of water source characterization and hydrodynamic responses in a karst aquifer. J. Hydrol. 2012, 450–451, 206–218.

42. Craig, H. Isotopic variations in meteoric waters. Science 1961, 133, 1702–1703.

43. Gat, J.R.; Carmi, I. Evolution of the isotopic composition of atmospheric waters in the Mediterranean Sea area. J. Geophys. Res. 1970, 75, 3039–3048.

44. Florea, L.J.; McGee, D.K. Isotopic and geochemical variability within shallow groundwater beneath a hardwood hammock and surface water in an adjoining slough (Everglades National Park, Florida, USA). Isot. Environ. Health Stud. 2010, 46, 190–209.

45. Fröhlich, K.; Gibson, J.J.; Aggarwal, P. Deuterium Excess in Precipitation and Its Climatological Significance. Available online: http://www.iaea.org/inis/collection/NCLCollectionStore/_Public/34/017/34017972.pdf (accessed on 12 March 2015).

46. Pang, Z.; Kong, Y.; Fröhlich, K.; Huang, T.; Yuan, L.; Li, Z.; Wang, F. Processes affecting isotopes in precipitation of an arid region. Tellus B 2011, 63, 352–359.

47. Dansgaard, W. Stable isotopes in precipitation. Tellus 1964, 16, 460–468.

48. Hamed, Y.; Dhahri, F. Hydro-geochemical and isotopic composition of groundwater, with emphasis on sources of salinity, in the aquifer system in northwestern Tunisia. J. Afr. Earth Sci. 2013, 83, 10–24.

49. Appelo, C.; Postma, D. Geochemistry, Groundwater and Pollution, 2nd ed.; Balkema: New York, NY, USA, 2005; p. 683.

50. Pierre, C. Isotopic evidence for the dynamic redox cycle of dissolved sulphur compounds between free and interstitial solutions in marine salt pans. Chem. Geol. 1985, 53, 191–196.

51. Mazor, I. Chemical and Isotopic Groundwater Hydrology, 3rd ed.; M Dekker: New York, NY, USA, 2004; p. 352.

52. Dafny, E.; Burg, A.; Gvirtzman, H. Deduction of groundwater flow regime in a basaltic aquifer using geochemical and isotopic data: The Golan Heights, Israel case study. J. Hydrol. 2006, 330, 506–524.

53. Kattan, Z. Environmental isotope study of the major karst springs in Damascus limestone aquifer systems: Case of the Figeh and Barada springs. J. Hydrol. 1997, 193, 161–182.

9
CHAPTER
SEASONAL RECHARGE AND MEAN RESIDENCE TIMES OF SOIL AND EPIKARST WATER IN A SMALL KARST CATCHMENT OF SOUTHWEST CHINA

Ke Hu[1,2,3], Hongsong Chen[1,2], Yunpeng Nie[1,2] & Kelin Wang[1,2]

[1]*Key Laboratory of Agro-ecological Processes in Subtropical Region, Institute of Subtropical Agriculture, ChineseAcademy of Sciences, Changsha, Hunan, 410125, China.*

[2]*Huanjiang Observation and Research Station for KarstEcosystems, Chinese Academy of Sciences, Huanjiang, Guangxi 547100, China.*

[3]*Graduate University of ChineseAcademy of Sciences, Beijing,*

ABSTRACT

Soil and epikarst play an important role in the hydrological cycle in karst regions. This paper focuses on investigating the seasonal recharge and mean residence time (MRT) of soil water and epikarst water in a small karst catchment of southwest China. The deuterium contents in precipitation, creek, soil baseflow (direct recharge of the saturated soil water to the stream), epikarst spring, and soil waters were monitored weekly for two years, and MRT was calculated by an exponential model (EM) and a dispersion model (DM). The obvious seasonal variation of deuterium in rainfall was buffered in epikarst water, indicating sufficient water mixing. Soil baseflow contained less rainy-season rainwater than epikarst spring discharge, reflecting the retarded effect of soil thickness on rainwater recharge. MRTs of all water bodies were 41-71 weeks, and soils in the depression extended those of shallow groundwater. This demonstrated that the deep soil layer played an important role in karst hydrological processes in the study catchment. The creek was recharged mostly by rainfall through epikarst, indicating its crucial role in water circulation. These results showed epikarst had a strong water-holding capacity and also delayed water contact time with dolomite.

Seasonal recharge and mean residence time (MRT) are important for a well-developed understanding of water movement and storage in subsurface systems[1,2,3,4]. Seasonal recharge reflects the relative importance of rainy-season versus dry-season precipitation and provides fundamental information to a catchment's water balance at annual scales. MRT is the average time water spends in a subsurface system before it emerges as surface flow. Knowledge of MRT is essential for evaluating the connectivity of a subsurface water system and for examining a catchment's response to land-use or environment changes[5,6,7].

Karst landscapes are found in carbonate bedrock sites which exhibit highly heterogeneous hydrological processes due to uneven distribution of dissolving rock, and they are characterized by sinkholes, sinking streams, closed depressions, subterranean drainage and caves[8,9,10,11,12]. Circulation and storage of groundwater in karst aquifers are obviously different from those in non-karst areas. Pockets of saturated soils occur atop karst (epikarst), and cracks and fissures in bedrock result in karst conduits[13,14]. Epikarst is a perched saturated zone above the groundwater table, and it stores part of the infiltrated rainwater and controls the water circulation dynamics[3,15]. The shallow and discontinuous soils and fissured epikarst comprise a double-porosity media. This makes water percolation and movement complex and difficult to be deciphered in karst regions.

Traditional hydrogeological investigation tools used in porous environments are not appropriate to identify the role of the epikarst in the complicated karst aquifers[16]. This circumstance results from the fact that the traditional methods cannot be used to track water infiltration and percolation under conditions of complex flow processes in an unsaturated zone[17,18]. Recently, natural tracers, such as oxygen-18 and deuterium concentrations in seasonal dynamics of soil water, stream, and spring water at the catchment scale, have become the method of choice for investigating hydrological processes[4,19,20,21]. This method is based upon the distinct seasonal changes of deuterium and oxygen-18 concentrations presented in rainfall and traced in output water[22,23,24]. Information about seasonal recharge and MRT under different subsurface systems can be explained more easily by use of stable isotopic methods than by use of other physical techniques. The seasonal recharge can be evaluated by long-term stable isotopic methods. Moreover, deuterium and oxygen-18 concentrations are nearly only affected by hydrologic flow paths and evaporation. Consequently, the different average annual deuterium and oxygen-18 values between rainfall and subsurface water could be seen as the seasonal recharge bias[1,2,25].

Besides seasonal recharge of groundwater, stable isotopes also can be used to reveal more complex hydrological issues. For example, these techniques have been used in recent research of MRT of groundwater to evaluate the storage and flux of subsurface water. The attenuation of precipitation signals by the study catchment has been used to evaluate the MRT for subsurface water using sinusoidal relationships between input and output waters and other lumped mathematical models that describe flow-path distributions[5,7]. Based on different flow paths, exponential model (EM) and dispersion model (DM) are applicable for unconfined and confined aquifers, respectively[26]. Previous studies have shown that, in the shallow soil area, DM approximates the MRT of tracers with high accuracy in fissured-porous aquifers, which are more similar to karst areas[5,27,28]. The exchange between immobile water in the micropores and mobile water in the fissures and fractures gives a steady-state condition to use DM[5,15]. Furthermore, the DM also can describe the realistic water exit time distribution[7,29]. Differently, EM is assumed to fit when there is no water exchange among different flow lines but well mixed in the outlet, and it has the exponential distribution of transit times[6]. If new water fraction has small effect on the output water resource, the EM perhaps has the good fit[6,7]. However, EM did not adapt for the soil water where there was frequent exchange between mobile and immobile water. Consequently, MRT calculated with EM might have a bias which did not meet the real flow patterns.

The hydrological functions of soil and epikarst probably have been underestimated because of poor understanding of hydrologic relationships to their structures and thickness[8,30,31]. Many previous studies had showed that catchment topography and aquifer structures controlled groundwater movement in sites with shallow soils[32,33]. Maloszewski *et al.*[5] reported that a karstic aquifer was characterized by two interconnected parallel flow systems: (a) a fissured-porous aquifer, and (b) karstic channels in the limestone. Moreover, they distinguished the fast and slow flow systems correspondingly[5]. However, this situation might be different in the experimental site with dolomite. Relatively deep soil cover and poor connectivity of aquifers existed in dolomitic karst areas, because there were different dissolution patterns in carbonate rocks of these types[9]. On the one hand, Soulsby *et al.*[34] claimed that the influence of catchment topography on MRT was largely mediated by its influence on soil cover and distribution in mountain catchment of Scotland. Moreover, the soil hydrology in a mountain forest catchment had a strong influence on MRT[33]. On the other hand, the longer vertical flow-path length for outlet water in mountainous regions might cause longer MRT relative to epikarst spring locations. This has been substantiated by McGuire *et al.*[6] in that flow paths from unsaturated soil to groundwater are shorter than vertical recharge. Assano *et al.*[35] also suggested that the length of vertical infiltration in an unsaturated zone controlled MRTs of soil water and transient groundwater. However, the influence of soil cover and thickness on MRT, seasonal recharge and water movement in epikarst was usually neglected in karst studies. Either only focusing on epikarst water movement or soil water movement would lead to mistakes. What's more, epikarst contacts between rock and soil typically consist of an unconsolidated layer of rock mixed with soil[30] and the hydrologic connection between footslopes and depression is continuous by means of lateral flow producing smaller responses to rainfall[36]. Consequently, the understanding of hydrological functions of both epikarst and the soil zone is urgent for knowing water movement in karst catchment.

The overall objectives of this paper are to evaluate the seasonal recharge and MRT of creek, soil baseflow (a flux of water going directly from the saturated soil to the stream), soil water and epikarst spring for better understanding the hydrological functions of soils and epikarst, based on seasonal variation of stable isotopes in a small karst catchment of southwest China. Moreover, the relationship among the above-mentioned water bodies is also discussed for investigating the influences of soil cover and thickness on water movement in underlying epikarst.

RESULTS

Rainfall amount and deuterium content

The total rainfall from May to October covered by about 70% of those during the study period (Fig. 1a). The δD (deuterium content) values in rainfall manifested strong seasonal variation and presented the characteristics of a sinusoidal wave with two sinusoidal cycles of 58‰ peak-to-peak variation during the sampling period. Although there was obvious seasonal variation for δD over a year, no clear relationship between event-based precipitation amount and δD was found. When rainfall amount was high, however, more depleted values of stable isotopes tended to be obtained.

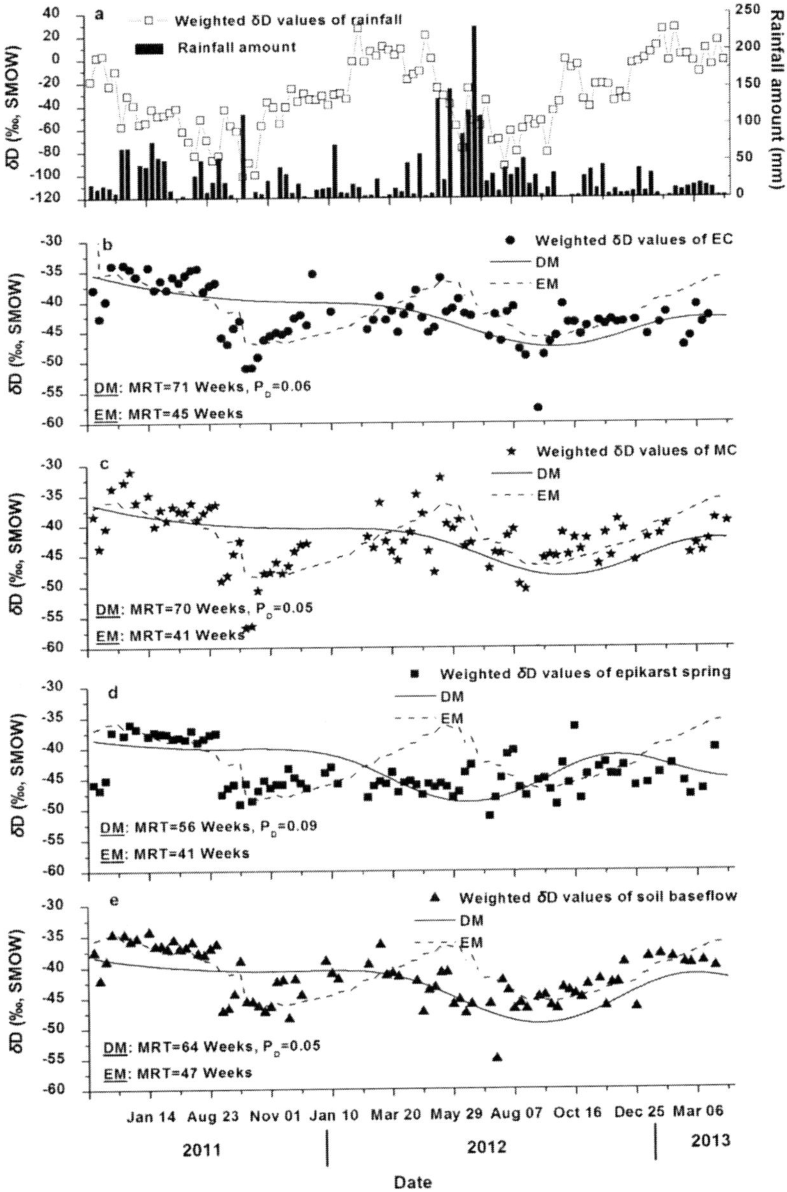

Figure 1. Seasonal variation of deuterium contents in rainfall (**A**), EC (**B**), MC (**C**), epikarst spring (**D**) and soil baseflow (**E**) and EM and DM fits in experimental site from April 10, 2011 to April 9, 2013.

Table 1 presents the number of weeks for different rainfall amounts during dry and rainy seasons from April 2011 to April 2013. The weeks with 0-10 mm rainfall were similar during dry and rainy seasons. However, the weeks with >50 mm rainfall during rainy season were more than those during dry season. This implied that rainy-season water had more opportunities to produce bypass flow in shallow soils and to recharge the deep soil layer as well as the underlying epikarst zone.

Table 1. The distribution of the different weekly amount of rainfall between rainy and dry seasons from april, 2011 to april, 2013.

Rainfall	0-10 mm (Weeks)	10-20 mm (Weeks)	20-30 mm (Weeks)	30-40 mm (Weeks)	40-50 mm (Weeks)	50-100 mm (Weeks)	>100 mm (Weeks)	Total (Weeks)
Dry season	22	18	3	6	3	2	0	54
Rainy season	21	4	6	4	4	9	6	54

Soil distribution

The distribution of soils was the premise for understanding water storage in soil zone. The results (Table 2) showed that the average soil depth and its standard deviation increased gradually with the increase of slope elevation. This indicated that the depression had deeper soil cover but higher uneven distribution than the slopes. The major soil texture was sandy soil in the top slopes, but clay loam soil in the depression and footslopes. The tendency of the above-mentioned soil distribution demonstrated that the water storage and hydrological function of soil at slopes were weaker than those in the depression. That is to say, the soil in the depression and footslopes, where soil thickness was deeper, played more important role in the water regulation relative to upslope soil zone.

Table 2. The distribution of soil depth in different slope positions in the study area.

Slope position	Number of plots	Average soil depth (cm)	Standard deviation (cm)
Upslope	38	20.8	9.8
Midslope	38	25.2	10.9
Downslope	38	36.9	20.0
Depression	37	63.8	35.0

Water response of creek and epikarst spring to rainfall

Water responses of the end of the creek (EC) and the epikarst spring to two typical rainfall events (rainfall amount were 50.1 and 123.8 mm, respectively) were different (Fig. 2). However, the changes of discharge between the EC and the epikarst spring were similar. The discharge increased rapidly with the first three hours and then decreased approximately 21 hours later. However, the portions of new water in the EC and the epikarst spring were different which were influenced by rainfall amount, and they were higher in the EC. During low rainfall events, the portions of new water in the EC and epikarst spring were all lower than 40% and 10%, respectively, even during the period with the highest discharge. However, during the high rainfall event, the portion of new water in the EC ranged from 25% to 100% and it was distinctly higher than that during the lowest rainfall event. Yet, epikarst spring also had lower than 30% new water most of the times. This is consistent with the supposition that the rainfall with high amounts tends to produce surface runoff, and fast flow channels are rare in the subsurface system of the study area.

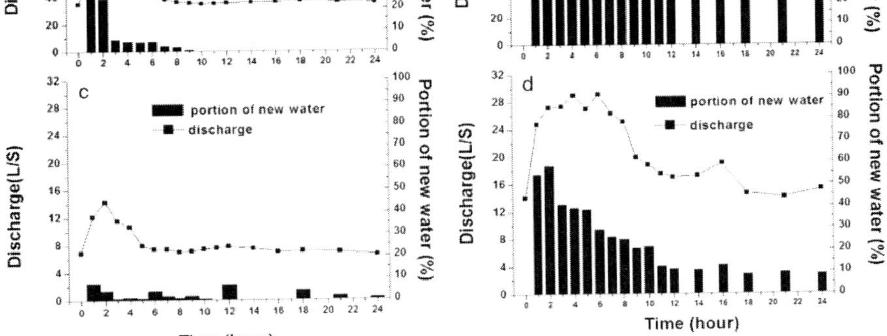

Figure 2. Water response of EC and epikarst spring to rainfall event on April 5 (**A**, **C**) and March 29 (**B**, **D**), 2012, respectively.

Seasonal recharges of different water bodies

In comparing with the change in rainfall isotopic content, the seasonal variations in deuterium contents of the middle of the creek (MC), EC, soil baseflow, and epikarst spring were sufficiently buffered as shown in Fig. 1. This indicated that each kind of water body sampled had small response to rainfall events, and the water was mixed sufficiently in the epikarst zone and deep soil zone.

Table 3 shows the average annual weighted δD values in the MC, EC, soil baseflow, and epikarst spring were close to those in rainfall during the rainy season. This demonstrated that major recharge of these water bodies occurred during the rainy season. Similar weighted δD values were found in the EC, MC and soil baseflow but they were more enriched than those in the epikarst spring. This suggested that the epikarst spring was recharged more by rainy-season rainwater than the other water bodies. Moreover, seasonal variation of δD in rainfall was strongly buffered in the creek, soil baseflow, epikarst spring and deep soil layer. This indicated that epikarst and soils controlled the rainfall recharge dynamics. The average annual weighted δD values of soil waters became depleted along with increase in soil depth. Deep soil waters (60-100 cm) in the depression had more enriched average weighted δD values than those at footslopes.

The recharge portion in the EC, MC, soil baseflow and epikarst spring was lower than the contribution of dry-season rainfall. This indicated that the recharge rate of rainy-season rainfall was higher than that of dry-season rainfall. The recharge portion in the epikarst spring was lower than those in the EC, MC and soil baseflow. In addition, soil waters had decreased recharge portion with increase in depth. Their recharge portions varied from 34.56% to 63.49% at 20 and 40 cm depths, but were lower than 26.19% at 60, 80 and 100 cm depths. These revealed that a large number of rainy-season rainfall did not recharge the shallow soil layers (0-40 cm) but produced bypass flow.

The highest coefficient of variation (CV) of δD (96.79%) was found in precipitation. The CV in δD of soil water decreased with increase in depth from 91.22% to 8.07% but it was sharply decreased in the SD-1 at 60 to 100 cm depth. The EC, MC, soil baseflow and epikarst spring had similar small CVs to the SD-1 at 60-100 cm depths, which were much lower than the others. This indicated that water mixed well in deep soil in the depression and epikarst zone. However, CV of δD in the SD-3 was special because of the existence of sandy soil, and fast infiltration might cause the less mixing.

Table 3. The average deuterium content, cv and recharge portion of entire year during dry-season in rainfall and different water bodies from april 10, 2011 to april 9, 2013.

Samples	Depth (cm)	δD (‰, SMOW)	CV (%)	δD_{dry} (‰, SMOW)	δD_{rainy} (‰, SMOW)	Portion (%)
Rainfall	—	−31.65	96.79	−11.39	−50.91	29.53
EC		−42.34	10.59	−42.91	−42.01	21.69
MC		−42.49	12.32	−42.87	−42.48	21.31
Soil baseflow	—	−42.03	9.97	−41.41	−42.29	22.47
Epikarst spring	—	−44.11	8.43	−45.20	−43.22	17.21
SS-1	20	−30.79	60.19			50.91
	40	−37.25	40.64			34.56
	60	−45.98	26.73			12.47
	80	−46.56	28.97			11.01
	100	−46.79	27.5			10.42
SS-3	20	−26.59	91.22			61.54
	40	−34.17	51.17			42.36
	60	−46.48	26.92			11.21
	80	−47.76	21.15			7.97
	100	−46.99	19.48			9.92
SD-1	20	−25.82	69.19			63.49
	40	−34.87	23.61			40.59
	60	−40.98	15.82			25.13
	80	−43.89	8.07			17.76
	100	−40.56	8.74			26.19
SD-3'	20	−29.55	55.27			54.05
	40	−33.56	43.34			43.90
	60	−44.31	35.23			16.70
	80	−43.24	31.54			19.41

MRTs of different water bodies

The MRTs of water in the MC, EC, soil baseflow, epikarst spring and soil zone, which were calculated by DM and EM (except soil waters), are illustrated in Table 4 and Fig. 1. The relative lower SIGMAs were found in the EC, MC, soil baseflow and epikarst spring, and the higher SIGMAs were

Table 4. The mrt of different water bodies calculated by dm and em, respectively from april 10, 2011 to april 9, 2013.

		DM			EM	
Samples	**Depth (cm)**	**T (weeks)**	**P_D**	**SIGMA**	**T (weeks)**	**SIGMA**
EC	—	71	0.06	0.4701	45	0.3919
MC		70	0.05	0.6431	41	0.5363
Soil baseflow	—	64	0.05	0.4093	47	0.4131
Epikarst spring	—	56	0.09	0.4550	41	0.5599
SS-1	20	10	2.00	0.8454		
	40	10	1.50	1.7860		
	60	24	0.03	2.7765		
	80	28	0.01	3.0080		
	100	31	0.02	2.3959		
SS-3	20	3	1.35	1.1120		
	40	9	2.40	1.7678		
	60	9	0.01	3.5505		
	80	9	0.04	3.6722		
	100	12	0.01	4.1931		
SD-1	20	6	1.20	1.3770		
	40	18	0.50	1.2942		
	60	57	0.12	1.0327		
	80	58	0.08	0.9067		
	100	60	0.11	0.6451		
SD-3*	20	10	2.50	1.1958		
	40	11	2.50	1.9297		
	60	10	0.01	4.3131		
	80	9	0.01	3.9500		

obtained in soil waters. Based on the fitting lines of DM and EM in Fig. 1, DM was found to be more appropriate than EM in epikarst spring. Conversely, both EM and DM had good fit in the EC, MC and soil baseflow. The MRTs calculated by EM were shorter than those calculated by DM. This resulted from that the MRT of soil water calculated by EM was invalid due to the existence of frequent water exchange in soils. Based on DM, EC and MC had similar MRTs of 71 and 70 weeks, respectively, and they were longer than those of soil baseflow (64 weeks) and epikarst spring (56 weeks). Moreover, soil baseflow had longer MRT than epikarst spring. The MRT of soil waters ranged between 3 to 60 weeks and increased with increase in soil depth except the SD-3. There were weathered sandy soil layers at depths of 60 and 80 cm in the SD-3, which made water movement more unusual than the other sites.

DISCUSSION

Epikarst water movement

In karst regions, the epikarst acts as a temporary storage and distribution system for water infiltrating into karst systems[8,15]. The MRT had some negative relationship with size and quantity of conduits in aquifer[41]. In limestone or rock catchment, the MRT of spring water largely depended on the size and quantity of conduits and ranged commonly from few weeks to several months[5,19,32,37]. However, in our study area (dolomite), the MRT of spring calculated with DM was 56 weeks (more than one year), which was similar to the results reported by other studies[5,6,19]. The long MRT of water in the epikarst and EC indicates this dolomite karst aquifer had less conduits and fractures to allow fast flow[5,38] than those in typical limestone karst aquifers. Micropores and fissures water, which was characterized by long MRTs[5,37], tended to be the major part of epikarst water storage in dolomite aquifers. Moreover, water mixing was sufficient due to diffusion exchange between mobile and immobile zones. Perrin *et al.*[3] found that if recharge exceeded a given threshold of epikarst, excess infiltrated water bypassed the soil and epikarst and reached the saturated zone as fresh flow with high CV and short MRT. The apparent threshold of epikarst in this study was high enough to prevent bypass flow and maintain slow flow. In addition, the quick and short response of outlet flow to rainfall commonly accompanied with the existence of conduits. However, such response was longer than anticipation. These supports our hypothesis that dolomitic epikarst would not have rapid hydrological processes, and rainwater infiltrated quickly into epikarst with bare or shallow soil cover but stayed for a long time rather than flowed out. The great numbers of pores and small fissures, which were characterized by slow flow, might be the reason[9]. The interaction among waters from different flows increased the MRT of fast water. Consequently, this type of epikarst has a large water storage capacity and contributes to delay the hydrological processes in the dolomitic karst groundwater system.

The MRT was influenced by multiple factors such as gradient, landscape organization, contributing area and flow-path length[32,33,34]. Similar to the argument reported by Soulsby *et al.*[34] the contributing area is less influential than that previously considered. One of the reasons in this study was that the landscape organization controlled the MRT of catchment. Another reason was that the depression mixed water well due to the deep soil. Dewalle *et al.*[22] and O'Driscoll *et al.*[2] found that annual average hydraulic diffusivity for water movement in the soil had positive relationship with soil depth. Well mixing water caused even distribution of MRT in the depression.

Model fitting

The fit of models reflected the water movement pattern of subsurface system to a certain degree[5,29,39]. The DM seemed to be much closer to the reality in the study area. This was because the EM assumed that there is no exchange of tracer between the flow lines. However, water movement in soils was difficult to satisfy this premise, unless the new water fraction contributed little on the output water resources[6,7]. Obviously, the soil in the slope was impossible, but soil in the depression was exception.

The low slope gradient, clay texture and deep soil thickness made the depression be a good water mixing reservoir. This was likely why the EM fit lines of creek, outlet and soil baseflow, which contained water from the depression, were better than that of the epikarst spring. However, this situation could not resolve the model error of EM for MRT of soil water. The slope epikarst water converged easily and dominated the water resources in the depression, because the recharge area and thickness of epikarst were larger than those of soils in the study area. These results were similar to the condition of no soil cover on the epikarst[15,19]. Consequently, the MRTs calculated by EM were shorter than those calculated by DM.

The impact of soil in the depression

Similar to previous studies, the major recharge of the creek and shallow groundwater occurred during the rainy season[1,2]. This was because most of the rainfall came from the rainy season and karst slopes had fast infiltration rate and rare overland flow[13,39]. However, there was a difference of recharge portion among creek, soil baseflow and epikarst spring. Compared with the soil baseflow, dry-season rainfall occupied a smaller recharge portion in epikarst spring. This is attributed to the difference between the soil baseflow and epikarst spring in that the former contained more deep soil water from the depression. The soil zone in the depression either intercepted input rainwater from the rainy season or increased the input rainwater from dry season. The former is the more likely reason. On the one hand, high infiltration rates and rare overland flow on the karst slopes in this study area were found and implied that such shallow soils facilitated the recharge of underlying epikarst for an entire hydrological year[39]. On the other hand, relative to the dry season, rainfall was more frequent during the rainy season. This made excess-infiltration runoff more common during rainy season. The hydraulic conductivity value for sand is several times greater than that for clay loam, indicating that the depression was the major place to intercept excess infiltrated rainy-season rainfall[41]. Also, the slope position had important influence. The depression which had lower slope gradient than upslopes, tended to yield runoff much easier[41,42].

Soils in the depression play an important role in water mixing due to the presence of capillary water storage[3]. Moreover, deep soil layer had higher water storage capacity than shallow soil layer because of fewer large apertures and more micropores[17,41,42]. This meant that the soil water storage had a positive relationship with soil thickness under the condition of similar soil texture. Soil in the depression decreased the recharge of underlying epikarst water and increased the overland flow at the yearly scale. In the combination of long MRT of the EC, it was found that the conduit water (fast water) was few. Consequently, the linear mixing model could be accepted in the study area, because rainwater mainly infiltrated into slow reservoir[43]. For example, Herrmann *et al.*[19] and Zarei *et al.*[44] also found that two-component hydrograph separation model got a good result of distinguishing old water and new water of the stream in karst basin.

The soil in the depression extended the MRT of subsurface water compared to upslope soil. This is because the topography and soil distribution in the depression was similar to riparian. Although there were no previous studies of this site, similarly free-draining riparian zone soils had an ability to dampen tracers more effectively[45]. Soulsby and Tetzlaff[4] also found that MRT was more closely correlated with riparian soil cover – even quite simply defined as that within 50 m of the river channel – than total cover of responsive soils. This relationship is consistent with the importance of the topography and soil distribution of catchment, particularly in relation to the connectivity between more responsive soils and the river channel network.

CONCLUSIONS

The seasonal recharge and mean residence time (MRT) of water in the end of creek (EC), the middle of creek (MC), soil baseflow, epikarst spring and soils were revealed by their seasonal variations of deuterium contents. The major recharge of subsurface and surface water occurred during the rainy season, and recharge rate of rainy-season rainwater was higher than that of dry-season rainwater. Deep soil (>60 cm) in the depression had higher water-holding capacity than shallow soil at the footslopes. The creek was recharged not by direct rainfall but by rainfall through the epikarst zone. This reflected that karst catchment had high infiltration rate. The MRT of all water bodies calculated with DM was longer than one year. This indicated that epikarst had poor connectivity and good water-holding capacity, and it could maintain the continuous recharge to major surface water. Furthermore, this also reflected the key role of epikarst in water circulation in the karst catchment. Soil baseflow had longer MRT than epikarst spring. This demonstrated that deep soil in the depression made input water stay in subsurface system for longer time and intercepted part of the rainy-season rainfall. Comparatively speaking, the contributing area had little impact on MRT than soil thickness in the small karst catchment. The results of this paper should improve our understanding of karst hydrological processes at catchment scales.

MATERIALS AND METHODS

Site descriptions

The study was conducted at the Huanjiang Observation and Research Station for Karst Ecosystems under the Chinese Academy of Sciences (24°43′–24°44′N, 108°18′–108°19′E) in Huanjiang County of northwest Guangxi, southwest China (Fig. 3). The experimental site is a typical peak-cluster depression area covering 1.01 km^2, and it is relatively closed area. About 60% of slope land has a slope gradient greater than 25° and the elevation ranges from 272 to 647 m above sea level. This region experiences a subtropical mountainous monsoon climate. The average annual rainfall is 1389.1 mm with 71% falling in the wet season from May to October. This period is defined as the rainy season. The average annual temperature is 18.5 °C. The soil depth in the depression and on the hillslope varies from 20 to 160 cm and from 0 to 50 cm, respectively. On most parts of the top hillslope, there is little soil cover but exposed bedrock. The shallow and discontinuous soils have been developed from dolomite and contain significant amounts of rock fragments. Soils are well-drained, gravelly and calcareous, and have a clay to clay loam texture (25–50% silt and 30–60% clay) with stable infiltration rate varying from 0.43 to 4.25 mm/min[39]. Soil organic matter content is relatively high ranging from 2.2% to 10.1%, and pH varies between 7.1 and 8.0. The cover of exposed bedrock range from 15% to 30%, and some rock outcrops covered mostly by deep-rooted trees are large (2–10 m in height).

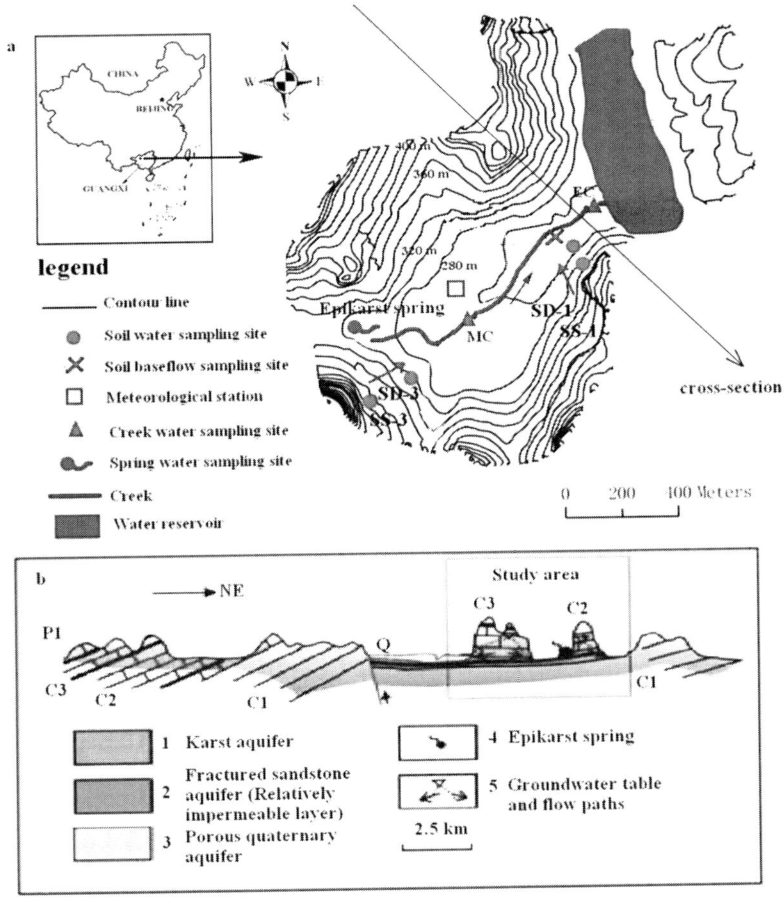

Figure 3. Schematic map of sampling sites (**A**) and vertical cross-section of the geohydrologic background (**B**) in small karst catchment of Huanjiang County of northwest Guangxi, China. Schematic map of sampling sites (**A**) was generated by software ArcGis and ArcView, ...

The study area is located at a wide and gradual edge of syncline from northwest to southeast. The exposed stratum is gray dolomite. There are gray and black limestones and yellow sandstone underneath the dolomite.

The gray dolomite is widely distributed in the area and characterized by good water-holding capacity. There is one continuous and two discontinuous epikarst springs locating at the bottom of the dolomite hillslope. A surface water reservoir lies on yellow sandstone in the northeast of the study area. The yellow sandstone is considered to be a relatively impermeable layer and water losses from the small catchment were limited. The groundwater table changed seasonally and was often below 1–3 m depth in the depression[46].

All residents have moved away and the cultivated lands have been abandoned since 1985. The vegetation consists of grassland and sparse shrubland. However, there are patches of zonal dense scrub and forest lands with a high cover of exposed bedrock, especially in the southwestern parts. Overland flow is low and the corresponding runoff coefficient is often less than 5% on the hillslopes under various land use types.

Water sampling

Rainfall and water samples from creek, soil baseflow and one epikarst spring (Fig. 3) were sequentially collected weekly within the study site from April 10, 2011 to April 9, 2013. The number of water samples from each water body was 99. In order to estimate the influence of contributing areas, two sampling sites were chosen in the creek. One located at the end of the creek (EC, outlet of the catchment), and one in the middle of the creek (MC). Soil baseflow was collected with a zero-tension PVC tube at the soil-bedrock interface (about 100-120 cm depth) near the EC in the depression. In order to avoid the impact of creek and depression, epikarst spring was chosen in the upper area of catchment. The precipitation sampling devices in the present study are similar to those used by IAEA and consist of a 150-mm-diameter funnel connected to a 0.5 L brown bottle[47]. An air outlet tube was welded at the lower part of the funnel. In order to avoid evaporation of water samples, paraffin was used at every joint. The weekly rainfall amount was obtained from a meteorological station of the study site. The water sampling sites are illustrated in Fig. 3. Weekly water samples were preserved at 4 °C to prevent evaporation.

Soil waters were sampled at 5 different depths from 20 to 100 cm in 4 sampling sites and collected weekly between April 10, 2011 and February 28, 2013. Two sites (SS-1 and SS-3) located in the slopes and the others (SD-1 and SD-3) in the depression. They formed two sampling lines, and one (SS-3 and SD-3) was near the epikarst spring (Fig. 3). This design was good for evaluating impact of slope position on soil water movement. Soil water was sampled by using tension lysimeters (produced by Institute of

Geographic Sciences and Nature Resources Research, CAS) which consist of suction lysimeters, ceramic porous cup, sampling bottle, tubes and pressure monitor[48]. A small hand pump was connected to the device and about an 80-Kpa vacuum was provided within the bottle.

In order to confirm high rainfall is more easily intercepted by soils in the depression than those on slopes and the corresponding surface runoff varies, the rainfall-runoff responses of the EC and epikarst spring were recorded hourly after two different typical rainfall events in 2012: April 5 (low rainfall) and March 29 (high rainfall). The discharges of the EC and epikarst spring were measured by PS 1000 automatic water level monitoring probe (Greenspan) and they were transformed to discharge by weir.

Deuterium and oxygen-18 analysis

Deuterium and oxygen-18 values of water samples were analyzed by the DLT-100 liquid water isotope analyzer (Los Gatos Research (LGR), Inc., model 908-0008), a liquid water isotope analyzer, at the Key Laboratory of Agro-ecological Processes in Subtropical Regions belonged to the Institute of Subtropical Agriculture under the Chinese Academy of Science. Results were reported in δ notation relative to V-SMOW as

$$\delta_{sample} = \left(R_{sample}/R_{standard} - 1\right) \times 1000 \qquad (1)$$

where R_{sample} and $R_{standard}$ are the ratio D/H or O^{18}/O^{16} of measured sample and standard sample, respectively. The standard deviation for repeated measurements was ±1‰ for δD and ±0.2‰ for $\delta^{18}O$.

Water recharge portion analysis

Water recharge portion is defined as recharge contribution of dry-season rainwater to different water bodies on year scale. It is calculated by two end-member model[52]. On a year scale, rainwaters from dry and rainy seasons are identified as two unique original water resources to surface and subsurface water in the study area. The relationship between input and output can be represented by as

$$C_{out} = PC_{dry-season} + (1-P)C_{rainy-season} \qquad (2)$$

where C_{out}, $C_{dry\text{-}season}$ and $C_{rainy\text{-}season}$ are δD or $\delta^{18}O$ value of water bodies, dry-season rainfall and rainy-season rainfall, respectively. P is recharge portion for year scale. The C_{out} is average weighted values with discharge during the stable discharge period. Because discharge of epikarst spring varies little under the baseflow condition, the C_{out} of epikarst spring is equal to unweighted value. $C_{dry\text{-}season}$ and $C_{rainy\text{-}season}$ are average weighted values with rainfall amount during dry and rainy seasons, respectively.

Additionally, the two end-member model also can be used to calculate the percentage of new water in storm-event. The relationship between input and output can be represented by as

$$C_{out} = PC_{new\ water} + (1 - P)C_{old\ water} \quad (3)$$

where C_{out}, $C_{new\ water}$ and $C_{old\ water}$ are weighted δD or $\delta^{18}O$ value of output water, new water and old water, respectively. P is percentage of new water. The weighted δD or $\delta^{18}O$ value of new water and old water is replaced by that of rainfall and baseflow water, respectively.

Mean water residence time analysis

Two lumped parameter mathematical models, i.e., DM and EM (version 3.1), were used to determine MRT of soil water at different depths and positions based on long-term isotope data[27]. A functional relationship between input and output can be represented as

$$C_{out}(t) = \Sigma C_{in}(t_t) g(T) \exp(-\lambda T) dt_t \quad (4)$$

where C_{out} and C_{in} are weighted δD and $\delta^{18}O$ values of different water bodies and rainfall, respectively; g(T) is system response function, which specifies residence time distribution of water within the system; t_t is the time of entry; and $T = t - t_t$ is the MRT of water, which can be calculated from system response function through calibrating models. Based on sufficient data of C_{out} and C_{in}, the g(T) can be calculated.

In the DM, the following uni-dimensional solution in the flux mode to the dispersion equation for a semi-infinite medium was used as the response function

$$g(t) = (4\pi P_D t/T)^{-1/2} t^{-1} \exp[-(1 - t/T)^2 (T/4P_D t)] \quad (5)$$

where P_D is the apparent dispersion parameter, which mainly depends on the distribution of travel times. The higher values reflect more inhomogeneity and greater width of transit time distributions. Consequently, the T can be obtained finally.

In the EM approximation, the flow lines are assumed to have the exponential distribution of transit times, i.e. the shortest line has the theoretical transit time close to zero, and the longest line has the transit time close to infinity. It is assumed that there is no exchange of tracer between the flow lines, and then the following response function is obtained:

$$g(t) = T^{-1}\exp(-t/T) \quad (6)$$

The accuracy of fit of simulations to the experimental data was computed using SIGMA function

$$SIGMA = \Sigma(S_i - M_i)^2/n \quad (7)$$

where S_i and M_i are simulated isotope values and measured isotope values, respectively; and n is the number of samples. The lower SIGMA value means a better fit.

In humid area, where evaporation is weak, the deuterium and oxygen-18 contents have strong linear correlations[19,27]. The isotopic enrichment due to fractionation processes is little during the evaporation. The results obtained from oxygen-18 and deuterium values met this situation and they were similar in our study area. However, in karst regions, Meißner *et al.*[50] found that oxygen-18 content is greatly affected by the soil carbonate while deuterium content is not. Obviously, this uncertainty needs to be avoided in our study area. Indeed, many previous studies only chose deuterium as stable isotope[27,50,51]. Consequently, the results obtained by the analyses of deuterium contents were introduced in this paper.

Soil depth survey

In order to get a better understanding of the role soil cover has in water circulation in the catchment and the interaction between soil water and epikarst water, the distribution of soil depth was investigated. In total, there were 151 soil plots, each with a dimension of 80 m × 80 m. For the purpose of equal distribution of soil plots in the study area, soil plots were designed by ArcGis. Soil depths were obtained by using surveying rod.

AUTHOR CONTRIBUTIONS

H.K. and C.-H.S. designed the study, collected and analyzed the data, and wrote the manuscript. N.-Y.P. and W.-K.L. discussed the experimental design and manuscript writing.

ADDITIONAL INFORMATION

HOW TO CITE THIS ARTICLE: Hu, K. *et al.* Seasonal recharge and mean residence times of soil and epikarst water in a small karst catchment of southwest China. *Sci. Rep.* **5**, 10215; doi: 10.1038/srep10215 (2015).

ACKNOWLEDGMENTS

This research was supported by the National Key Basic Research Program of China (2015CB452703), the Action Plan for the Development of Western China of Chinese Academy of Sciences (KZCX2-XB3-10), and the National Natural Science Foundation of China (51379205 and 41171187). The authors wish to thank Dr. Ken Mix (Texas State University) for his help in improving the language of this article.

REFERENCES

1. Lee K. S., Kim J. M., Lee D. R., Kim Y. & Lee D. Analysis of water movement through an unsaturated soil zone in Jeju Island, Korea using stable oxygen and hydrogen isotopes. J. Hydrol. 345, 199–211 (2007).

2. O'Driscoll M. A., DeWalle D. R., McGuire K. J. & Gburek W. J. Seasonal ^{18}O variations and groundwater recharge for three landscape types in central Pennsylvania, USA. J. Hydrol. 303, 108–124 (2005).

3. Perrin J., Jeannin P. Y. & Zwahlen F. Epikarst storage in a karst aquifer: a conceptual model based on isotopic data, Milandre test site, Switzerland. J. Hydrol. 279, 106–124 (2003).

4. Soulsby C. & Tetzlaff D. Towards simple approaches for mean residence time estimation in ungauged basins using tracers and soil distributions. J. Hydrol. 363, 60–74 (2008).

5. Maloszewski P., Stichler W., Zuber A. & Rank D. Identifying the flow systems in a karstic-fissured-porous aquifer, the Schneealpe, Austria, by modelling of environmental ^{18}O and ^{3}H isotopes. J. Hydrol. 256, 48–59 (2002).

6. McGuire K., DeWalle D. & Gburek W. Evaluation of mean residence time in subsurface waters using oxygen-18 fluctuations during drought conditions in the mid-Appalachians. J. Hydrol. 261, 132–149 (2002).

7. Vitvar T., Burns D. A., Lawrence G. B., McDonnell J. J. & Wolock D. M. Estimation of baseflow residence times in watersheds from the

runoff hydrograph recession: method and application in the Neversink watershed, Catskill Mountains, New York. Hydrol. Process. 16, 1871–1877 (2002).

8. Aquilina L., Ladouche B. & Dörfliger N. Water storage and transfer in the epikarst of karstic systems during high flow periods. J. Hydrol. 327, 472–485 (2006).

9. Bonacci O., Pipan T. & Culver D. C. A framework for karst ecohydrology. Environ. Geol. 56, 891–900 (2009).

10. Field M. S. A lexicon of cave and karst terminology with special reference to environmental karst hydrology. 214 US EPA: Washington DC, 2002).

11. Ford D. Carbonate karst in Encyclopedia of cave and karst science (ed. Gunn J.) 184–186 Fitzroy Dearborn, New York, 2004).

12. Yuan D. X. On the karst ecosystem. Acta. Geol. Sin. 75, 336–338 (2001).

13. Chen H. S., Liu J. W., Wang K. L. & Zhang W. Spatial distribution of rock fragments on steep hillslopes in karst region of northwest Guangxi, China. Catena. 84, 21–28 (2011).

14. Kendall C. & McDonnell J. J. Isotope tracers in catchment hydrology. 245–247 Elsevier Science: Amsterdam, 1999).

15. Hartmann A., Lange J., Weiler M., Arbel Y. & Greenbaum N. A new approach to model the spatial and temporal variability of recharge to karst aquifers. Hydrol. Earth Syst. Sci. 16, 2219–2231 (2012).

16. Smart C. Artificial tracer techniques for the determination of the structure of conduit aquifers. Ground Water. 26, 445–453 (1988).

17. Gazis C. & Feng X. A stable isotope study of soil water: evidence for mixing and preferential flow paths. Geoderma. 119, 97–111 (2004).

18. Robertson J. A. & Gazis C. A. An oxygen isotope study of seasonal trends in soil water fluxes at two sites along a climate gradient in Washington State (USA). J. Hydrol. 328, 375–387 (2006).

19. Herrmann A., Koll J., Maloszewski P., Rauert W. & Stichler W. Water balance studies in a small catchment area of paleozoic rock using environmental isotope tracer techniques. Conjunctive Water Use (Proceedings of the Budapest Symposium, July 1986). IAHS Publ. no. 156, 111–124 (1986).

20. Hu K., Chen H. S., Nie Y. P., Yang J. & Wang K. L. Characteristics of Seasonal Variation of Deuterium and Oxygen-18 Isotope Composition of Precipitation in Karst Peak-cluster Depression Area, Northwest Guangxi of China. Transactions of the Chinese Society of Agricultural Engineering (China) 5, 53–62 (2013).

21. Tetzlaff D., Soulsby C., Hrachowitz M. & Speed M. Relative influence of upland and lowland headwaters on the isotope hydrology and transit times of larger catchments. J. Hydrol. 400, 438–447 (2011).

22. DeWalle D. R., Edwards P. J., Swistock B. R., Drimmie R. J. & Aravena R. Seasonal isotope hydrology of three Appalachian forest catchments. Hydrol. Process. 11, 1895–1906 (1997).

23. Fenicia F. *et al.* Assessing the impact of mixing assumptions on the estimation of stream water mean residence time. Hydrol. Process. 24, 1730–1741 (2010).

24. Schwarz K., Barth J. A., Postigo-Rebollo C. & Grathwohl P. Mixing and transport of water in a karst catchment: a case study from precipitation via seepage to the spring. Hydrol. Earth. Syst. Sci. 13, 285–292 (2009).

25. Sharma M. L., Bari M. & Byrne J. Dynamics of seasonal recharge beneath a semiarid vegetation on the Gnangara Mound, Western Australia. Hydrol. Process. 5, 383–398 (1991).

26. Maloszewski P. Lumped-parameter models as a tool for determining the hydrological parameters of some groundwater systems based on isotope data. IAHS Publ. 262, 271–276 (2000).

27. Maloszewski P., Rauert W., Stichler W. & Herrmann A. Application of flow models in an alpine catchment area using tritium and deuterium data. J. Hydrol. 66, 319–330 (1983).

28. Maloszewski P., Stichler W. & Zuber A. Interpretation of environmental tracers in groundwater systems with stagnant water zones. Isotopes Environ. Health Stud. 40, 21–33 (2004). [PubMed]

29. Maloszewski P. & Zuber A. Determining the turnover time of groundwater systems with the aid of environmental tracers: 1. Models and their applicability. J. Hydrol. 57, 207–231 (1982).

30. Pipan T. & Culver D. Forty years of epikarst: what biology have we learned? Int. J. Speleol. 42, 215–223 (2013).

31. Williams P. W. The role of the epikarst in karst and cave hydrogeology: a review. Int. J. Speleol. 37, 1–10 (2008).

32. McGuire K. J. *et al.* The role of topography on catchment-scale water residence time. Water Resour. Res. 41, W05002 (2005).

33. Tetzlaff D. *et al.* How does landscape structure influence catchment transit time across different geomorphic provinces? Hydrol. Process. 23, 945–953 (2009).

34. Soulsby C., Tetzlaff D., Rodgers P., Dunn S. & Waldron S. Runoff processes, stream water residence times and controlling landscape characteristics in a mesoscale catchment: An initial evaluation. J. Hydrol. 325, 197–221 (2006).

35. Asano Y., Uchida T. & Ohte N. Residence times and flow paths of water in steep unchannelled catchments, Tanakami, Japan. J. Hydrol. 261, 173–192 (2002).

36. Detty J. M. & McGuire K. J. Topographic controls on shallow groundwater dynamics: implications of hydrologic connectivity between hillslopes and riparian zones in a till mantled catchment. Hydrol. Process 24, 2222–2236 (2010).

37. Bottrell S. H. & Atkinson T. C. Tracer study and storage in the unsaturated zone of a karstic limestone aquifer in Tracer hydrology. (eds. Hotzl H. *et al.*) 207–211 Balkema, 1992).

38. Maloszewski P. & Zuber A. Principles and practice of calibration and validation of mathematical models for the interpretation of environmental tracer data in aquifers. Adv. Water Res. 16, 173–190 (1993).

39. Chen H. S., Liu J. W., Zhang W. & Wang K. L. Soil hydraulic properties on the steep karst hillslopes in northwest Guangxi, China. Environ. Earth Sci. 66, 371–379 (2012).

40. Rawls W. J., Brakensiek D. L. & Saxton K. E. Estimation of soil water properties. Trans. ASAE. 25, 1316–1320 (1982).

41. Calvo-Cases A., Boix-Fayos C. & Imeson A. C. Runoff generation, sediment movement and soil water behaviour on calcareous (limestone) slopes of some Mediterranean environments in southeast Spain. Geomorphology 50, 269–291 (2003).

42. Cantón Y. *et al.* A review of runoff generation and soil erosion across scales in semiarid south-eastern Spain. J. Arid. Environ. 75, 1254–1261 (2011).

43. Kiraly L. Karstification and groundwater flow. Speleogenesis Evol. Karst Aquifers 1 3: 1–26 (2003).

44. Zarei H., Akhondali A. M., Mohammadzadeh H., Radmanesh F. & Laudon H. Runoff generation processes during the wet-up phase in a semi-arid basin in Iran. Hydrol.Earth Syst. Sci. Discuss. 11 4: 3787–3810 (2014).

45. Laudon H., Sjöblom V., Buffam I., Seibert J. & Mörth M. The role of catchment scale and landscape characteristics for runoff generation of boreal streams. J. Hydrol. 344, 198–209 (2007).

46. Yang J., Chen H. S., Nie Y. P. & Wang K. L. Variation of Precipitation Characteristics and Shallow Groundwater Depth in the Typical Karst Peak-cluster Depression Areas. J. Soil Water Conserv. (China). 26, 239–243 (2012).

47. Schotterer U., KFröhlich K. & Oldfield F. Global Network for Isotopes in Precipitation: To Joel Gat and Hans Oeschger (GNIP)[Z]. IAEA (1996).

48. Wang S. Q. *et al.* Shallow groundwater dynamics and origin of salinity at two sites in salinated and water-deficient region of North China Plain, China. Environ. Earth Sci. 66, 729–739 (2012).

49. Meißner M., Köhler M., Schwendenmann L., Hölscher Dirk. & Dyckmans J. Soil water uptake by trees using water stable isotopes (δ^2H and $\delta^{18}O$)– a method test regarding soil moisture, texture and carbonate. Plant Soil. 376, 327–335 (2014).

50. Sprenger M., Volkmann T. H. M., Blume T. & Weiler M. Estimating flow and transport parameters in the unsaturated zone with pore water stable isotopes. Hydrol. Earth Syst. Sci. Discuss. 11 10): 11203–11245 (2014).

51. Stumpp C., Maloszewski P., Stichler W. & Maciejewski S. Quantification of the heterogeneity of the unsaturated zone based on environmental deuterium observed in lysimeter experiments. Hydrol. Sci. J. 52 4): 748–762 (2007).

52. Long A. J. & Valder J. F. Multivariate analyses with end-member mixing to characterize groundwater flow: Wind Cave and associated aquifers. J. Hydrol. 409, 315–327 (2011).

10 CHAPTER

KARST AQUIFER RECHARGE: A CASE HISTORY OF OVER SIMPLIFICATION FROM THE ULEY SOUTH BASIN, SOUTH AUSTRALIA

Nara Somaratne

South Australian Water Corporation, 250 Victoria Square, Adelaide, South Australia 5000, Australia;

ABSTRACT

The article "Karst aquifer recharge: Comments on 'Characteristics of Point Recharge in Karst Aquifers', by Adrian D. Werner, 2014, Water 6, doi:10.3390/w6123727" provides misrepresentation in some parts of Somaratne [1]. The description of Uley South Quaternary Limestone (QL) as unconsolidated or poorly consolidated aeolianite sediments with the presence of well-mixed groundwater in Uley South [2] appears unsubstantiated. Examination of 98 lithological descriptions with corresponding drillers' logs show only two wells containing bands of unconsolidated sediments. In Uley South basin, about 70% of salinity profiles obtained by electrical conductivity (EC) logging from monitoring wells show stratification. The central and north central areas of the basin receive leakage from the Tertiary Sand (TS) aquifer thereby influencing QL groundwater characteristics, such as chemistry, age and isotope composition. The presence of conduit pathways is evident in salinity profiles taken away from TS water affected areas. Pumping tests derived aquifer parameters show strong heterogeneity, a typical characteristic of karst aquifers. Uley South QL aquifer recharge is derived from three sources; diffuse recharge, point recharge from sinkholes and continuous leakage of TS water. This limits application of recharge estimation methods, such as the conventional chloride mass balance (CMB) as the basic premise of the CMB is violated. The conventional CMB is not suitable for accounting chloride mass balance in groundwater systems displaying extreme range of chloride concentrations and complex mixing [3]. Over simplification of karst aquifer systems to suit application of the conventional CMB or 1-D unsaturated modelling as described in Werner [2], is not suitable use of these recharge estimation methods.

Keywords: karst aquifers; point recharge; sinkholes; recharge in semi-arid; Australia

1. INTRODUCTION

The article of Werner [2], "Karst Aquifer Recharge: Comments on Characteristics of Point Recharge in Karst Aquifers", contains inappropriate over-simplifications of the hydrogeology and recharge processes in Uley South basin. Werner [2], using the conventional chloride mass balance method (CMB) and one-dimensional unsaturated flow modelling, arrives at distorted characteristics of the Uley South QL aquifer. The simplistic conceptual model in Werner [2], appears "fit-for-purpose" for demonstrating the general application of the conventional CMB and 1-D unsaturated flow modelling, but is far removed from the reality of the basin. Werner [2] describes the conceptual model and well-mixed groundwater system in the QL aquifer without providing supporting evidence, although substantial lithological, hydrogeological, hydrochemical and salinity profile data are available. For reader convenience, this reply is sub-divided into two sections as in Werner [2]: (1) Uley South Recharge Processes; and (2) Uley South Recharge: Methods and Estimates.

2. ULEY SOUTH RECHARGE PROCESSES

2.1. Basin Features: Sand Dunes, Vegetation Cover and Groundwater Flow Direction in Uley South

Since a more detailed description of Uley South basin features is given in Somaratne [1], Somaratne et al. [4] and Somaratne and Frizenschaf [5], only a succinct summary is provided here, focusing in features relevant to the points of contention [2]. The basin is topographically closed due to surrounding sand dunes and basement high areas except in the landward direction where dry limestone is found (Figure 1a). Vegetation is sparse in Uley South central basin, except on dunes, which generally occur 30 m above mean sea level (Figure 1a). Groundwater flow direction is from north-east to south-west (Figure 1b), as the basement high area between Coffin Bay and Uley South restricts groundwater movement to the west. There is a basement low area of 700 m length (gap) near the north-western corner of the basin (Figure 1a), however, no investigation well has been drilled in this area. Groundwater flow towards the "gap" is restricted due to basement high areas and dry limestone in the landward direction. During intense rain events, runoff of neighboring limestone dry areas and basement high areas drain to karstic features in Uley South that recharge the aquifer. Thus the area beyond the Uley South landward boundary forms an allogenic recharge zone.

2.2. Characteristics of QL Aquifer

Werner [2] describes the Uley South Quaternary Limestone as mainly comprising aeolian sediments, which are generally either unconsolidated or loosely aggregated [2]. Unconsolidated sediments are nonlithified having no mineral cement or matrix binding such as $CaCO_3$ in their grains. Technically, an aquifer comprised of such sediments would not be called "Quaternary Limestone" because limestone is one type of sediment that has undergone lithification and formed into a sedimentary rock.

The description of the basin geology in Werner [2] is based on a previous work [6] rather than examination of available geological information. In their study, McKee and Ward [7] survey the pertinent literature, and show that no "one term" has had exclusive adoption and that approximately 20 names or variants have been applied to the calcareous dune deposits and

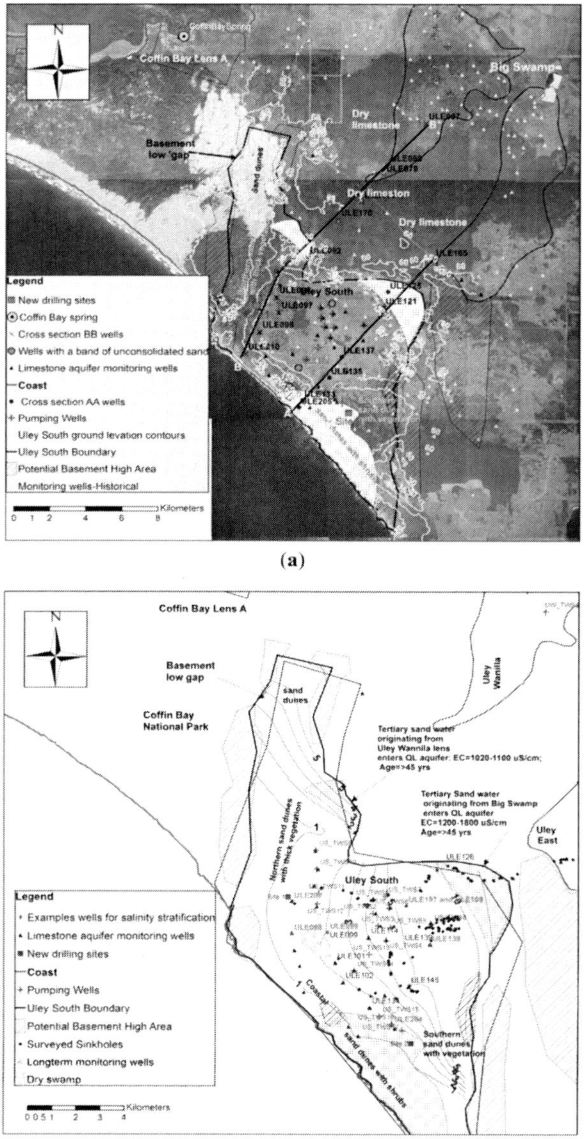

Figure 1. Uley South basin features **(a)** Topographic features **(b)** Groundwater elevations contours.

consolidated rock derived from them. The feature of $CaCO_3$ composition has been variously indicated in the sediment name or rock name by use of such terms as calcareous, carbonate, calcarinite, sandstone or limestone [7]. The agent of deposition is mostly indicated by the name aeolian, wind-blown, dune sand or dune rock. Only modern dunes are generally unconsolidated, mostly active, and becoming cemented and stabilized over time, even though there is a great variation in the amount of time required for lithification [7]. Thus QL lithology in Uley South is variously described over the 1960 to 2014 period of well drilling, but the primary composition is calcarenite-albeit with varying degrees of cementation.

With the above background, and in an effort to more accurately describe basin characteristics, 98 available lithological descriptions with corresponding drillers' logs spanning the period 1961–2003 were re-examined. Two lithological descriptions contain "Aeolian unconsolidated bands" (ULE186 at 6–11.5 m depth and, ULE102 at 4–13 m depth in Figure 1a), three lithological descriptions of historical wells show "Aeolinite-Hard" (ULE115, ULE117, ULE113) and one well describes the lithology as "Aeolinite-buff, fine to medium consolidated" (ULE101). The remaining wells' lithological logs describe either Limestone (LST) or Sandstone (SST) with common descriptions such as: "LST: cream, buff, fine, chalky, hard; LST: grey boulders, SAND at places strongly cemented (or moderately and places with weakly cemented) bars; LST: off white, chalky indurated; LST: off white moderately cemented; LST: sand hard bars; SST cream, dense (or fine) calcarenite". The lithology of wells adjacent to sand dunes mostly fitted the description as shown in ULE202 "Calcrete: aeolinite, medium grained, very strongly cemented, whitish brown".

The occurrence of near-vertical cliffs over 100 m height, along the entire coastline (about 18 km) (Figure 2) suggests that the QL aquifer comprise of consolidated sediment. All of the above evidence appears to contradict the conception, as shown in the schematic diagram of Werner [2] of upper subsurface conditions with unconsolidated or poorly consolidated sediments. This schematic diagram also shows a continuous subsurface calcrete horizon [2]. Not all wells show a calcrete layer in the lithological descriptions. Whilst some wells show the presence of a calcrete layer in vadose zone, in some other wells, the calcrete layer only appears in the saturated zone. It is also common that limestone or sandstone "hard bars" occur at multiple horizons in the profile even in a single well. However, no evidence is found in the lithological descriptions, that such horizons form continuous layers.

During drilling, it is noticeable when intersecting a large cavity, but not so noticeable if intersecting a small conduit, unless core samples are collected. No core samples have been collected from Uley South drillholes making it difficult to identify conduit porosity zones, particularly of small-pipe conduits. However, in several wells "lost circulation" (no drill cutting returns to surface) has been reported, indicating the presence of large cavities. For example in ULE208 well (Figure 1b) drilled in 2003, at 24–42 m depth interval, "lost circulation, no returns, no samples, hard bar at 38 m" is reported. Depth to water of ULE208 is about 28 m and therefore "lost circulation" begins in the unsaturated zone indicating the presence of conduit porosity in both saturated and unsaturated zones. The new investigation well drilled in 2014 at Site 1 (Figure 1a) which is about 240 m west of ULE208 and 15.5 m higher elevation shows the presence of lost circulation and hard bars at the same elevation as that in ULE208 (Figure 3). The well at Site 1 is 0.1 m diameter with an open hole production zone from 54.9 to 63 m. The depth to standing water level is about 44 m. The occurrence of conduit porosity at about the same elevation in the Site 1 well and ULE208 is an indication of possible inter-linking of conduit porosity in both unsaturated and saturated zones. The lithological description of the well at Site 1 (Figure 3a) shows the occurrence of calcarenite with varying degrees of cementation.

Figure 2. Section of the coastal cliff.

Depth (m)	Description
0–4	Calcarenite; light brown off white
4–14	Calcarenite, buff, very weakly cementing
14–16	Calcarenite;no returns; injecting form
16–63	Calcarenite; very weak cementing to moderate cementing and mostly no returns; water at 44 m depth. At 41 m, 51 m, and 52–53.5 m very hard bars; Cavity from 53.5–55 m; Hard from 58–60 m

(a)

(b)

Figure 3. Geology and topographic features at Site 1 (**a**) Summary of lithological description (**b**) Drilling at Site 1 (note the surface calcrete in foreground and coastal dunes with vegetation cover in the background).

Investigation drilling at Site 2 (Figure 1a), located at the foot of the coastal dunes reveal similar lithological facies. At 28–30 m depth, a cavity is encountered and at 79–93.5 m depth "lost circulation" is reported. The well is 0.25 m diameter and cased to 69 m with an open hole production zone from 69 to 84 m. Depth to water is about 60 m from ground level.

Further evidence of the presence of conduit porosity emerges from the hydraulic parameters of the QL aquifer. Pump tested hydraulic conductivity of water supply wells in Uley South basin is 300–1300 md^{-1} and specific yield is 0.03–0.72 indicating large heterogeneity. This is typical of karstic aquifer but not of unconsolidated sediments. For example, town water supply well number 1, TWS 1, (Figure 1b) had been pump tested at 126 Ls^{-1} for 16 h and produced only 2.1 m drawdown. The saturated thickness of the QL aquifer at TWS 1 well is about 10 m. The TWS 1 was constructed in 1964; TWS 2-8 constructed during 1969–1975 and TWS 9-17 constructed in 1999. There have been no reports of collapse or instability since their construction and are still functioning as water supply wells. The TWS wells are 0.25 m diameter, apart from TWS 9, 11 and 12, all the other 14 water supply wells are constructed with open hole production zones indicating stable saturated zones. This would not be possible in unconsolidated or poorly consolidated sediments.

Similar to the Uley South basin, Werner [2] describes the Coffin Bay aquifer as "surface calcrete overlying unconsolidated sediment". Located adjacent to Uley South basin, the Coffin Bay Lens A aquifer is geologically similar to Uley South, comprised of calcarenite. However, there are no sinkholes evident in Coffin Bay and therefore no point recharge sources as in Uley South. Despite this, conduit porosity does exist as evidenced in spring discharges along the coastline (Figure 1a). It is thus concluded that the primary recharge mechanism in the Coffin Bay Lens A aquifer is diffuse recharge.

2.3. Solution Sinkholes in Uley South

In Werner [2], sinkholes are described as "predominantly solution features within the calcrete capping" and "the limited depth of sinkholes over the majority of Uley South is such that there are no known dissolution features that reach the watertable within the basin". In order to guide the reader on recharge through sinkholes, a brief description of sinkhole classification and recharge process is provided below. Out of six sinkhole types (solution, collapse, caprock, dropout, suffosion and buried) [8], solution (Figure 4a) and buried (Figure 4b) sinkholes are evidenced in Uley South. Solution sinkholes occur in areas where limestone is exposed at land surface or is covered by thin layers of soil [8]. Solution of carbonate rock is most active along joints, fractures, cavities, bedding planes or other openings in the limestone that permit water to move easily into the subsurface. Thus solution sinkholes develop along these openings. Dissolved limestone and insoluble residue are carried downward by percolating water along enlarged openings as solution of limestone progresses.

Circulation of water through karstified rocks is affected by three concentration mechanisms [9]: Overland flow; through flow, that is, flow through a layer of soil above limestone; and subcutaneous flow. Each of the above flows differs depending on the level of development of the karst process. Similarly, the input mechanism from sinkhole to groundwater can also be divided into three mechanisms [9]: Shaft flow from predominantly vertical shafts (connected sinkholes, Somaratne [1]); flow through enlarged joints, cavities and fractures of the vadose zone (connected sinkholes, Somaratne [1]); and vadose seepage or slow sinking through small fissures (macro pores in unconnected sinkholes, Somaratne [1]). Point recharge at sinkholes feeds karst conduits to the saturated zone, bypassing the surrounding matrix and thus largely escaping evapotranspiration. It is expected that a small component of recharge around sinkholes also seeps slowly into the granular porosity around the recharge point.

(a) (b)

Figure 4. Sinkhole types in Uley South (**a**) Large active solution sinkhole showing upper calcrete layer and below consolidated sediment (calcarenite); (**b**) Large buried sinkhole showing upper calcrete layer and infill materials.

Groundwater flow in karst aquifers is shared between the granular porosity and karst conduits, but groundwater flow through even a porous limestone aquifer is generally mostly through the conduits, because groundwater in the conduits flows several orders of magnitude more quickly than in the granular porosity. In contrast to solution sinkholes, buried sinkholes are formed due to subsequent deposition of material washed into sinkholes from sides. However, these sinkholes become active once the new lateral dissolution of limestone occurs as shown in Figure 4b.

2.4. Sources of Groundwater Mixing in the QL Aquifer and Salinity Stratifications

Werner [2] and Ordens et al. [10] posit diffuse recharge as the primary recharge mechanism in Uley South. The different recharge mechanisms and their influence on salinity stratification and hydrochemical changes, evident in Uley South QL aquifer is described below. The reader is directed to Somaratne and Frizenschaf [5] for more detailed description of the different water types and mixing in the Uley South basin.

Rainfall in Uley South is seasonal, occurring predominantly during winter months. Uley South QL aquifer groundwater is derived from three sources, two of which are rainfall driven. Diffuse recharge from rainfall takes place through surface exposed epikarst, buried and unconnected sinkholes and pockets of soil that occur in the basin; point recharge of runoff through connected solution sinkholes which is intermittent and highly transient; and finally, leakage of TS water into QL aquifer at the landward boundary where Tertiary Clay is absent (Figure 1b and Figure 5). This leakage is independent of rainfall and occurs throughout the year. The TS water quality has influenced salinity stratification, hydrochemical and isotope composition of QL aquifer.

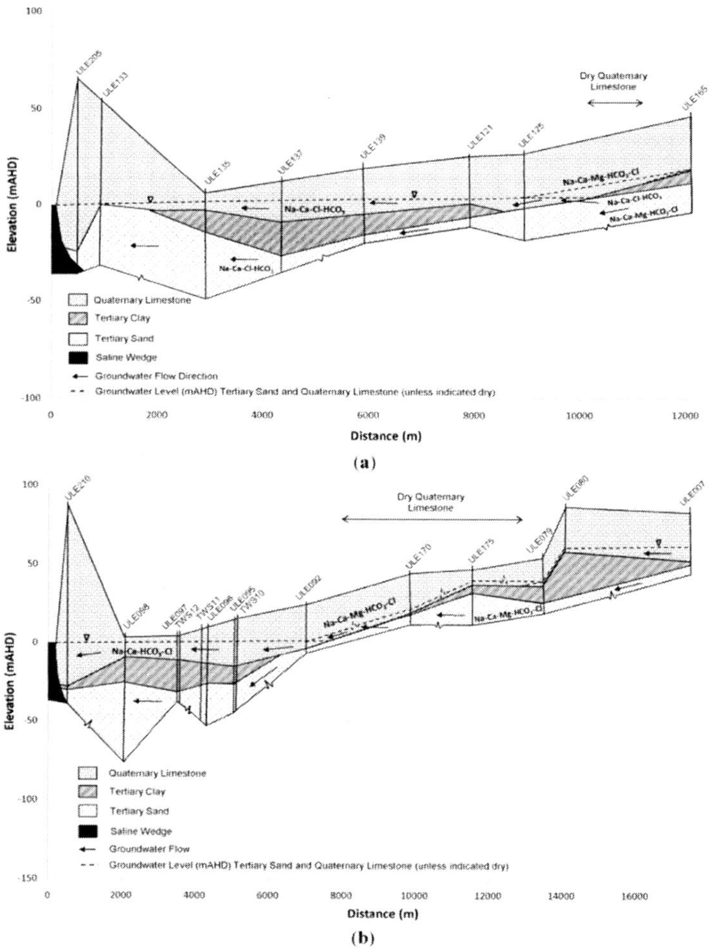

Figure 5. Geological cross section: (**a**) Cross section AA-Uley South to Uley East lens [1,5]; and (**b**) Cross section BB-Uley South to Uley Wanilla lens.

Harrington et al. [11] and Somaratne and Frizenschaf [5] show that the high-salinity plume of groundwater in the TS aquifer enters Uley South at the central part of the landward boundary where Tertiary Clay is absent (Figure 1b and Figure 5a). The plume originates from the Big Swamp (Figure 1b) area located 18 km north-east of Uley South as a result of downward leakage of Big Swamp's surface water through the clay aquitard [5]. As the plume moves down gradient along its main flow path, the western edge of the Uley East basin, it is diluted from electrical conductivity (EC) 9000–10,000 μS/cm at Big Swamp to EC approximately 1800 μS/cm at the southern boundary of the Uley East basin. In Uley South in the area where the aquitard is absent, the TS aquifer receives direct recharge, resulting in further dilution of EC to about 1200 μS/cm [5]. Similarly, Tertiary sand water originating from Uley Wanilla lens with EC up to 1100 μS/cm enters the north-central part of the basin where Tertiary Clay is absent at the landward boundary (Figure 1b and Figure 5b).

In order to track the possible pathway of freshwater from point recharge sources, salinity profiles were obtained according to the method described by Somaratne [1]. Salinity profiles are considered to be superior to water sampling as the sonde records EC of water at 0.05 m interval, thus enabling identification of EC changes at short depth intervals. Out of 32 monitoring wells in QL aquifer, 22 were sonded and about 70% of sonded wells showed salinity stratification. Where EC measurements were available from drilling undertaken in 1961, these were presented for comparison (Figure 6, Figure 7 and Figure 8).

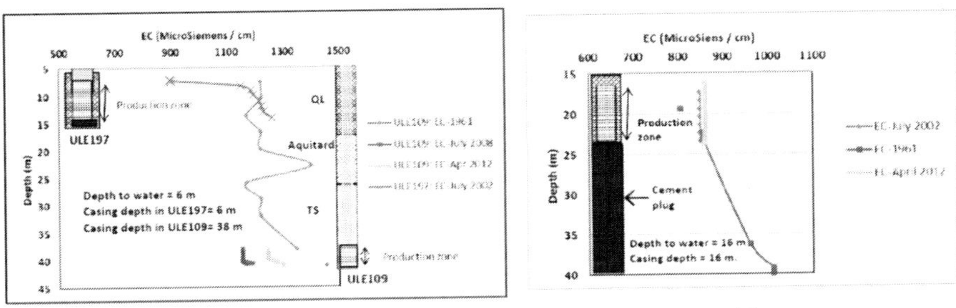

(a) (b)

Figure 6. Effect of Tertiary Sand water leaking into QL aquifer on salinity stratification (**a**) ULE109 and ULE 197; (**b**) ULE139.

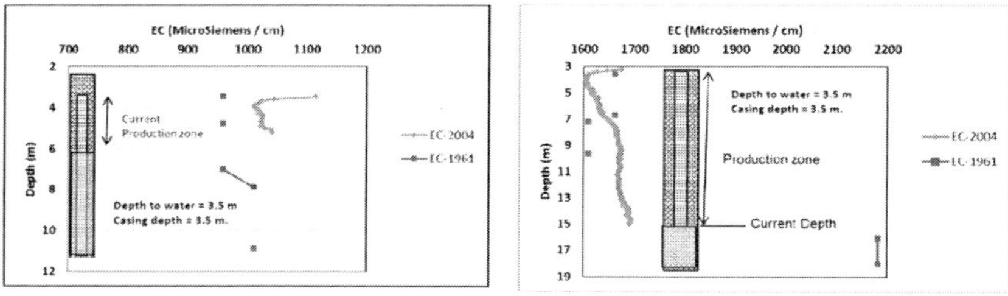

(a) (b)

Figure 7. Salinity Stratification influenced by evaporation in QL aquifer (**a**) ULE099 and (**b**) ULE098.

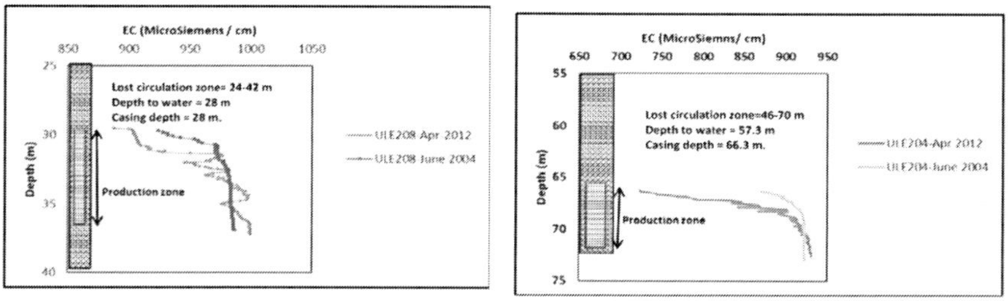

(a) (b)

Figure 8. Salinity stratification influenced by conduit porosity in QL aquifer (**a**) ULE208 and (**b**) ULE 204.

As a result of higher salinity water leaking to QL aquifer, pronounced salinity stratification is found in QL aquifer monitoring well ULE197, located 2 km down-gradient to the Tertiary Clay absence area (Figure 6a). In the ULE197 well, salinity of the QL base is identical to that in TS aquifer monitoring well ULE109, located at the same site. In contrast, the southern half of the QL aquifer has no link to the Big Swamp area through either QL or TS aquifers due to the occurrence of the basement high at the Uley South boundary (Figure 1). Therefore, QL water in the southern part of Uley South is derived primarily from direct rainfall recharge within Uley South. ULE139 (Figure 6b) drilled as an investigation well in 1961 and completed as a QL aquifer monitoring well shows no evidence of salinity stratification suggesting no influence by TS water or connection to sinkholes through conduit. Comparing ULE197 and ULE139 salinity profiles of QL aquifer, leakage of Tertiary Sand water has increased EC of ULE197 by 400 μS/cm at the base of the QL aquifer and this disperse upward towards the piezometric surface (Figure 6a) creating a non-uniform mixing zone.

The influence of TS water entering Uley South is evidenced in all monitoring wells in the central northern parts of the basin. In shallow wells ULE099 and ULE098 (Figure 7), the upper part of the EC profile is slightly higher, possibly due to the effect of evaporation. The difference of EC in the upper and lower parts of the QL aquifer has declined due to dispersion of higher salinity water. ULE098 is located downgradient to a dry swamp (Figure 1b), and EC is higher than upgradient well ULE099 due to evaporation enriched swamp water leaking into groundwater that is possibly intercept by ULE098.

EC stratification due to point recharge is evident in monitoring wells located away from the influence of TS water in the QL aquifer. ULE208 drilled in 2003, is located at the northern end of the central basin (Figure 1b) with a depth to water of about 28 m. The well is located at the foot of a limestone slope. Since the surface is calcretised, it is expected that little direct recharge is received from the surface. The low EC zone (Figure 8a) corresponds to the “lost circulation zone” described previously. The nearest known sinkhole is located down-slope approximately 200 m to the east. It is therefore possible that conduit carrying point recharge from the sinkholes is intercepted by the “lost circulation” zone of ULE208.

Drilled in 1999, monitoring well ULE204 is located on limestone slopes in the southern side of the basin (Figure 1b). Depth to water is about 57.3 m and the well is cased up to 66.3 m. During drilling, circulation was lost between 46 and 70 m depths. The well is constructed as “open hole” in the production zone. Similar to ULE208, low salinity in the upper “lost circulation” area is due to conduit porosity containing low salinity water (Figure 8b) which is thought to originate from cluster of sinkholes in an up-gradient valley, about 1 km to the north-east. This suggests that in Uley South, high flow pathways are evident. This evidence is removed from wells in the central basin by the input of higher salinity water continuously received from the TS.

In similar settings, Allison et al. [12] investigated rates and mechanisms of recharge beneath two major landscape settings in the Murray Basin in South Australia; these are calcrete flats with sinkholes and sand dunes in an adjacent landscape. Allison et al. [12] suggest that recharge varies from in excess of 100 mm·year^{-1} for sinkhole areas to less than 0.1 mm·year^{-1} beneath sand dunes with native vegetation.

The groundwater age in Big Swamp and Uley Wanilla is >45 years [11]. Mixing of old waters originating from Uley Wanilla and Big Swamp with younger water (<20 years) in Uley South has important ramifications on applicable recharge estimation methods. When mixing occurs between different groundwater systems of different ages and different chlorides concentrations, this results in unreliable recharge estimates using groundwater age and the conventional CMB method [4]. The combination of aquifer heterogeneity, salinity stratification, recharge through solution features, leakage of Tertiary Sand water and non-uniform mixing of different water types and modification along flow paths, results in Uley South being more hydrochemically complex than Werner [2] suggests.

3. ULEY SOUTH RECHARGE: METHODS AND ESTIMATES

Werner [2] discusses various approaches and data limitations on application of recharge estimate methods in karst aquifer systems, and in particular for the Uley South basin. Numerous studies on recharge estimation methods and their limitations are available. For details, the reader is directed to: Simmers [13], Sharma [14], Healy and Cook [15], Scanlon et al. [16], and a summary of selected methods in Somaratne et al. [4].

With regard to obtaining consistent results and greater reliability of recharge, Scanlon et al. [16] hold the view that it is highly beneficial to apply multiple methods but Healey and Cook [15] consider that consistency by itself should not be taken as an indication of accuracy [4]. Concurring both with Scanlon et al. [16] and Healey and Cook [15], Somaratne et al. [4] introduce a simple technique to assess reliability of recharge based on three distinct criteria: applicability of the method, reliability of data, and spatial coverage of the basin. Somaratne et al. [4] show that if the theory or method of recharge estimation is not valid, it always produces a "Low" reliable recharge value. This "Low" level will stand unless the theory or method is improved. Reliability of data and spatial coverage can be improved, if the method is applicable [4].

With this background, underpinning assumptions of the conventional CMB method are provided below. Following Ericksson and Khunakasen [17], the CMB equation can be expressed as:

$$R = \frac{Pc_{p+D}}{c_R} \tag{1}$$

where R is recharge (LT^{-1}); c_{p+D} (ML^{-3}) is the representative mean chloride concentration in rainwater including contributions from dry deposition;

and c_R is chloride concentration in recharge (ML^{-3}). For Equation (1) to hold after replacing c_R with groundwater chloride (c_g), c_R must be in equilibrium with c_g (saturated version of CMB) [3]. This is not the case with point recharge through sinkholes as Werner [2] anticipates, as point recharge occupies mostly conduit porosity areas. Zhu et al. [18] highlights that for the saturated version of CMB to be applicable; groundwater movement in both unsaturated and saturated zones should be approximated as one-dimensional piston flow. This essentially means that chloride concentration of the mass flux crossing the piezometric surface (c_R) is at equilibrium with groundwater chloride (c_g) [3]. When chloride concentrations of groundwater samples (c_g) are used as c_R values in Equation (1) for saturated zone application, the recharge rates determined from the conventional CMB apply to locations in the catchment where the samples are recharged, not where the samples are collected [3]. This requires that hydrodynamic dispersion of chloride between the recharge point and sampling location be small, which essentially requires the piston flow requirement in the unsaturated and saturated zones of Zhu et al. [18]; and the necessity of minimal mixing for less variability within the aquifer (valid under steady-state condition), otherwise the conventional CMB equation must be viewed as a gross simplification as indicated by Subyani and Sen [19]. Clearly, when point recharge is a contributing recharge source, fundamental assumptions of the conventional CMB are violated [3]. It is noted here that the conventional CMB is the simplest form of the mass balance equation in hydrology, and is not applicable to account for mass balance in systems displaying complex groundwater mixing caused by point recharge [3]. In particular, groundwater basins, such as Uley South, where the chloride in the aquifer originates from several sources and not just from precipitation directly on the aquifer, such application of the conventional CMB method is likely to provide a distorted recharge value. This is why Somaratne [1] suggest incorporating duality of the recharge mechanism into the conventional CMB, as in the generalised CMB method [3], if it is to be used for karst aquifers. The method must be valid in the first instance, and then parameterisation of the model is a different issue.

In the article of Somaratne [1], point recharge is described as "highly transient and may occur in relatively short-time periods, yet is capable of recharging large volume of water, even from a single extreme rainfall event. Preferential groundwater flows are observed in karst aquifers with local fresher water pockets of low salinity that develop around point recharge sources". Werner [2] comments that "hence, the seemingly stable freshwater pockets in the vicinity of point recharge" but the word "stable" is not given in Somaratne [1] nor is it implied.

Referring to Herczeg et al. [20], Werner [2] states water chemistry and isotope interpretations indicated that point sources contribute less than 10% of total recharge, with diffuse recharge providing the remainder. This is used to justify a diffuse recharge estimate using the conventional CMB method in Uley South basin. For a valid comparison, it is prudent to consider the catchment area and number of sinkholes considered in the Herczeg et al. [20] study. Herczeg et al. [20] considered three large sinkholes (including the Poocher Swamp sinkholes), one swamp and a drainage bore in the > 500 km^2 Tatiara catchment, whereas hundreds of sinkholes (surveyed 161) are present in the Uley South basin (113 km^2), and 400 drainage wells and three sinkholes are present in the Mount Gambier capture zone (26.5 km^2). The diffuse recharge rate of the Tatiara catchment is 1–15 mm per year [20] and in Uley South is 52.7 mm per year [1].

Werner [2] and Ordens et al. [10] interpret the gap in the chloride vs. $\delta^{18}O$ relation as evidence that the sinkhole is not directly recharging the aquifer. Somaratne contests [1] that the gap is a sampling bias because of the inability to obtain water samples under solution sinkholes due to their complex architecture and inaccessibility. Sampling to reduce this bias is possible in drainage wells in Mount Gambier as Herczeg et al. [20] use a drainage well for their study. Average catchment areas contributing to a sinkhole in Uley South (0.07 km^2) [1] and the drainage well in Mount Gambier (0.03–0.12 km^2) are similar. These small catchments generate small runoff volumes that occur mostly in conduit porosity.

Even though 400 drainage wells are spread across the capture zone in Mount Gambier, a gap of 43 mg·L^{-1} exits between rainfall and monitoring well chlorides in the chloride vs. $\delta^{18}O$ relation (Figure 9). Any water samples taken from a well is generally considered to be mixture of waters from all flow lines reaching the well. The gap in chloride vs. $\delta^{18}O$ relation does not mean drainage wells are not directly recharging the aquifer. This gap could only be filled by taking water samples from drainage wells themselves.

In previous recharge estimation efforts, Ward et al. [21] adopt a conceptual model of recharge for the Uley South basin that comprised the addition of runoff to the watertable via numerous sinkholes. This produced an average annual total recharge of 75 mm, whereas Ordens et al. [10] apply the conventional CMB method to the Uley South basin resulting in an average annual total recharge of 52–63 mm. The three criteria reliability analysis of Somaratne et al. [4] results in the total recharge to the basin calculations of both Ward et al. [21] and Ordens et al. [10] as "Low" reliability estimates [4] as Ward et al. [21] ignored the diffuse recharge and Ordens et al. [10] ignored the effects of point recharge and leakage of TS water into the QL aquifer. However, the estimation of Ward et al. [21] of point recharge and that of Ordens et al. [10] of diffuse recharge yield moderately reliable recharge.

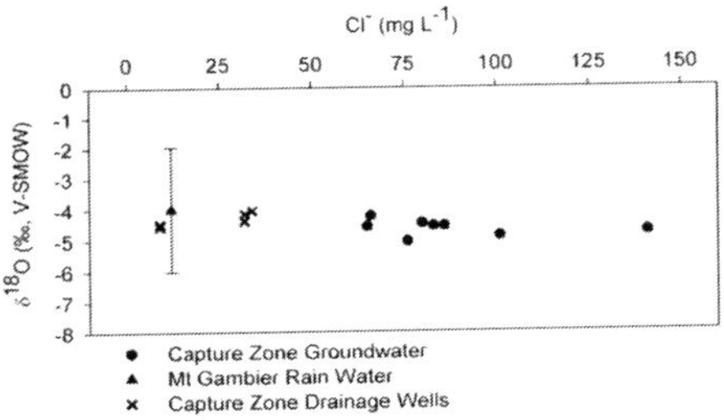

(a)

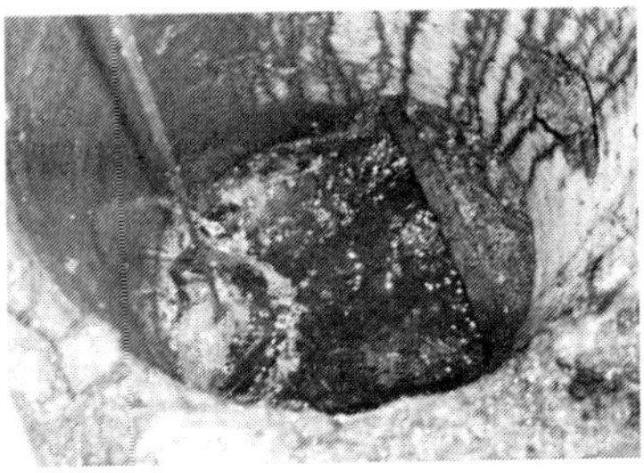

(b)

Figure 9. Characteristics of chloride vs. $\delta^{18}O$ relation, Mount Gambier Blue Lake capture zone (**a**) Chloride vs. $\delta^{18}O$ relation [1]; (**b**) Direct recharge through a drainage well.

The implications of reliance on the conventional CMB estimated recharge in Uley South basin is clear in the long-term water level trends. Average annual pumping from the basin over the past 14 years (2000–2014) is about 60 mm per year (6.8×10^6 m^3 over 113 km^2), which is equivalent to the CMB estimated recharge range of Ordens et al. [10]. Based on extraction of 100% of estimated recharge, groundwater levels should be falling. However, the water level trend observed in the basin since 1999, is characterized by stable or slight water level rise (Figure 10); the coastal saline wedge is stable [5] and as shown before, groundwater salinities in both QL and TS aquifers are equivalent to 1961 levels. The rise and fall of basin water levels is primarily climate driven as indicated in cumulative rainfall deviation from the mean. Declining water levels from 1968 to 1977 have no relation to groundwater extraction as the extraction began in 1977. The year 1968 recorded the highest rainfall (917.4 mm) at the Big Swamp station (Station: 18017) in the last century and 1977 recorded the lowest rainfall of 414 mm during 1960–2014 period. With the average groundwater extraction during the 1977–1992 period of 5×10^6 m^3 (basin equivalent depth of 44.2 mm), water levels generally increased due to average annual rainfall being slightly higher (592.4 mm) than the mean annual average of 550 mm. The period from 1993 to 1999 is mostly dry, with an average 15% reduction in rainfall compared to the mean annual average. During this period water levels declined even though extraction remained unchanged. Years 2000 and 2001 featured up to about 20% higher annual rainfall and since then there has been generally average annual rainfall. During this period, the basin water levels were stable and steadily rising even though the average annual extraction increased to 6.8 × 10^6 m^3 (basin equivalent depth of 60 mm).

Monitoring wells ULE101 and ULE126 are not considered representative wells for the main limestone aquifer, even though they have long-term data sets. ULE101 is located at the edge of a swamp and terminates 5 m into the Tertiary Clay layer, and ULE126 is at the basin's outer margin (Figure 1b) of wet and dry limestone [4]. The current water level of the most up-gradient well within the basin, ULE139, is the same as the pre-development water level of 1961. The most down gradient well, ULE102 is about 0.7 m lower than the 1961 level but steadily rising towards a new equilibrium level. Therefore, recharge estimates based on a conventional CMB are unrealistically low in this karstic setting. Similarly, the minimum average annual recharge value, 47 mm, reported by Ordens et al. [10] using the watertable fluctuation method (47–129 mm) is well below the annual extraction. This highlights the danger of presenting recharge values without assessing reliability levels of estimations.

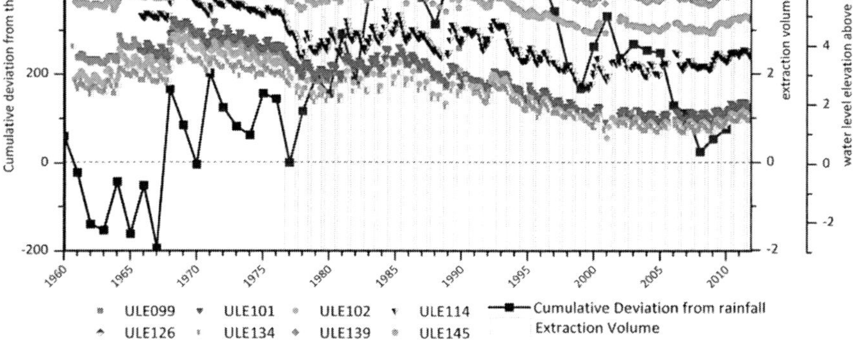

Figure 10. Water level trends in long-term monitoring wells, cumulative deviation from mean rainfall and groundwater extraction in Uley South basin.

4. CONCLUSIONS

Understanding recharge mechanisms and reliable recharge estimates is critical to evaluation and management of water resources. An accurate estimate of recharge to aquifers in arid and semi-arid areas still remains a challenge; especially where point recharge (recharge from discrete locations) dominate. The effectiveness of a groundwater basin management plan depends largely on the reliability of estimated recharge and the monitoring to assess the basin's status. Recognising all recharge estimation methods have some degree of uncertainty, a precautionary approach is recommended as one of the guiding principles in utilising groundwater resources. Groundwater allocation based on over-estimated recharge leads to depletion of the resource, damaging the aquifer and the environment that supports it [4]. Similarly management plans based on under estimated recharge result in economic and social costs to the community in the form of development of alternative resources or limitation on development [4].

Reliability of estimated recharge depends on applicability of the particular method employed, availability of reliable data and adequate spatial coverage of data for the basin [4]. This reliability can be compromised by overly simplifying the characteristics of the karst aquifers, recharge processes or the use of invalid methods for recharge estimation; thus contradicting the precautionary principle. The recharge estimation of karst aquifers demands a significant research effort. This could be achieved through prudent analyses and interpretation of data along with modification and development of new methods to capture the recharge processes. As Somaratne [1] suggests, one area of further development of karst aquifer recharge estimation is the incorporation of duality of recharge regimes into the CMB as featured in the generalized CMB method [3].

ACKNOWLEDGMENTS

The helpful comments of two anonymous reviewers are gratefully acknowledged. Glyn Ashman is thanked for review of the manuscript.

CONFLICTS OF INTEREST

The author declares no conflict of interest.

REFERENCES

1. Somaratne, N. Characteristics of point recharge in karst aquifers. Water **2014**, 6, 2782–2807.
2. Werner, A.D. Karst aquifer recharge: Comments on Somaratne, N. Characteristics of point recharge in karst aquifers. Water **2014**, 6, 3727–3738.
3. Somaratne, N. Pitfalls in application of the conventional chloride mass balance (CMB) in karst aquifers and use of the generalised CMB method. Environ. Earth Sci. **2015**.
4. Somaratne, N.; Smettem, K.; Frizenschaf, J. Three criteria reliability analyses for groundwater recharge estimations. Environ. Earth Sci. **2014**, 72, 2141–2151.
5. Somaratne, N.; Frizenschaf, J. Geological control upon groundwater flow and major ion chemistry with influence on basin management in a coastal aquifer, South Australia. J. Water Resour. Prot. **2013**, 5, 1170–1177.
6. Evans, S.L. Estimating Long-Term Recharge to Thin, Unconfined Carbonate Aquifers Using Conventional and Environmental Isotopes Techniques: Eyre Peninsula, South Australia. Master's Thesis, Flinders University of South Australia, Adelaide, Australia, 1997.
7. McKee, E.D.; Ward, W.C. Eolian Environment, Chapter 3. In Carbonate Depositional Environment: AAPG Memoir 33; Scholle, P.A., Jebout, D.G., Moore, C.H., Eds.; The American Association of Petroleum Geologists: Tulsa, OK, USA, 1983.
8. Waltham, T.; Bell, F.D.; Culshaw, M. Sinkholes and Subsidence: Karst and Cavernous Rocks in Engineering and Construction; Springer-Verlag: Chichester, UK, 2005.
9. Gunn, J. Point-recharge of limestone aquifers-a model from New Zealand karst. J. Hydrol. **1983**, 61, 19–29.
10. Ordens, C.M.; Werner, A.D.; Post, V.E.A.; Hutson, J.L.; Simmons, C.T.; Irvine, B.M. Groundwater recharge to a sedimentary aquifer in the topographically closed Uley South Basin, South Australia. Hydrogeol. J. **2012**, 20, 61–72.
11. Harrinngton, N.; Zulfic, D.; Wohling, D. Uley Basin Groundwater Modelling Project, Volume 1, Project Overview and Conceptual

Model Development; DWLBC Report 2006/01. Government of South Australia: Adelaide, Australia, 2006.

12. Allison, G.; Stone, W.; Hughes, M. Recharge in karst and dune elements of a semi-arid landscape as indicated by natural isotopes and chloride. J. Hydrol. **1985**, 76, 1–25.

13. Simmers, I., Ed.; Estimation of Natural Groundwater Recharge; Reidel: Boston, MA, USA, 1988.

14. Sharma, M.L., Ed.; Groundwater Recharge; A.A. Balkema: Rotterdam, The Netherlands, 1989.

15. Healy, R.W.; Cook, P.G. Using groundwater levels to estimate recharge. Hydrogeol. J. **2002**, 10, 91–109.

16. Scanlon, B.R.; Healy, R.W.; Cook, P.G. Choosing appropriate techniques for quantifying groundwater recharge. Hydrogeol. J. **2002**, 10, 18–39.

17. Eriksson, E.; Khunakasem, V. Chloride concentration in groundwater, recharge rate and rate of deposition of chloride in the Israel Coastal Plain. J. Hydrol. **1969**, 7, 178–197

18. Zhu, C.; Winterle, J.R.; Love, E.I. Late Pleistocene and Holocene groundwater recharge from the chloride mass balance method and chlorine-36 data. Water Resour. Res. **2003**, 39.

19. Subyani, A.; Sen, Z. Refined chloride mass-balance method and its application in Saudi Arabia. Hydrol. Process. **2006**, 20, 4373–4380.

20. Herczeg, A.L.; Leaney, F.W.J.; Stadter, M.F.; Allan, G.L.; Fifeld, L.K. Chemical and isotope indicators of point source recharge to karst aquifers, South Australia. J. Hydrol. **1997**, 192, 271–299.

21. Ward, J.D.; Hutson, J.; Howe, B.; Fildes, S.; Werner, A.D.; Ewenz, C. A modelling framework for the assessment of recharge processes and climate change. Report Developed through the Eyre Peninsula Groundwater allocation and Planning Project; Eyre Peninsula Natural Resources Management Board, Government of South Australia: Adelaide, Australia, 2009.

11
CHAPTER

PEATLANDS AS FILTERS FOR POLLUTED MINE WATER?—A CASE STUDY FROM AN URANIUM-CONTAMINATED KARST SYSTEM IN SOUTH AFRICA PART I: HYDROGEOLOGICAL SETTING AND U FLUXES

Frank Winde [1,*] and Ewald Erasmus [2]

[1] *North-West University, School of Environmental Sciences and Development, Private Bag X6001, Potchefstroom, 2520, South Africa*

[2] *Geotechnical Environmental Specialists, 108 Wenning Street, Groenkloof, Pretoria, 0181, South Africa*

ABSTRACT

Located downstream of goldfields of the Witwatersrand basin, the Gerhard Minnebron (GMB) Eye—as major water source for downstream community of some 300,000 people—may be impacted on by mining-related water pollution especially with uranium (U). Containing up to 5 m-thick deposits of peat that is frequently reported to act as a filter for U and other heavy metals, this paper is the first part of a series that aims to quantify the ability of the GMB peatland to act as buffer against current and future U pollution.

In a first step, this paper outlines the geohydrological conditions and discusses how deep–level gold mining impacted on the dolomitic aquifers. Subsequently, the potential influx of U into the wetland is estimated and associated sources and pathways analyzed. Finally, a model is proposed explaining the significant differences in degree and dynamics of U observed within a single groundwater compartment.

Keywords: peat; uranium; gold mining; dolomite; karst; compartments; Gerhard Minnebron; South Africa

1. INTRODUCTION

Peat consists of partially decomposed wetland plants accumulating in waterlogged environments where the surplus of dead organic matter results in anaerobic conditions under which complete decomposition cannot be achieved. With 99% of all known peat deposits being located in humid regions of the northern hemisphere, peat in southern Africa is a generally scarce resource [1,2]. This is particular true for the semiarid interior plateau of South Africa where the studied peatland is located. The peatland at the Gerhard Minnebron (GMB) wetland owes its existence mainly to a strong perennial discharge of groundwater from a karst spring known as the 'Gerhard Minnebron Eye' and can thus be classified as a karst fen [3,4]. Currently, the spring water is mainly used for the domestic water supply system of the downstream municipality of Potchefstroom as well as by local farmers.

The GMB eye is fed by groundwater emanating from an extensive system of discrete, interlinked dolomitic karst aquifers (so-called 'compartments'). Four of the nine compartments that affect the fluvial system of the Wonderfonteinspruit (WFS) that runs upstream of the GMB eye are impacted on by deep level gold mining. Apart from large-scale dewatering, which lowered the groundwater table by up to 1,000 m in places, this also includes pollution through the filling of caves and sinkholes with uraniferous tailings, the discharge of polluted effluents into the WFS as well as through significant volumes of seepage flowing directly from tailings deposits into the underlying karst aquifer. While the associated water pollution of the GMB spring has been discussed in previous studies [5,6,7,8], the actual extent and the specific sources as well as the exact pathways and mechanisms are still largely unknown.

Since 1993, an estimated 60% of the peat has been extracted from the wetland mainly for mushroom production (casing substrate) and for use in horticultural soil enhancing products. In view of the rapid destruction of a potentially beneficial resource, in 2006, the Department of Water Affairs (DWA) commissioned a study to assess associated impacts on the hydrological system [9]. A major focal point of the study is to investigate to what extent the remaining peat deposits may act as a buffer between upstream mining pollution, especially with waterborne uranium (U), and the downstream water supply of Potchefstroom. Such buffer function was inferred from the frequently reported ability of peat to remove uranium and other contaminants from water [10,11,12,13,14,15] and could be vital in a possible (worst case) post-mining scenario where the GMB eye is expected to be one of three major outflow points through which large volumes of highly contaminated mine water may be discharged from flooded underground mine voids [16].

In part one of this paper the hydrogeological setting of the GMB peatland and its significance for the formation of the peatland at GMB are discussed. Furthermore, hydraulic links between the peatland and adjacent fluvial systems and aquifers, on surface and underground, are analyzed and associated water fluxes quantified including a preliminary quantification of the total contribution of the peatland system to the downstream water supply system. In order to estimate the pollution potential posed by the upstream gold mines, the total annual load of dissolved U entering the Wonderfonteinspruit as well as a karst aquifer associated with the GMB peatland is determined.

The second part of the paper concentrates on quantifying the buffer/filter function of the peatland for waterborne uranium. This includes a brief overview on possible mechanisms of U removal and accumulation as reported in the literature as well as a series of batch experiments to assess to what extent peat, under local conditions, is able to trap and retain U from typical mine waters. U concentration in different types of waters of the peatland as well as sediments and peat samples from various depths are determined to identify possible effects of mining-related pollution.

The effectiveness of peat as a potential filter is not only determined by its ability to remove U from the water phase but also by the extent of surface and groundwater water fluxes moving through the peat. In order to obtain a first order approximation the hydraulic conductivity of peat is determined in a column experiment as well as in situ. The latter is based on quasi-continuous datalogger-controlled measurements in undisturbed peat. Based on the above, a conceptual model will be presented on the formation and hydrological significance of the peatland and its ability to act as buffer against current and future uranium pollution emanating from upstream mining areas.

2. HYDROLOGICAL AND HYDROGEOLOGICAL CONDITIONS

2.1. Regional Overview

The GBM eye together with the upper and lower Turffontein eyes constitutes the major outflow point of dolomitic groundwater from a large karst aquifer called 'Boskop-Turffontein Compartment' (BTC). This compartment is the largest and lowest lying in a succession of several others located further up in the Wonderfonteinspruit (WFS) catchment [17,18].

Originating south of the sub-continental divide near Krugersdorp (now Mogale City) the 90 km-long WFS runs over approximately 80 km across several dolomitic compartments to finally join the upper Mooi River. The GMB peatland, however, falls outside the (surface) catchment of the WFS and feeds via an unnamed stream (in this study called 'GMB-stream') also into the upper Mooi River, some 4 km upstream of Boskop Dam as the main reservoir for the water supply of some 250,000 people of the Potchefstroom municipality (now Tlokwe) (Figure 1).

Separated from each other by near impervious, approximately north-south trending syenite and dolorite dykes the dolomitic compartments, perennially, feed large volumes of dolomitic groundwater via so-called eyes (karst springs) into the WFS as reflected in its Afrikaans name ('Miraculous Fountain Stream') (Figure 2).

Most of the groundwater is stored in the upper 40–100 m (below the original water table prior to dewatering) of the outcropping Malmani dolomite in what is termed the 'cavernous zone' [19,20,21]. This zone consists of a network of caves and cavities interconnected by solutions slots, underground channels and fractures, totaling a storage capacity which exceeds that of the full Vaal Dam (with 2,536 million m^3 at full capacity the second largest dam in South Africa) by several times. The 2.65–2.47 billion year-old dolomite [22] has been subjected to extensive karstification resulting in the five longest caves in southern Africa being present in the WFS catchment as well as a number of karst springs (locally termed 'eye' or 'oog' in Afrikaans) which are amongst the strongest in the country [23]. Owing to their significant storage capacity and associated spring flow, the dolomitic compartments in the WFS catchment and other karst aquifers in the Transvaal (a former province in south Africa) have historically been the subject of many water-related investigations [6,17,18,19,24,25,26,27].

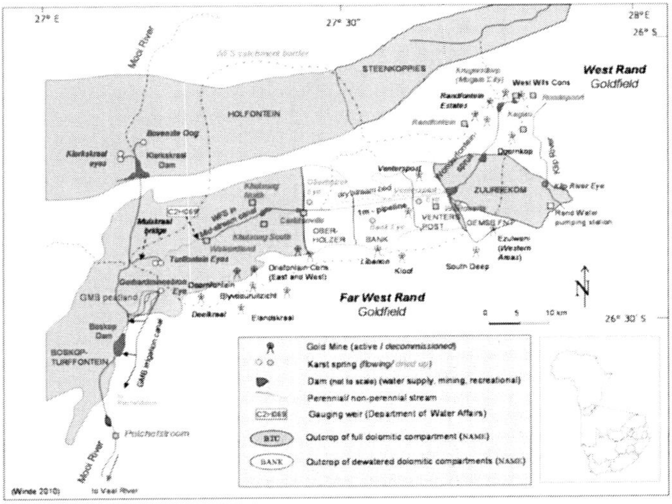

Figure 1. Location of the Gerhard Minnebron wetland in relation to dolomitic compartments and gold mines in the adjacent catchment of the Wonderfonteinspruit.

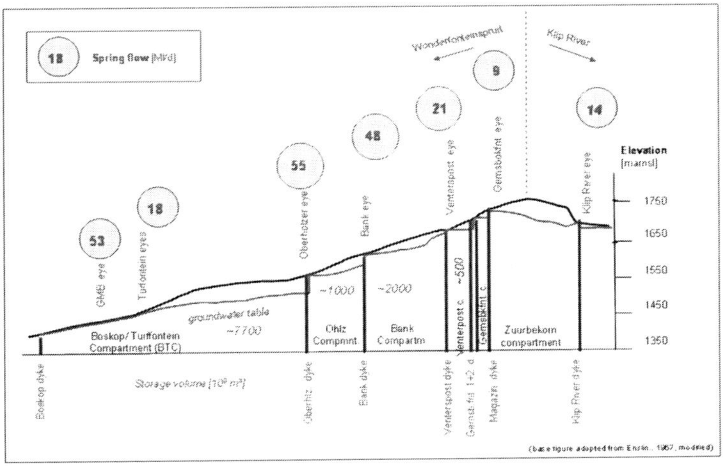

Figure 2. Schematic cross section of the dolomitic compartments associated with the Wonderfonteinspruit and the Gerhard Minnebron eye indicating pre-mining discharge rates from dolomitic eyes, groundwater storage volumes in associated compartments as well as differences in water

table elevation between the individual compartments (based on data and a figure in Enslin, 1967, [27] the storage volume for the Boskop-Turffontein compartment is estimated).

The degree of karstification is not homogenous across the up to 1.5 km-thick dolomite in the study area but varies in accordance with a number of parameters, the most significant being the percentage of chert (layers) in the various dolomitic formations. Generally, chert-rich formations display a higher degree of karstification and thus contain more groundwater than chert-poor or chert-free dolomite [6].

Table 1 lists the relevant different dolomitic formations and the associated chert- and water content in relation to the general stratigraphy of the study area.

Table 1. Stratigraphic position of different dolomitic formations in study area.

Sequence/ Supergroup (SG)	Group	Subgroup	Formation	Sub-formation	Chert-/water contents of dolomite * (Yield [L/s]) **	Type of rocks	Av. thickness in study area [m] *
Transvaal sequence	Pretoria		Rooihoogte			shales	~300
	Chuniespoort	Malmani	Eccles		high (11)	dolomite	~380
			Lyttleton		Low (3)	dolomite	~50
			Monte Christo	upper	high (12)	dolomite	~260
				middle	low	dolomite	~160
				lower	high	dolomite	~270
			Oak Tree		Low (6)	dolomite	~200
			Black Reef			quartzite, shales	~10
Ventersdorp SG			Ventersdorp lava			lavas	~1,800
Venterspost conglomerate				Ventersdorp Contact Reef		quartzite	0...~3
Witwatersrand SG	Central Rand		Various gold reefs (incl. carbon leader)			quartzite, shales	~3,000
Basement						granites, gneiss	

* [28]; ** average borehole yield according to [29] as observed in a total of 950 boreholes in the Schoonspruit compartment

Table 2 displays the hydraulic transmissivity for dolomite at different depths as determined by [6] through pumping test conducted in boreholes of the dewatered Bank compartment.

Approximately 95% of the water overlying the mine voids is stored in the upper 5% of the rock column. Based on an average porosity of some 10 volume%, the dewatered part of the outcropping dolomite underlying the WFS (365 km^2) has a storage capacity of 3,500 million m^3 (estimated based on data in [6,21,30]. This is in addition to the storage capacity of the downstream Boskop–Turffontein compartment that comprises an outcrop area of an estimated 800 km^2. At a similar porosity and weathering depth as found in the upstream compartments that would add some 7,700 million m^3 of underground storage capacity (Figure 2). However, the actual storage capacity depends to a large extent on the chert-contents of the most prominent outcropping formation. Since the dolomite dips toward the south at an angle of approximately 7 degrees older, that is stratigraphically lower, formations are also exposed to the surface. Figure 3 illustrates the areas the different dolomitic formations occupy in the WFS catchment as well as a vertical cross section.

Table 2. Hydraulic transmissivity for dolomite in the Bank compartment [6].

Depth (below groundwater rest level*)	Transmissivity [m^3/d × m^2]
0–2 m	>7,000 --> highest transmissivity found next to the Bank Dyke and the Wonderfonteinspruit >1,000 Southern flank of the Wonderfontein Valley
2–12 m	1,000–100
>12 m	<100

(* 'Groundwater rest level'—the elevation at which the groundwater table stabilizes after a pumping test. Since boreholes first need to intersect water bearing karst channels before water is struck the 'rest level' is frequently several tens of meters higher than the bottom of the drilled borehole owing to water equilibrating with the level in higher lying karst reservoir connected to the struck karst channel).

Figure 3 illustrates that the stream channel of the WFS is mainly eroded into highly karstified (chert-rich) dolomite of the upper and lower Monte Christo formation that also covers much of the BTC (the dolomitic outcrop area to the west of the Oberholzer dyke). In contrast, chert-poor and thus water free formations such as Oak Tree (at the northern rim of the dolomite outcrop) and Lyttleton (between the lower Monte Christo and the Eccles formation at the southern rim of the dolomite outcrop) cover only a comparably small proportion of the outcrop area. Since the proportions between water-rich and water-poor formations are similar across all compartments it is assumed that the Boskop-Turffontein displays a similar high storage capacity as developed in the three upstream dewatered compartments.

In this context, it is worth noting that the upper part of the BTC is also affected by increased sinkhole formation which, amongst others, necessitate the (expensive) relocation of some 18,000 households from Khutsong North, which was established on chert-rich dolomites, to an (also dolomitic, but chert poor) area further south [31]. Reasons for the increased ground instability include the area being underlain by chert rich dolomite formations in a zone of the BTC that is naturally dewatered, and urban induced concentrated surface ingress. The latter is caused by a combination of urban disturbances of a naturally occurring soil seal that acts as a dispersing mechanism for ingressing surface water, and that the large proportion of sealed surface areas commonly found in urban settlements tend to concentrate stormwater run off, which in the upstream dewatered compartments was found to be a major trigger of sinkhole formation. (This is mainly through subterraneous erosion by infiltrating water that removes soil and other filling material from pre-existing, underground cavities into even deeper lying karst receptacles that fell dry after the groundwater table and thus the erosion base was lowered. The continued removal of fine fill material results in a growing hollow cavity, the arch of which finally collapses due to a lack of support resulting in the sudden, and therefore often catastrophic, appearance of sinkholes. Large sinkholes occurring in the wake of dewatering measured up to 100 m in diameter and several tens of meters in depth [20]). In the case of the BTC, such lowering of the groundwater table has occurred due to erosion in the lower (western) part of the compartment, lowering the topographic surface and water table intersection plane and thus the associated groundwater table. In the somewhat steeper upper (eastern) part of the BTC, this may have resulted in a water level drop that dried up a previously water bearing karst channel system that now acts as a receptacle for the eroded overburden. Such natural drops in groundwater level over long periods of time (since the deposition of the dolomite) is the reason for the frequent presence of sinkholes that were subsequently filled with sediment and soil—paleo-sinkholes—and are now reactivated in and around the Khutsong area.

Apart from the increased volume of storm water drainage, disturbances of natural seals such as clay layers and increased stress on building ground that is associated with the continued growth of the settlement and urban development, the observed rise in ground instability may also relate to a more recent drop in the elevation of the groundwater table. Since the compartment is not actively dewatered, this is regarded as largely unrelated to the nearby deep-level mining activities. However, localized groundwater depression cones surrounding pumping shafts at the Blyvooruitzicht GM may indicate that some groundwater is abstracted by mining (and subsequently used for mining purposes) and may thus somewhat contribute to the overall drop in groundwater levels. Since Khutsong North is located in close proximity to the dyke, where the groundwater level is currently some 60 m below surface, this may have contributed to accelerated ground instability.

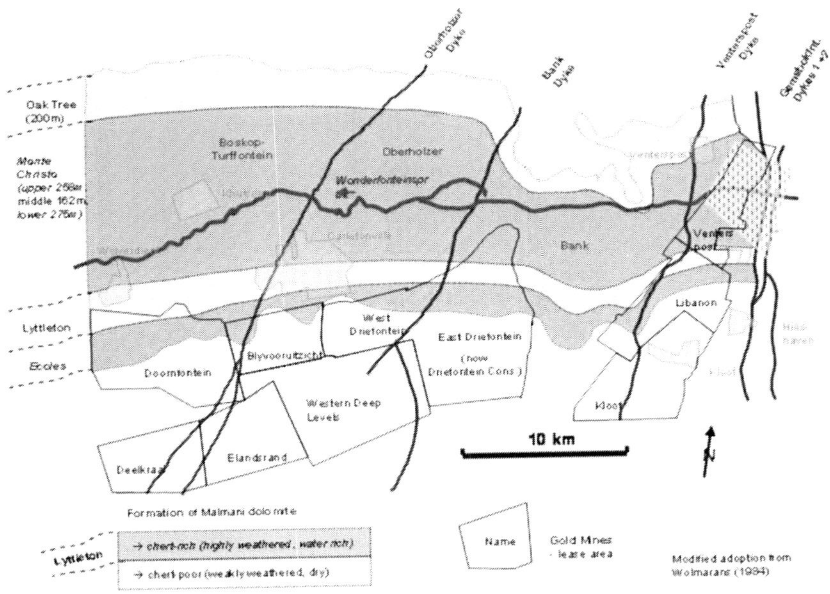

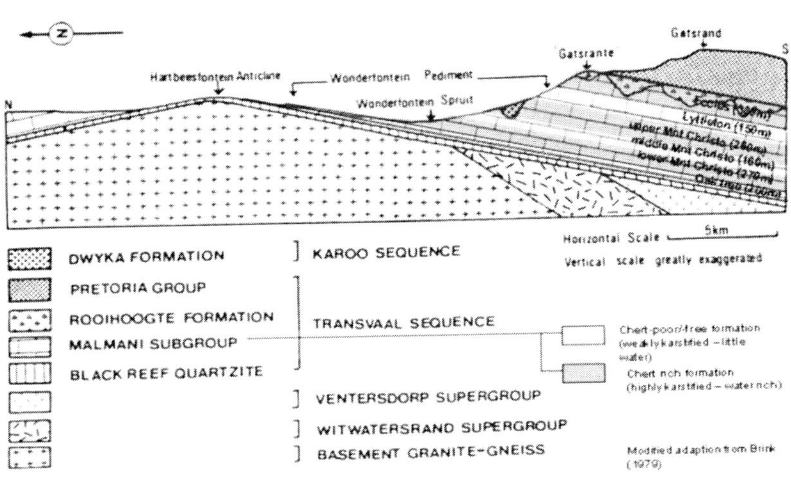

Figure 3. Horizontal and vertical location of water-rich and water poor dolomitic formation in the study area (upper map based on [6] (modified); lower cross section adopted from [28]).

It should perhaps also be investigated if possible abstraction of groundwater for irrigation is taking place in the area and if so to what extent it may contribute to the lowering of the groundwater level in this area.

2.2. Type and Formation of Karst Springs

Dolomitic eyes in the Wonderfonteinspruit catchment

Most of the dolomitic eyes in the WFS catchment are 'dammed springs' that owe their existence to the dykes which dam groundwater in the underground karst aquifer resulting in rising groundwater levels [32]. The actual spring or eye occurs where the groundwater table intersects the topographic surface commonly close to the dyke and the WFS stream channel in low lying parts of the floodplain that are connected to the underground karst channel system. Fed by a continuous stream of up-welling groundwater overflowing the lowest point of the perimeter of the eye the flow from these springs is commonly slow and steady even though large volumes may be discharged. Spring discharge is particularly high where the associated dyke cuts across the entire compartment forcing all the water of the karst aquifer to drain towards this spring. Typical examples for such high-volume, dammed springs are the Venterspost-, Bank- and Oberholzer-eye, which all dried up soon after dewatering in the associated compartments commenced. They followed the Klip River eye in the Zuurbekom compartment that was the first spring that fell dry in 1913 due to sustained groundwater abstraction by the Rand Water pumping station [33]. The latter is still in operation. Where dykes do not cut through the entire compartment but remain localized barriers within a larger aquifer, some of the groundwater may be able to bypass the dyke rendering flow from the associated eye less strong. Examples in the study area include the upper and the lower Turffontein eye.

The Gerhard Minnebron Eye

Compared to the dammed springs in the WFS catchment, the GMB eye appears to be different in nature as the spring is not related to a dyke. The early travelers reported that water was 'gushing out' from the spring, which presents an image of a free-draining spring type [24]. Situated at the foot of a 10–15 m high cliff that is located at the head of a narrow valley less than

100 m wide, the spring appears to be largely morphology-controlled with the relief dropping below the upstream groundwater table. There is no major upstream drainage line or large surface catchment that allowed for the carving out of such valley by fluvial erosion. The sudden drop in land surface elevation, therefore, most probably relates to the collapse of an underground karst channel possibly in the format of what is described as a 'polje' structure. Although the present morphological configuration of the GMB peatland conforms more to a semi-polje structure, being filled up with peat and having an outflow point in contrast with swallow holes of a classic polje structure, it is assumed that the original structure had indeed swallow-hole-type outflow points as the peat consists predominantly of sedge reed material, and it is observed that such vegetation is not established in water deeper than 50–70 cm [1]. The resulting linear depression would have been permanently filled with spring water and thus allowed for the subsequent accumulation of plant debris and sediment that finally made up to 5 m-thick peat deposits even under prevailing semi-arid conditions. Figure 4 depicts a 3-D model of the wetland illustrating the abrupt change in relief at the location of the eye and the linear nature of the associated wetland downstream.

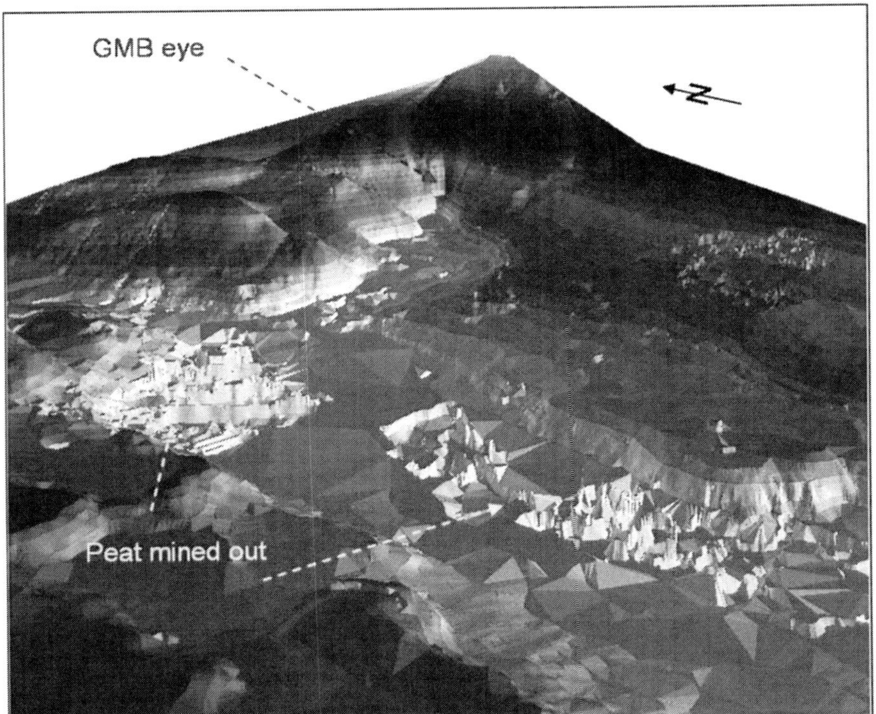

Figure 4. Digital elevation model (DEM) of the GMB wetland.

Since the age of the oldest peat sampled at a depth of 4.5 m below surface in the lower part of the wetland was determined by the radio-carbon method at 13,910 years before present (a bp) (this corresponds to 11,310 a bp for a 5.5 m deep sample taken at GMB by Smuts [1]) this collapse must have occurred before the last Pleistocene cold period ended. Even though South Africa was not covered by ice at the time, as was much of Europe and North America, the climate was generally colder and drier than today (much of the precipitation water was taken out of the global circulation and bound in N-hemisphere glaciers and ice sheets) [34,35]. The fact that peat forming vegetation appeared in the GMB wetland (mainly consisting of common reeds, Phragmaties australia, which according to pollen analyses of peat deposits at the nearby Schoonspruit eye did not change much since then, [36]) confirms that the spring flow must have provided sufficient water throughout the year despite the generally drier conditions fed by a comparable large catchment with a high recharge rate within the Boskop Turffontein compartment.

A slightly modified explanation for the origin of the peat deposit would be that the polje collapse was triggered by lowering of the BTC water table associated with the most recent cold period that culminated about 18,000 years ago, lowering precipitation to less than 200 mm/a in the central parts of Southern Africa (thus a natural dewatering event). Reed growth and thus the onset of peat accumulation only commenced 14,000 years ago as presumably the climate started to warm up at that period in time and precipitation increased sufficiently for the GMB eye to decant. Evidence for a brief warming period starting around 15,000 a before present is reported by [34] based on pollen analyses from Aliwal North (Eastern Cape, South Africa).

In how far the polje development may relate to tectonic movements or seismic activity associated with movements along the 'Graben' structure in the area (known as 'Gerhard Minnebron Graben') and so-called 'thrusts faults' created by a meteorite impact at the nearby Vredefort dome is uncertain. It is, however, of possible relevance that the eye lies very close or even on top of a regional SE-NW running thrust fault that crosses the area and the peatland's polje structure is intimately associated with it [37]. There appears to be an inter-relationship of cave systems on the Highveld with Vredefort-related thrust faulting ramp zones including the Gatsrand, Sterkfontein and Kromdraai caves [38].

Given the fault width (approximately 500 m), and that the fault material consists largely of impervious, metamorphosed (re-crystallized) dolomite and chert, it probably acts as a localized aquiclude creating a dammed spring similar to the dykes. The reported 'gushing out' of free flowing water at the foot of a cliff would thus be indicative of water released under some hydrostatic pressure from a possible reservoir dammed up behind the cliff wall.

Since the GMB eye was dammed by a concrete wall in the early 1960s to divert a large proportion of the spring water (some 80%) around the wetland into a newly built irrigation canal to Potchefstroom (opened April 1964) the original configuration of the multiple eyes and the reported 'gushing out of water' are no longer visible. It is, currently, difficult to establish whether the phrase 'gushing out' was indeed an adequate description of the original water flow. A historic document somewhat differs in its description of the nature of the spring from the one cited earlier. Water from the eye had been used for irrigation since 1865 and later for driving a water mill. A dispute concerning water rights was triggered by the drying up of the Mooi River and WFS in 1911/12. The resulting judgment contains a detailed description of the original appearance of the eye [39]. According to evidence given before Judge Jeppe by Mr Davis on behalf of the landowners Messrs. Levis and Marks, the GMB eye originally consisted of an (unspecified) number of springs arranged at the 'foot of a cliff' in a 'semi-circle with a diameter of perhaps 60 or 70 yards' (54–63 m). The southern most of these springs was significant enough to be named the 'Mark's spring' after the owner of that portion of the farm Gerhardminnebron. Located some 20 yards (18 m) south of the nearest eye, its water level was reportedly 1.2 feet (~40 cm) higher than that of the center eye [39]. The latter 'fact' suggests that the southern eye constituted an artesian or 'confined' type of spring [32] implying the presence of a layer that confines the outflow of groundwater creating artesian pressure. This layer could well be formed by one of the many horizontal chert bands under which typically well developed karst systems are found especially in chert-rich dolomite such as the Eccles formations in which the GMB eye is located. This layer function could, however, just also be fulfilled by unweathered dolomite stretching between enlarged fractures, sinkholes and solution slots through which the pressurized groundwater escapes to form artesian springs. The continued artesian outflow at the springs suggests that the hydraulic conductivity of the existing conduits is too small to equalize the pressure difference created by the higher upstream groundwater table (i.e., the water behind the 'cliff').

It is likely that the hydraulic gradient that drives the artesian flow in the eye area is comparatively steep as a significant rise in water level at the spring that was associated with the construction of a 2 m-high dam wall in the early 1960s did not appear to impede the discharge from the GMB eye. Expecting declining flow rates, the constructing DWA engineer at the time,

Mr. Fischer, raised the dam gradually over a period of two years by approximately 140 cm from the some 60 cm older mason dam to 2 m above ground level of the newly constructed concrete weir while monitoring the spring flow volume at the same time. Over this period, 1962 to 1964, there was no reduction in spring flow volumes according to the observations [33]. The 140-cm-rise in water level would have been significant in most parts of the rather flat topography surrounding the area, yet it did not affect the discharge of the GMB eye. This suggests that the upstream groundwater table at the time must have been at least 1.4 m above the original spring water level. Since artesian outflow of groundwater continues to this day despite the pressure exerted by the overlying 2-m-high water column in the artificial pool, it is assumed that the water level in upstream karst aquifer that feeds the GMB eye remains considerably higher than the increased water level at the spring.

For a later argument regarding recharge of the BTC, it should be noted that the water table at the spring has risen over the past 3–4 years to a level where the spring water now permanently overflows the top of the weir. This is in addition to the relatively large volume of water that flows directly and mostly unrestricted, via a sluice in fixed position, from the pool into the GMB irrigation canal. Owing to the continuous flow into the canal and over the weir, as well as leakage around the weir that almost immediately equalizes possibly increased inflow volumes, water level responses to rain events can no longer be observed. To quantify the responses of the eye to rainfall events, records from the flow gauging station in the irrigation canal a few meters from the outflow point were used as a proxy since rising discharge volumes of the eye increase the flow rate in the canal via a increasing flow velocity even though water levels may remain almost constant or are too small to be detected.

3. IMPACTS OF DEEP LEVEL GOLD MINING ON HYDROLOGICAL AND GEOHYDROLOGICAL CONDITIONS

Owing to large volumes of dolomitic groundwater pushing into underlying mine workings, deep level gold mines soon embarked on the 'dewatering' of dolomitic compartments in order to reduce the excessive costs for pumping the water from depths of up to 3 km back to the surface. Dewatering was effected by pumping out more water from the receiving mine void than the karst aquifer could naturally be recharged with, and disposing of the groundwater outside the aquifer, lowering the groundwater table by up to 1,000 m in places (normally at pumping shafts as centers of the dewatering cones). Consequences of the large-scale dewatering included the drying up of four springs with a total discharge of approximately 133 ML/d as well as irrigation boreholes as predicted but also some unforeseen effects such as the wide-spread occurrence of sinkholes and dolines often with disastrous effects on people's live and infrastructure [21]. During an above average wet period in the mid 1970s, many sinkholes had formed right in the stream channel of the WFS diverting large volumes of stream water directly into the underlying mine void partly defeating the original purpose of dewatering. In order to counteract the increased groundwater recharge, in 1977, the stream was diverted into a 30 km-long pipeline (colloquially known as '1 m-pipeline' referring to the diameter of the pipe) that carries the water across the three dewatered compartments (Venterspost, Bank and Oberholzer) to the non-dewatered BTC. From here on the WFS runs for the last 35 km or so in its original stream bed (Figure 1). In order to accommodate the drastically increased stream flow that resulted from the discharge of pumped groundwater from the mines during the active draw down in the initial phase of the dewatering (when pumping rates reached well over 400 ML/d) and to prevent adjacent farm land that was often cultivated right up to the river bank from being flooded, the stream channel on the BTC had to be extended. This was effected by constructing an earthen (unlined) canal within the natural stream bed (the so-called 'Mid-stream canal', sometimes also referred to as 'Westcott canal' after the constructing engineer) through deepening of the natural stream bed and using the dug-out sediments for raising the elevation of the stream banks. Owing to the earthen nature of the canal and initially high loads of suspended solids in much of the pumped groundwater, the mid-stream canal silted up frequently requiring the regular removal of deposited sediments in order to maintain the needed flow capacity. This service was provided by the mining industry via the Far West Rand Dolomitic Water Association, a body originally created in conjunction with governmental departments to deal with compensation claims by farmers adversely affected by dewatering [40]. Currently gold mines pump approximately 113 ML/d back into the WFS while using some of the abstracted groundwater for internal purposes such as tailings disposal, irrigation of vegetated slimes dams and domestic needs.

Apart from severe impacts on the surface water system through the drying up of springs, diversion of stream flow and changing natural runoff characteristic, deep level gold mining also changed the hydrogeological system by mining through the dykes that compartmentalize the aquifer. With more than 43 km of tunnels, through fares, etc., running through dykes (estimate derived from [20]), deep level mining hydraulically linked previously separate compartments. This also applies to the full (non-watered) BTC, which is now connected to three dewatered compartments upstream forming one large, single 'Mega-compartment'. In a possible* future rewatering scenario, this results in the final water level being controlled by the elevation of the lowest lying springs i.e., the two Turffontein eyes, the GMB eye and possibly some smaller eyes near Boskop dam (Figure 5).

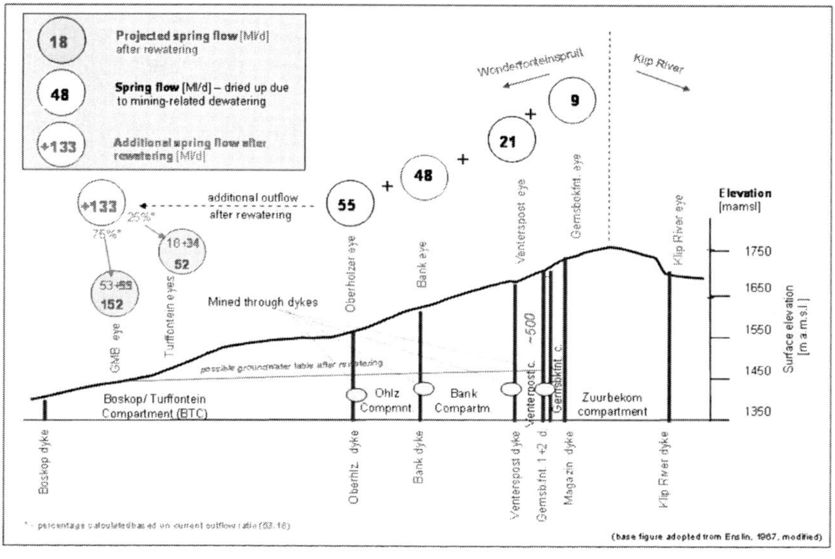

Figure 5. Simplified E-W cross section of the dolomitic compartments upstream of the Gerhard Minnebron Eye. Owing to the penetration of dykes a kind of 'Mega-compartment' could be created consisting of the 4 dewatered compartments and the Boskop-Turffontein compartments (figure adopted from [26]).

In the Mega-compartment scenario, the final groundwater table after rewatering would remain up to more than 100 meters below the pre-mining water table resulting in most springs upstream of the BTC remaining dry while the associated recharge contributes to a flow increase in the lowest lying eyes of the BTC estimated to be in the order of 133 ML/d. Allocated to the Tfnt. and GMB springs according to their current discharge ratio of 25%:75%, their possible post-rewatering flow rates are 52 ML/d and 152 ML/d respectively, (Figure 5). After rewatering, the GMB spring would form a major decant point for polluted mine water emanating from the vast network of flooded mine voids. Judged by the extremely poor water quality of water decanting from the flooded mine void in the head water region of the WFS (known as 'Western Basin'), where pH values <2 and U concentrations of 16 mg/L have been measured, such concentrated outflow could have severe consequences for downstream water users as well as the receiving environment.

Analyses of recent satellite imagery as supplied by Google Earth have revealed that the WFS over the past few years (since April 2003) did not reach the Mooi River but dried up completely well upstream of where the Turffontein eyes feed into the river and re-establish stream flow. After passing through the 1 m–pipeline and receiving mining effluents from Driefontein GM, the river enters the BTC with an average flow rate of approximately 100 ML/d (1,157 L/s) of which only 44 ML/d (long-term average) used to pass gauging station C2H069 some 8 km downstream. In the absence of large-scale abstraction, this reduction indicates that the WFS always lost a significant volume in the upper part of the BTC (on average 56 ML/d). Between October 2002 and April 2003 (i.e., somewhat paradoxically over the wet season), the flow at C2H069 dropped to only about 10 ML/d. Provided that the mining discharge and flow from the 1 m-pipeline remained constant, this drop equals a loss of some 100 ML/d (36.5 million m^3/a, nearly double the long-term average loss) of stream water over a total flow distance of just under 10 km. Since no water is abstracted from the stream for irrigation or other purposes and calculated evapotranspiration losses from wetlands were found to be negligible (and are over-compensated by additional discharges of sewage effluents from several municipalities such as Carletonville, Khutsong and Welverdiend along the way amounting to approximately 10 ML/d) it is assumed that nearly all of the stream water is lost to the underlying cavernous karst. However, a possible reduction in discharge volumes of the 1 m-pipeline and the downstream mines also needs to be explored.

This bed loss is particularly likely to occur along stretches of the now silted Midstream canal. After the regular removal of sediments was abandoned by the mines in the late 1990s, the accumulating sediments increasingly force water out of the canal into the adjacent floodplain. In some instances this resulted in water now flowing directly from the inundated floodplain into low-lying karst openings such as cave entrances, sinkholes and dolines that feature prominently along the stream and could explain the high volume of bed loss (Figure 6).

Figure 6. Satellite image (dated 4 December 2009) depicting the silted 'Midstream canal' of the Wonderfonteinspruit and the resulting inundation of adjacent dolomitic land allowing stream water to directly flow underground via openings such as cave entrances, sinkholes, etc. (source of image: Google Earth, 2010).

(Although the described water losses are a relative new phenomenon since dewatering commenced, it was also the case in the pre-mining situation where the WFS only flowed past Welverdiend during exceptionally wet periods [40]. However, during those times irrigation schemes abstracted most of the decanting eye's water, thus removing the river baseflow element. However, the irrigation water allocation was regulated by a court decision in such way that—at normal flow conditions—no water flowed past Welverdiend.)

Whether the increased recharge rate of the BTC caused by the bed-loss will lead to rising discharge volume at the three springs is still uncertain. A possible indication for that being indeed the case is the fact that an increase of water levels was observed at the GMB eye where water, since approximately 2006/7, permanently overflows the dam wall, which previously was a rare occurrence. This coincides with farmers downstream of Welverdiend, who for the first time in over 30 years experienced a few years earlier (2003/4) that their dams dried up [41]. Both phenomena are most likely consequences of the fact that currently much more water is lost directly to the underlying karst aquifer that feeds the GMB spring leaving dams downstream of the ingress area dry.

In this context, it may also be relevant that the wetland that accompanies the lowest reach of the WFS downstream of the Turffontein springs continues to expand in width, possibly indicating increased baseflow volumes in this reach (i.e., groundwater exfiltrating from the BTC). This, too, could be in response to increased recharge rates associated with the silting of the Midstream canal.

4. URANIUM FLUX INTO THE GMB PEATLAND

4.1. Sources and Extent of U Pollution

Based on close to 3,400 data readings on U concentration in water samples from the WFS catchment generated by a multitude of studies and monitoring programs between 1997 and 2008, [42] estimated the total load of dissolved U moving down the WFS. For the critical reach at the upper BTC (between the end of the 1 m-pipeline and the dams downstream of Welverdiend, Figure 1) where a large stream loss has been detected, [42] determined an annual U load of close to 3,500 kg transported in the WFS. Assuming that almost all of the stream water is lost to the underlying karst aquifer, the groundwater in the BTC receives nearly 3.5 t of the dissolved radioactive metal. While the U load in the lower WFS remained more or less constant over the observed 12-year period, a sharp increase was found for dolomitic springs in the BTC as well as the lower Mooi River. At the Turffontein eye the annual U load rose over 13-times (7–95 kg/a), while at the GMB eye and the Mooi river at Boskop Dam the U load quadrupled to 43 kg/a and 803 kg/a, respectively [42], probably as a result of the U influx associated with losing contaminated stream water to the BTC. While the absolute amount of U reaching Boskop Dam is still relatively small compared to the load transported in the WFS, it is the consistent and significant increase that constitutes a reason for concern. Should this trend continue over the next five decades where gold mining operations are still expected to be active in the area, ultimately U loads of several tons per year may finally flow into the Boskop Dam as the main reservoir of Potchefstroom's water supply.

However, analyzing the temporal distribution of U concentrations found in the springs of the BTC it appears that there is no continuous, steady rise in U concentration. The increased average U levels are mainly caused by higher U peak concentrations between which the spring water often returns to almost pristine conditions [42]· Increased levels of EC and dropping pH-values observed by quasi-continuous measurements of water quality in the GMB eye indicate that polluted water possibly of mining origin (high EC, low pH is typical for acid mine drainage) arrives at the eye approximately 5–6 days after rain events occurred in the area [43]. Since slimes dams are deposited throughout the region, polluted stormwater run off from the uraniferous tailings deposits is a potential source of groundwater pollution. This is especially because some of the run off from the slimes dams in the lowest part of the WFS (Blyvooruitzicht and Doornfontein GMs) is captured in a canal that frequently spills over onto adjacent dolomitic land where, consequently, a range of sinkholes developed dotted along the natural drainage line (Figure 7).

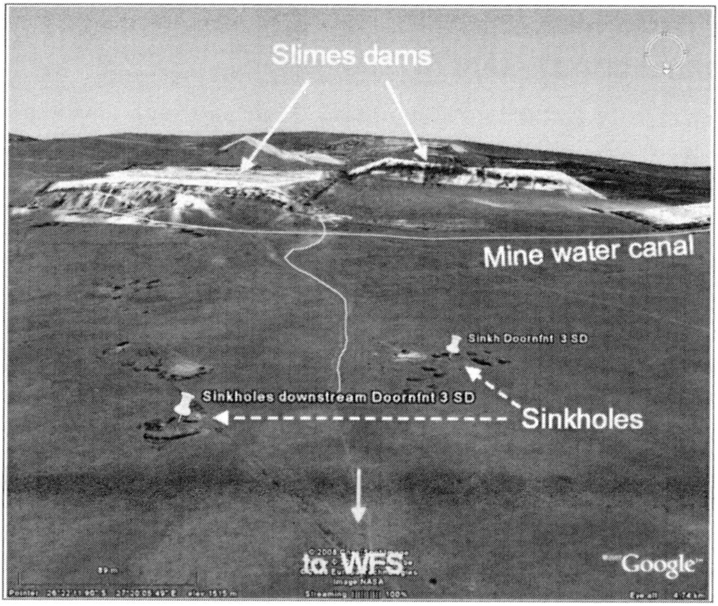

Figure 7. 3-D view of a satellite image depicting sinkholes formed in a drainage line down-gradient from uraniferous slimes dams at Blyvooruitzicht Gold Mine caused by frequent spillages from a mine water canal especially after rain events (source of image: Google Earth, 2009).

Since these sinkholes now intercept much of the spilled and highly polluted stormwater run off and divert it directly into the underlying karst aquifer, this may explain the deteriorating spring water quality in response to rain events. With a distance of some 18 km and a time lag of six days in the case of GMB (for Turffontein no high resolution time series are available) this would require an average flow speed of some 3 km/d (3.3×10^{-2} m/s), which does not seem unreasonable for flow in a multiporosity karst system, given the transmissivity values determined in the Bank Compartment (Table 2). A direct link between slimes dam runoff and springs via sinkholes would also explain the frequent U peaks observed at Turffontein (assuming that the secondary data used are accurate).

Apart from polluted storm water run off and stream loss, a third source of water contaminating the GMB wetland may exist. This could relate to polluted mine water currently filling the Deelkraal mine void where underground operations ceased in 2006/7. Being the closest of all mine voids in the Far West Rand (FWR) to the GMB wetland and located upstream of it, escaping mine water may reach the wetland via underground karst channels. Since the mine displays a considerably elevated U level of 2,450 µg/L (n = 2) it could well be the source of the U pollution detected in the GMB wetland [44]. Although the water was comparatively warm when sampled in the mine void (42 °C) it is unlikely that a temperature signal may still be used as a tracer at the GMB wetland after more than 10 km underground transport and possible mixing with groundwater along the way.

Long-term water monitoring data suggest that mining-related impacts on both springs are not a recent phenomenon but have been ongoing for several decades. This is illustrated, amongst others, by elevated levels of sulfate (SO_4). Originating mainly from the oxidation of sulfides in the mined gold reefs, sulfate concentration in mine waters are generally high rendering SO_4 a good indicator for mining-related impacts. Since elevated sulfate levels lead to a rising electrical conductivity of the water, this easier and quicker determination of EC-values is often used as a proxy to detect mining-related water pollution. However, relying exclusively on the lump parameter EC (which measures the contribution of all dissolved solids to conducting an electrical current through the water) may be misleading in instances where, e.g., decreasing sulfate levels are compensated by increased concentration of other salts from non-mining sources such as sewage effluents or others. In this paper, therefore, not the absolute EC or the corresponding TDS (concentration of total dissolved solids) are used as indicators, but the percentage of sulfate on the TDS. For the upper Turffontein eye this ratio indicates a significant mining impact on the spring water that started as early as 1967 and reached its peak in the mid 1980s when sulfate from the mines accounted for close to 40% of the total salt concentration (Figure 8).

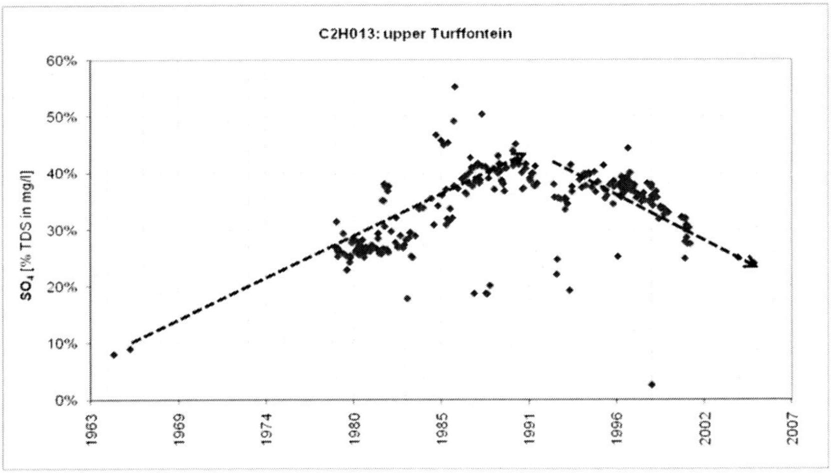

Figure 8. Changes in the proportion of sulfate at the Turffontein upper eye as indication of mining-related impacts on groundwater quality in the BTC (Data source: [45]).

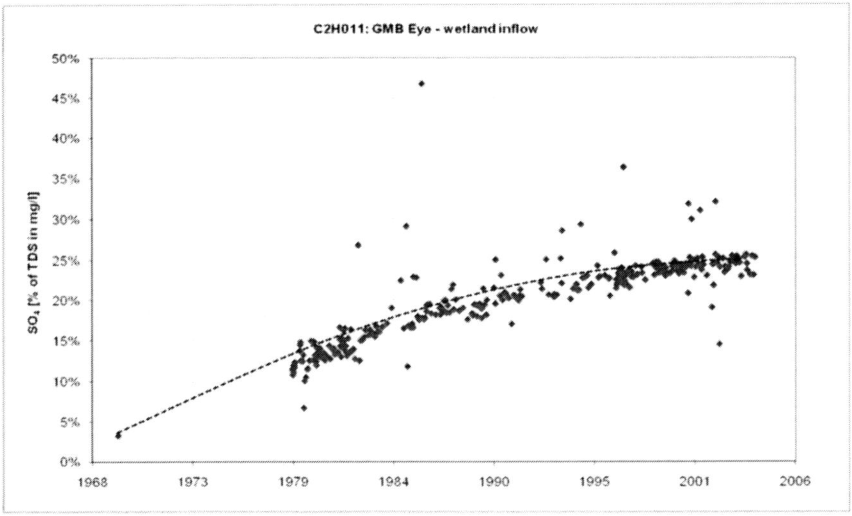

Figure 9. Proportion of sulfate at the GMB eye as indication of mining-related impacts on groundwater quality in the BTC (data source: [46]).

Since then the water quality improved possibly due to the decommissioning of a nearby slimes dam at the Doornfontein GM shortly before the improvement started [40]. In contrast to the Turffontein eye at GMB, the proportion of sulfate on TDS is still rising (Figure 9)

With 25% of the TDS, SO_4-levels at GMB are now approaching the current level at Turffontein eye, which has decreased from over 40% reached more than 20 years ago (Figure 8). Since both springs are located in the same dolomitic compartment, the question arises why their water quality differs significantly, a fact that was noted as early as 1905, i.e., well before any mining commenced in the catchment [24]. Regarding the U pollution of groundwater of the BTC it is interesting to note that most boreholes, springs and groundwater-fed streams show distinct peaks of U concentrations while very low concentrations (i.e., pristine conditions) between these peaks indicate the absence of a permanent polluting U source (Figure 10).

One exception is the WFS at Muiskraal bridge (second diagram from the bottom in Figure 10) where the stream not only displays an average U concentration that is an order of magnitude higher than in the upstream Tfnt. spring (the max. U concentration is even two-orders of magnitude higher) but also shows elevated minima that point to continuous U-input. Since the data were gathered before the WFS dried up well upstream of Muiskraal bridge (following the silting of the midstream canal further upstream) the elevated U-level there was caused by polluted stream water which, at the time, still reached this sampling point on the surface.

Regarding the temporal pattern of the U peaks at the different sites many corresponding peaks were detected. However, this is not consistent at all boreholes but differs from peak to peak and site to site. For example, the relatively high U peak observed at the GMB eye on 18 February 1997 (21 µg/L) corresponds with an U peak in the WFS at the Muiskraal bridge (55 µg/L) and a very slight increase in borehole 1 in Welverdiend (A Coetzee farm, 1.8 µg/L) while no corresponding response can be detected at the Turffontein eye, for example (Figure 10). Overall, the U concentration in groundwater of the BTC displays a rather complex temporal pattern where U peaks correspond at sites that are spatially remote from each other while no such correspondence occurs in sites of close proximity (Figure 10).

For the BTC it can be concluded that not only the absolute level of U but also its dynamics vary considerably within this compartment. Possible explanations are discussed in the next section, exploring possible pathways of U pollution.

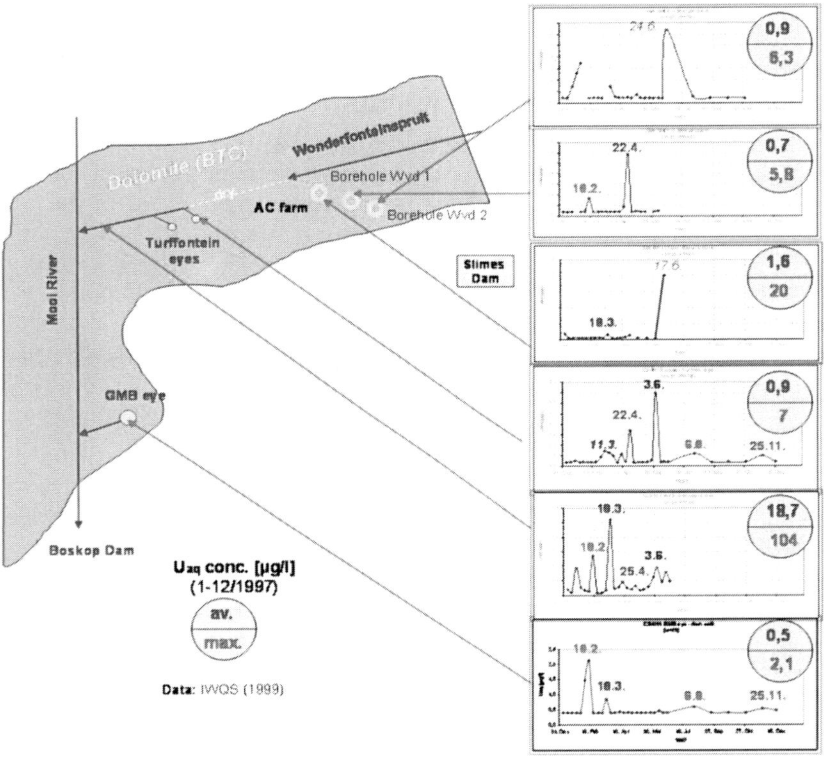

Figure 10. The charts display U concentrations as observed in stream water, groundwater from boreholes and springs of the BTC during 1997. Differently colored dates indicate U peaks that occurred at more than 1 site (Data source: [47]).

4.2. Pathways of U Pollution

While dolomitic compartments are often treated as a kind of closed hydraulic entity, especially when compared with each other they are by no means to be regarded as a homogenous, underground reservoir that contains water in much the same way as, for example, a dam does. In contrast to a single large water body of a dam, the karst aquifer consists of a multitude of water filled cavities ranging from small fractures to channels with free flowing water and vast caves spanning ten to hundreds of meters in diameter. While many may be hydraulically interconnected other may be not or only to a limited extent, resulting in a multiporosity system with the associated hydraulic conductivity spanning several orders of magnitude. Apart from horizontal water flow also vertical flow between karst levels at different depths may occur. For the WFS at least three distinct karst levels have been identified, each relating to a period of extended chemical weathering at a relative constant groundwater level as indicated by exposed cave systems [23].

It is furthermore important to note that the outcropping dolomites consist of different formations with profound differences regarding the associated water storage capacity. While chert-rich formations tend to be extensively weathered and therefore contain large volumes of groundwater, chert-poor or chert-free dolomite may not be karstified at all and effectively act as an aquiclude.

In the Far West Rand area the outcropping dolomite consists of alternating bands of chert-free/poor and chert-rich formations that run nearly parallel to each other due to a shallow dip component, forming the southern limb of the Rand anticline (Figure 3). In theory, this should result in parallel zones of aquifers separated by aquifuges (the chert-free dolomites). For a number of reasons, beyond the scope of this article, this was not the case in the dewatered, hence best-studied compartments. However, most of these reasons do not seem to apply to the BTC where, as illustrated above, the various eyes display different chemical signatures. Evaluation of exploration-derived cross sections (Figure 11) indicates that the BTF compartment, at least west of Welverdiend (the area covered by drilling), is not only divided in the different lithological zones, but even further subdivided into smaller 'sub-compartments' by strike-parallel (Vredefort event-related) faulting.

Boreholes in the BTC monitored in the past by the DWA were analyzed in terms of groundwater levels, and although the borehole coverage is limited, at least one borehole per identified 'sub-compartment' exists. For the dewatered compartments it was established that the groundwater levels follow topography and are not absolutely flat as thought earlier, displaying a gradient of 1:1,250 [6]. From borehole information it became clear that each sub-compartment has a unique groundwater level, and furthermore, approximately halfway downstream the BTC compartment these levels were found to be well below the (chert-free) dolomite 'walls' separating neighboring sub-compartments. That means that the exchange of groundwater between different sub-compartments, at least for the lower half of the BTC, is unlikely to take place under normal conditions.

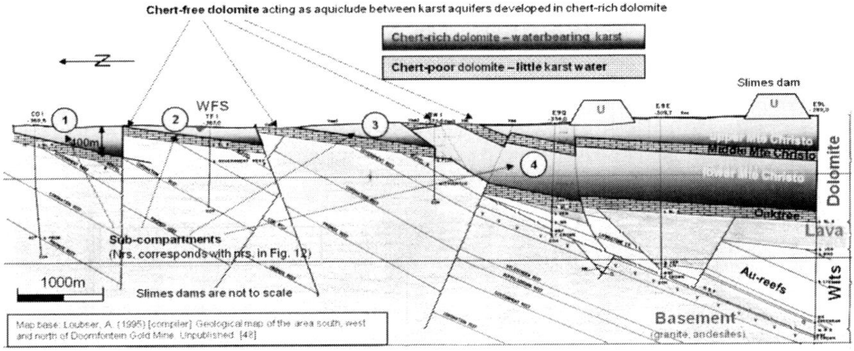

Figure 11. Geological N-S cross section through the floodplain of the Wonderfonteinspruit at Carletonville indicating chert-poor/-free dolomitic formations acting as aquifuges between karstified and water bearing chert-rich dolomite (termed 'sub-compartments, small print in figure is not relevant to the argument) (Map base of Figure: [48]).

That would mean that the BTC consists of a number of discrete 'sub-compartments' between which little mixing of water may occur. Since each sub-compartment has its own surface and underground catchment area with specific lithology, soil and other relevant features the water quality in the different sub-compartments may differ (Figure 12).

It is therefore likely that the Turffontein eyes are fed by a different sub-compartment than the GMB eye explaining the long-known differences in water quality between the two springs. Mining later may have amplified these differences by placing sources of water pollution such as slimes dams, rock dumps, pipe discharging mine effluents, etc., to different extent and size in the respective catchments (Figure 13).

For the Tfnt. eye the density of slimes dams deposited in its surface catchment resulted in a higher degree of water pollution compared to the GMB eye (Figure 13). Apart from (horizontal) differences in the extent of U pollution in the surface catchments of the various sub-compartments there are indications that vertical differences in water quality exists within sub-

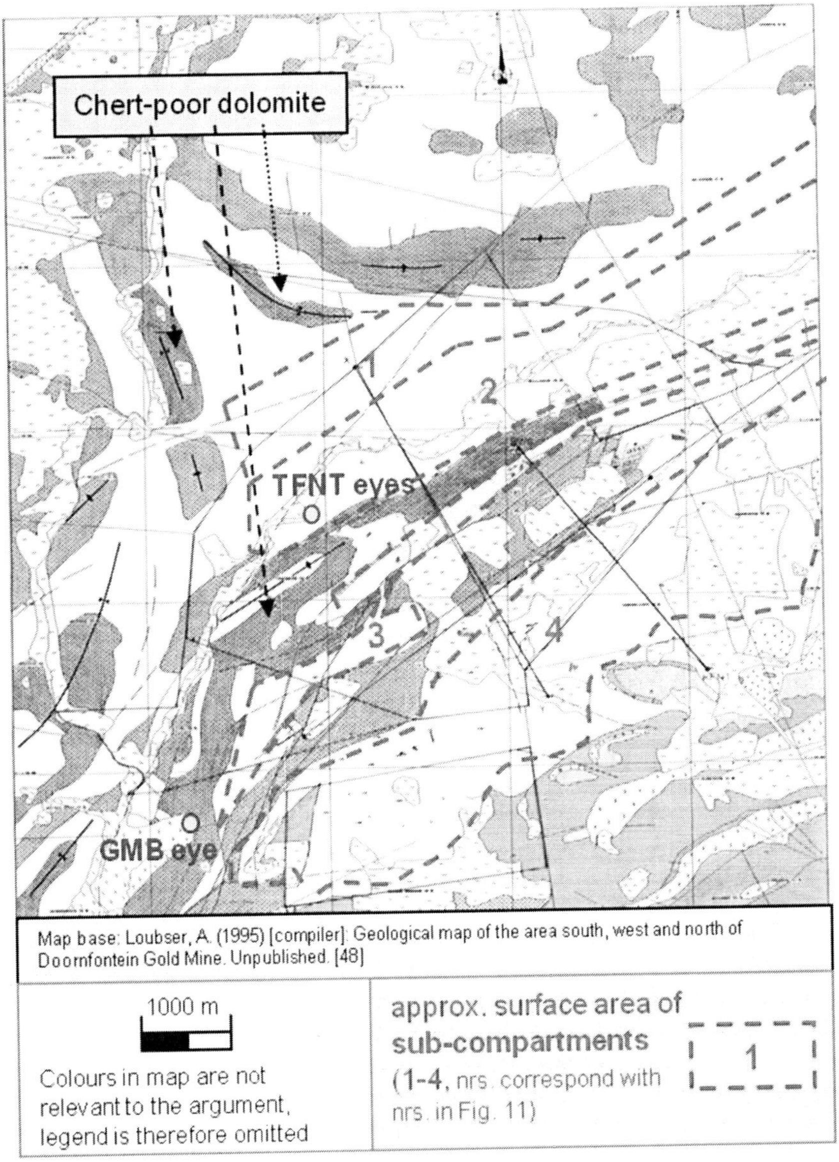

Figure 12. Extent of the surface area outcrop of different sub-compartments in the Boskop-Turffontein Compartment. (Map base of Figure: [48]).

Figure 13. Satellite image indicating the presence of slimes dams in surface catchments associated with sub-compartments that feed the Turffontein and GMB eyes (approximate delineation) (image source: Google Earth, 12 June 2007).

compartments. An example is sub-compartment number 2 that comprises much of the WFS as well as the Tfnt. eyes (Figure 13). U levels in the upper Tfnt. eye (1,426 m amsl) were found to be significant lower (20-times) than those in the WFS at Muiskraal (1,412 m amsl) even after the WFS dried up well upstream of this point in April 2003. Since polluted stream water after April 2003 could no longer reach the Muiskraal bridge and other surface tributaries are not existing, the high U-level found at the site (average 13 µg/L, max. 172 µg/L; measured by the Blyvooruitzicht GM between February 2003 and March 2006; 105 samples [42]) can only be derived from polluted groundwater diffusely entering the stream downstream of the Turffontein eyes. The occurrence of a wide and comparatively large wetland covering this stretch is indicative of significant baseflow feeding the stream in this low lying part of the BTC where the groundwater table frequently intersects the stream channel. In contrast to the somewhat higher lying Turffontein eyes where elevated U levels exclusively appear in form of sporadic, isolated peaks, the exfiltrating groundwater feeding the WFS upstream of Muiskraal appears to carry a permanently elevated U load as indicated by U minima a being above the regional natural background level of 0.2 μg/L). Unfortunately, the Blyvooruitzicht GM stopped monitoring this part of the stream in March 2006. Samples taken thereafter as part of the weekly monitoring program of the Potchefstroom municipality as well as grab samples by the authors showed significantly reduced U levels for the period 2007–2010 (1–2 μg/L), even though in December 2010 a U-peak of nearly 16 μg/L was detected following an intense rain period that allowed the upper WFS to reach the Muiskraal site on surface again.

In order to explain these differences in the degree and temporal pattern of U pollution occurring within the same sub-compartment, a vertical differentiation of the underlying karst system is assumed. Such differentiation could relate to the observed multitude of cave levels with each level acting as conduit to separate and deliver groundwater of different qualities over relatively long distances (Figure 14).

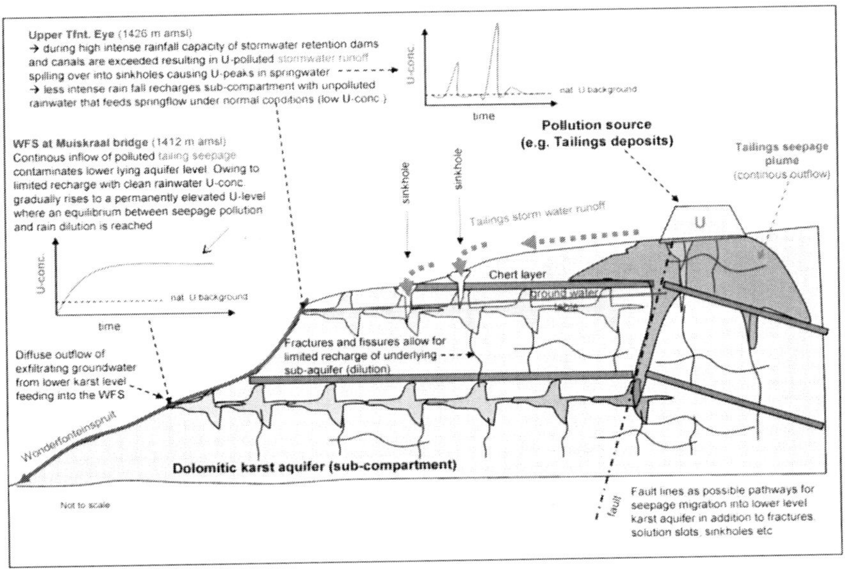

Figure 14. Schematic sketch depicting how a possible vertical differentiation of sub-compartments (different karst levels) may explain the different degree and temporal patterns of U concentration observed in the Turffontein eye and the Wonderfonteinspruit at the Muiskraal bridge sampling site.

In the proposed model, the sub-compartment consists of multiple karst levels (in this case two were used as an example) between which hydraulic interactions are limited to some degree of downward flow along fractures and fissures. The flow from the upper to the lower levels allows for limited recharge of the lower aquifers with mostly unpolluted rain/groundwater. In the upper level aquifer, which feeds the Turffontein eye, U polluted water can only enter after intense rain events when contaminated stormwater run off from nearby slimes dams exceeds the carrying capacity of dams and stormwater canals and spills into adjacent sinkholes that are directly linked to the upper karst level. This short-term input of U polluted runoff results in the sporadic U peaks that are typical for the eye. The lower karst level provides baseflow that diffusely enters the WFS at the wetland near Muiskraal bridge (Figure 14). Being directly hydraulically linked via fractures and possibly even sinkholes at the base of slimes dams, the lower karst level permanently receives U polluted seepage that more or less continuously flows out of the slimes dam. This, in turn, results in a gradually rising U concentration in the receiving aquifer up to a level where the U input is balanced by the influx of unpolluted groundwater from the upper aquifer (Figure 14). This would explain that the U level at the Muiskraal bridge is significantly higher than at the Turffontein eye (even in times when there is no surface flow of polluted water from the upper WFS arriving as it was the case since approximately April 2003) as well as the fact that the U level appears to be constant over the past decade or so. The fact, that after 2006, U levels appear to decrease again may relate to the disappearance of a major U source that used to feed U-polluted seepage into the lower sub-compartment level. Such disappearing source could relate to the reclaiming of a large slimes dams complex at the Blyvooruitzicht GM, which was recently completed. However, this all remains speculative at the time of writing as so far no investigations into verifying the model have been conducted.

5. SUMMARY AND CONCLUSIONS

The project reported on in this paper aims to determine the role that peat deposits at the GMB wetland can play to mitigate impacts of U pollution arising from an upstream gold mining area on the water supply of a municipality with some 250,000 inhabitants. As the first part of a series, this paper outlines the hydrogeological conditions in the wider study area and characterizes some of the impacts associated with decades of deep level gold mining taking place in the area. The latter include a first order quantification of the potential U load that may impact on the GMB wetland as well as a discussion of possible pathways along which such impact may be facilitated. This forms the base for the second part of the paper that explores in how far the peat deposits at the wetland may indeed act as a filter for U and thus protect the water reservoir of the downstream community.

In contrast to the other major karst springs present in the region, which owe their existence to near vertical dykes that force groundwater to the surface, the GMB spring is not a dammed spring in the classical sense. Its formation is probably associated with the collapse of a karst channel and the presence of a regional thrust fault that crosses the area. The collapsed, polje-type of karst system was subsequently filled with up to 5 m thick peat which began to accumulate some 4,000 years before the end of the last Pleistocene cold period.

Deep level gold mining has significantly altered the hydrogeological conditions in the area by adopting a dewatering policy. Apart from lowering the groundwater table and the associated drying up of some of the largest karst springs in South Africa it also caused unpredicted ground movements in form of sinkholes which not only had partly catastrophic impacts on human life and infrastructure but also presumably increased groundwater recharge rates of the area by diverting surface run off, including stream flow, underground. By penetrating through previously impervious dykes in geological strata below the dolomites, gold mines could have irreversibly (it is regarded uneconomical to close the pierced dykes) changed geohydrological conditions underground and—in a worst case scenario—hydraulically linked several dolomitic compartments together creating a single 'Mega-compartment'. As one of the three main outflow points of this 'Mega-compartment' the GMB eye may receive drastically increased volumes of highly polluted mine water decanting from a systems of flooded mine voids years after mines closed in the area. High U levels in mine water currently decanting from a flooded void system in the mined out West Rand area located in the headwater region of the WFS indicate that the radioactive heavy metal may also pose a future threat. This is in addition to current U pollution from diffuse sources like tailings and rock dumps that mainly affects the WFS. Based on an analysis of several thousands of U concentration data from the catchment, Winde calculated an annual U load of some 3.5 t for the lower WFS [42]. Since most of the polluted stream water is lost to the subjacent karst aquifer via either stream bed loss or water directly flowing into karst openings in the flooded area next to the stream, a significant amount of U is annually injected into the BTC. This is reflected by significant increases in U levels observed in the two main (and monitored) springs of the compartment namely the GMB and Tfnt. eye, where the average U level since 1997 rose 4- and 13-times, respectively [42].

Distinct differences in the degree and temporal pattern of U pollution between the two springs as well as other associated water bodies can only be explained by hydraulically largely disconnected karst systems within the compartment (termed 'sub-compartments'). These sub-compartments result from the different degree of karstification of chert-rich and chert-poor dolomitic formations with only the former allowing for the formation of extensive water bearing karst aquifers, while chert-poor or -free dolomite may in fact act as aquifuge given the hydrological conditions prevailing in the BTC. Owing to differing concentrations of U sources placed by mining in the associated surface and underground catchments of these sub-compartments, their water quality, especially U levels, may differ significantly. However, significant differences in U levels and U dynamics also occurred within the same sub-compartment. A model assuming a vertical differentiation of the sub-compartment in at least two or more karst aquifer levels is able to explain the observed differences but is still in need of verification.

Generally, it can be concluded that a significant U pollution potential already exists under current conditions that may impact on the GMB wetland. Major U sources relate all to mining and include stormwater run off from tailings deposits that was found to directly impact on karst aquifers via sinkholes, contaminated seepage continuously emanating from the base of slimes dams which were often placed on cavernous dolomite with pre-existing sinkholes to facilitate better drainage of porewater and enhance stability, and lastly, direct discharge of U polluted mine effluents via pipelines and canals into the WFS rendering stream water a (secondary) source of U pollution.

While the influx of U polluted stream water into the GMB wetland during extreme flood events cannot be ruled out it is assumed that most of the U arrives in the wetland via underground pathways. So far, it appears that such U influx is confined to a few isolated rain events with water quality recovering to pristine conditions (<0.4 μg U/L) for the overwhelming majority of the time in between.

ACKNOWLEDGEMENTS

The study is funded by the Department of Water Affairs of South Africa (project no. 213-2006) which is gratefully acknowledged.

REFERENCES

1. Smuts, W.J. Characteristics of South African Peats and Their Potential Exploitation; PhD Thesis, Faculty of Science, University of Pretoria, Pretoria, South Africa, unpublished. 1997; p. 222.

2. Grundling, P.L.; Marneweck, G.C. Mapping, Characterisation and Monitoring of Highveld Peatlands. Compilation of Existing Data and Evaluation of Inventory Methodology; Wetland Consulting Services Report No. 28/99; Wetland Consulting Services: Pretoria, South Africa, unpublished; 1999.

3. Dini, J.; Cowan, G.I.; Goodman, P. Proposed Wetland Classification System for South Africa (First Draft); Water Research Commission: Pretoria, South Africa, 1998.

4. Grundling, P.L. The karst fens of South Africa. IMCG Newsletter 2001, 2, 15–16.

5. Fleischer, J.N.E. Artificial Recharge and Pollution Aspects in the Dolomitic Aquifers, Turffontein Area; Technical report GH 3109; DEPARTMENT OF WATER AFFAIRS AND FORESTRY OF SOUTH AFRICA: Pretoria, South Africa, unpublished; 1979.

6. Wolmarans, J.F. Ontwatering Van Die Dolomietgebied Aan Die Verre Wes-Rand: Gebeure in Perspektief (Dewatering of the Dolomitic Area at the Far West Rand: Events in Perspective). DSc thesis, University of Pretoria, Pretoria, South Africa, 1984; p. 205.

7. Grundling, P.L. Environmental Impact Assessment for Stander Veen Cc. (Middleground) Expansion to Portions 4,8 & 9; Environmental Impact Assessment Report No. IHL 2002/4/001; Ihlaphosi Enviro Services: Pretoria, South Africa, unpublished; 2002.

8. Bredenkamp, D.B. Use of Natural Isotopes and Groundwater Quality for Improved Recharge and Flow Estimates in Dolomitic Aquifers. Water SA 2007, 33, 87–92.

9. Winde, F. A Hydrological Study on the Gerhard Minnebron Wetland to Determine how the Wetland System Functions and Fits into the Wonderfonteinspruit Catchment due to Licence Applications to Harvest Peat and Existing Peat Harvesting Operations. DWAF Project No. 2006-231,Report on Reconnaissance Phase (1 November 2006–30 April 2007); DEPARTMENT OF WATER

AFFAIRS AND FORESTRY OF SOUTH AFRICA: Pretoria, South Africa, 2007; p. 52.

10. Owen, D.E.; Otton, J.K. Mountain wetlands: efficient uranium filters—potential impacts. Ecol. Eng. 1995, 5, 77–93.

11. Brown, P.A.; Gill, S.A.; Allen, S.J. Metal removal from wastewater using peat. Water Res. 2000, 34, 3907–3916.

12. Ringquist, L.; Oeborn, I. Copper and zinc adsorption onto poorly humified sphagnum and carex Peat. Water Res. 2001, 36, 2233–2242.

13. Ringquist, L.; Holmgren, A.; Oeborn, I. Poorly humified peat as an adsorbent for metals in wastewater. Water Res. 2001, 36, 2394–2404.

14. Coggins, A.M.; Jennings, S.G.; Ebinghaus, R. Accumulation rates of the heavy metals lead, mercury and cadmium in ombrotrophic peatlands in the west of Ireland. Atmosph. Environmt. 2005, 40, 260–278.

15. van Roy, S.; Vanbroekhoven, K.; Dejonghe, W.; Diels, L. Immobilization of heavy metals in the saturated zone by sorption and in situ bioprecipitation processes. Hydrometall 2006, 83, 195–203.

16. Winde, F.; Stoch, E.J.; Erasmus, E. Identification and Quantification of Water Ingress into Mine Voids of the West Rand and Far West Rand Goldfields (Witwatersrand Basin) with a View to Long-Term Sustainable Reduction Thereof; Final Report to Council for Geoscience, Project No. 5512; Council for Geosciences: Pretoria, South Africa, unpublished; 2006; p. 261.

17. Fleischer, J.N.E. Report on the Investigation of the Far West Rand Dolomite Aquifers; DWAF interim report; Department for Water Affairs: Pretoria, South Africa, unpublished; 1979.

18. Fleischer, J.N.E. An Analysis of Springs in the Dolomitic Aquifer South-Western Transvaal; Technical report GH 3121; Department for Water Affairs: Pretoria, South Africa, unpublished; 1979.

19. Fleischer, J.N.E. The Geohydrology of the Dolomite Aquifers of the Malmani Subgroup in the South-Western Transvaal; Technical

Report No. GH 3169; Department for Water Affairs: Pretoria, South Africa, 1979.

20. Swart, C.J.U.; James, A.R.; Kleywegt, R.J.; Stoch, E.J. The future of the dolomitic springs after mine closure on the far west rand, Gauteng, RSA. Environ. Geol. 2003, 44, 751–770.

21. Swart, C.J.U.; Stoch, E.J.; van Jaarsveld, C.F.; Brink, A.B.A. The lower Wonderfontein Spruit: An exposé. Environ. Geol. 2003, 43, 635–653.

22. Walraven, F.; Martini, J. Zirkon Pb-evaporation age determination of the Oak Tree formation, Chuniespoort group Transvaal sequence: implications for Transvaal-Griqualand West Basin correlations. S. Afr. J. Geol. 1995, 98, 58–67.

23. Martini, J.E.J.; Kavalieris, I. The karst of the Transvaal (South Africa). Int. J. Speleology 1976, 8, 229–251.

24. Middleton, J.L.; Hunter, W.; Duff, M.B. Report on the Transvaal-Rand Water Supply. Civil Engineers; private archive EJ Stoch: Welverdiend, Pretoria, unpublished; 1904.

25. Enslin, J.F.; Kriel, J.P. Some Results Obtained from a Hydrological Study of a Dolomitic Catchment Area in the Transvaal; Union of South Africa: Department of Water Affairs, Pretoria, South Africa, 1960.

26. Jordaan, J.M.; Enslin, J.F.; Kriel, J.P.; Havemann, A.R.; Kent, L.E.; Cable, W.H. Final Report of the Interdepartmental Committee on Dolomitic Mine Water: Far West Rand; Department for Water Affairs: Pretoria, South Africa, 1960.

27. Enslin, J.F. Dolomitic Water Supplies in the Transvaal, Republic of South Africa; Technical report No. GH 1317; Department for Water Affairs: Pretoria, South Africa, 1967.

28. Brink, A.B.A. Engineering Geology of Southern Africa; Building Publications Pretoria: Silverton, South Africa, 1979.

29. Polivka, J. Geohydrological Investigation of the Schoonspruit Compartment in the Dolomitic Area of Ventersdorp; Technical Report No. GH 3524; Department for Water Affairs: Pretoria, South Africa, 1987.

30. Schwartz, H.I.; Midgley, D.C. Evaluation of geo-hydrology constants in the far west rand dolomitic formations. Trans. S. Mr. Instn. Civ. Engrs. 1975, 17, 31–36.

31. Stoch, E.J.; Winde, F. Threats and opportunities for post-mining development in dolomitic gold mining areas of South Africa—An hydraulic approach. Part III: Planning and uncertainty—lessons from history. Water SA 2010, 36, 83–88.

32. Ford, D.; Williams, P. Karst Hydrogeology and Geomorphology, 1st ed.; Wiley: Chichester, UK, 2007.

33. Liebenberg, W.; Head of regional office, Department of Water Affairs, Potchefstroom, South Africa. Personal communication, 2007.

34. Lewis, C.A. Later quaternary climatic changes, and associated human responses, during the last ~45 000 yr in the Eastern and Adjoining Western Cape, South Africa. Earth Sci. Rev. 2008, 88, 167–187.

35. Chase, B. Evaluating the use of Dune Sediments as a Proxy for Paleo-Aridity: A Southern African Case Study. Earth Sci. Rev. 2009, 93, 31–45.

36. Richer, L. The Hydrology and Geomorphology of the Wetland of the Oog Van Schoonspruit, North-West Province, South Africa; Honours Thesis, University of the Witwatersrand, Johannesburg, South Africa. undated.

37. Basson, I.J. The Structural Geology of the Northern Portion of the Potchefstroom Fault. Private archived by Erasmus, E.. Pretoria, South Africa, unpublished. 1994.

38. Brink, M.C.; Waanders, F.B.; Bischoff, A.A. Katdoornbosch—Witpoortjie fault: A ring thrust of Vredefort event age. S. Afr. J. Geol. 2000, 103, 15–31.

39. Jeppe (Judge). Judgement on Water Allocation on the Farms Gerhard Minnebron and Katdoornbosch. Private archived by Stoch, E.J.; Welverdiend, South Africa, unpublished; 1915.

40. Stoch, E.J.; Extraordinary professor of the North-West University, Potchefstroom Campus, Faculty of Natural Sciences, Resident in

study area, farmer representative and consulting scientist, Welverdiend, South Africa. Personal communication, 2007.

41. Visser, G.; Local farmer, Welverdiend, South Africa. Personal communication, 2007.

42. Winde, F. Uranium pollution of the wonderfonteinspruit: 1997–2008. Part II: U in water—concentrations, loads and associated risks. Water SA 2010, 36, 257–278.

43. Winde, F. Uranium Pollution of Water Resources in Mined-Out and Active Goldfields of South Africa—A Case Study in the Wonderfonteinspruit Catchment on Extent and Sources of U Pollution and Associated Health Risks. In International Mine Water Conference, Pretoria, South Africa, 19–23 October 2009.

44. Pranic, Z. Groundwater Sample Taken from Flooded Stopes of the Deelkraal Gold Mine at 7 July 2009 and submitted for analysis. 2009.

45. Hydrological Information System: Water Quality Data for Station C2H013, Upper Turffontein Eye; Department of Water Affairs and Forestry of South Africa (DWAF): Pretoria, South Africa, 2005.

46. Hydrological Information System: Water Quality Data for Station C2H011, Gerhard Minnebron Eye; Department of Water Affairs and Forestry of South Africa (DWAF): Pretoria, South Africa, 2005.

47. Report on the Radioactivity Monitoring Programme in the Mooi River (Wonderfonteinspruit) Catchment; Report No. N/C200/00/RPQ/2399; Institute for Water Quality Studies (IWQS): Pretoria, South Africa, 1999.

48. Loubser, A. [compiler]: Geological map of the area south, west and north of Doornfontein Gold Mine. Private archive by Erasmus, E.; Pretoria, South Africa, unpublished; 1995.

12

CHAPTER

PEATLANDS AS FILTERS FOR POLLUTED MINE WATER?—A CASE STUDY FROM AN URANIUM-CONTAMINATED KARST SYSTEM IN SOUTH AFRICA PART II: EXAMPLES FROM LITERATURE AND A CONCEPTUAL FILTER MODEL

Frank Winde

North-West University, School of Environmental Sciences and Development, Private Bag X6001, Potchefstroom, 2520, South Africa;

ABSTRACT

As the second part of a series of four, this paper reviews a number of case studies of natural uranium attenuation in peat, as well as underlying chemical mechanisms reported in literature. Based on this review, a generic, conceptual, model for peat to act as filter for dissolved uranium (U) is developed for guiding subsequent field investigations. The model consists of a chemical and an hydraulic component which is derived largely from data reported in literature as well as from limited field observations. For the chemical model component 10 different processes, each controlled by factors relating to water chemistry, have been identified to govern the attenuation of U in peat via a net balance of immobilization and remobilization. For the hydraulic aspect of the filter model, five different principal modes of U polluted water coming in contact with peat are discussed, focusing on the associated peat-water contact time as a crucial parameter controlling chemical U attenuation. Moreover, links between the two model components are discussed and, based on the integrated conceptual model, possible effects of natural and anthropogenic events on U attenuation in peatlands are outlined. Guided by the model, various site-specific field and laboratory investigations are finally designed to verify how far the identified generic factors and processes are indeed applicable to the Gerhard Minnebron Peatland.

Keywords: uranium; peat; natural attenuation; conceptual model; immobilization; remobilization; water chemistry; hydraulic processes; Gerhard Minnebron; dolomitic karst

1. INTRODUCTION

1.1. Project Background

This paper is the second part of a four paper series. Part I outlined the regional hydrological and geohydrological settings in the study area including impacts of deep level gold mining and the large-scale dewatering of dolomitic aquifers. The latter also quantifies the amount of waterborne U transported in the fluvial system which may impact on the Gerhard Minnebron (GMB) wetland. The findings indicate that well over three tons of waterborne U is annually lost from the Wonderfonteinspruit (WFS) into underlying karst aquifers that feed the GMB peatland. Since much of the water used by the downstream community of Potchefstroom is derived from the GMB wetland this U-influx is reason for concern. Furthermore, as one of three potential outflow points in a future rewatering scenario, the GMB wetland may be subjected to massive influx of highly U-polluted water [1]. Following reports of peat being able to act as an efficient filter for U and other dissolved heavy metals (e.g., [2,3,4,5,6,7,8]), the aim of this paper is to develop a model to assess the ability of local peat to fulfill a filter function for U.

Currently two peat mining operations excavate peat from the GMB wetland. Results of the study are expected to inform decisions by governmental departments regarding applications to extend existing operations.

Furthermore, the study also bears relevance to a current court case where one of the peat mining companies claims compensation from an upstream gold mine for polluting the peat reserve with U.

1.2. Case Studies of Natural U-Accumulation in Peat

Peat consists of partially decomposed wetland plants accumulating in waterlogged environments where the sustained influx of organic matter results in the depletion of dissolved oxygen through bacteria-aided decomposition of dead plant material. The resultant anaerobic conditions under which a complete mineralization of the plant material cannot be achieved leads in the long-term to the preservation and subsequent accumulation of organic matter as peat.

Sampling some 145 wetlands in the Rocky Mountains, Sierra Nevada, and other mountainous regions of the western United States, researchers from the U.S. Geological Survey report a maximum U-level of 10,000 ppm (mg/kg dry matter) detected in woody peat and organic rich sediments at the Flodelle Creek wetland (Washington) [2]. Similar to localized U-peaks (1,000–3,000 ppm) in other wetlands, this peak occurs near the input of naturally U-enriched groundwater. The majority of peat samples, however, contained less than 150 ppm, frequently displaying significant variations over short distances within the same wetland [2]. U-peaks of nearly 1% (10,000 ppm) were also reported by Schöner [3] for peat horizons in wetlands of the Wismut region (Germany) contaminated by seepage from adjacent tailings of former U mills. Based on 48 samples from South African peatlands, Smuts (1997) reports an average U concentration in peat of 18 mg/kg [4].

Almost half (46%) of the wetlands sampled by Owen and Otton [2] showed peat with maximum U-levels of at least 20 ppm, i.e., close to ten times the natural background concentration in the continental crusts. In 15% of the investigated wetlands the highest U-concentration found in peat are comparable to those in commercially mined low-grade U-ore (>100–1,000 ppm).

Reviewing pertinent literature, Owen and Otton (1995) list a number of other examples of reported natural uranium enrichment in peat including areas where the U-content of the water did not exceed natural background levels [3,5]. Quantifying the degree of enrichment as ratio between the U-concentration in the peat and the U-level in the water phase (termed 'Geochemical enrichment factor', GEF) a GEF of 2×10^6 indicates that U-levels in peat could exceed groundwater concentrations by a factor of two million [5]. Lower U enrichment factors were found for natural peat deposits ranging from 500 to 10,000 [6]. Examples from the western U.S. show enrichment factors ranging from 10,000 to 20,000 for wetland sediments in California and Colorado [2]. Based on batch experiments, Szalay (1974) found a similar high value (10,000) and concluded that "peat absorbs uranium almost perfectly, even from very dilute solutions occurring in nature." [7] quoted in [2]. This is in agreement with earlier results from Moore (1954) who found peat able to remove 98% of the dissolved U from an applied uranyl sulphate solution ([7] cited in [2]).

Apart from uranium, peat tends to accumulate other heavy metals such as mercury, lead and cadmium (including airborne) and is therefore frequently used as a geo-archive aiding the reconstruction of pollution levels in paleo-environments [9,10].

2. CONCEPTUAL MODEL FOR PEAT AS A URANIUM FILTER

The potential of peat to act as a filter for dissolved U is determined by two major components namely (i) its sorption capacity for the radioactive heavy metal ('chemical component') and (ii) the volume of contaminated water that comes into contact with the peat ('hydraulic component') per time unit.

2.1. The Chemical Filter Component

In order to characterize the chemical filter component of peat, an attempt is made to categorize the different (often competing) mechanisms responsible for the removal of dissolved U from the water phase ('immobilization'), based on a review of relevant literature. Since the net filter function of peat not only depends on the amount of U initially removed, but also on the amount that may subsequently be released ('remobilisation'), the mechanisms responsible for the remobilization of U are also discussed.

2.1.1. Mechanisms of U-Immobilization

Sorption onto organic material: Despite the long-term application of U-accumulation in peat and organic-rich wetland sediment for exploration purposes and, increasingly, since the early 1990s for passive water treatment in artificial wetlands, Schöner (2006) remarks that mechanisms responsible for the removal and fixation of dissolved uranium from water are still not sufficiently understood [3].

Generally, however, it appears that sorption is the major mechanism for immobilizing dissolved U from the water phase. Partially decomposed plant fibers, as the main constituent of peat, contain cellulose and hemicellulose, humic and fulvic acids, waxes, resins, and other substances which display a range of polar functional groups such as alcohol, aldehydes, ketones, and ethers. Facilitating chemical bonding and ion exchange, such functional groups favor polyvalent cations with high atomic weight, such as hexavalent U (U^{6+}) [11,12]. Metals are initially only loosely adsorbed to peat and may subsequently be reduced to more stable forms [13]. Since H^+ ions compete with other cations for sorption sites, the rate of metal removal from the water phase is strongly pH-dependent and appears to be at a maximum between 3.5 and 6.5 [14]. At values above 8.5, metal sorption decreases since peat starts to degrade [14]. Sorption is further strengthened by the fact that peat, as the first stage of coal formation, displays some properties of activated carbon including micro-porosity and the associated high reactive surface area as well as a high organic carbon content (a minimum of 30 weight-percent is required to be classified as peat). While confirming that a high content of organic carbon is a prerequisite for U accumulation in wetland sediments, Schöner (2006) only found correlations between the two parameters above a certain threshold content of organic carbon [3]. Winde (2003) reports somewhat lower U-levels for the same area and measured the highest U levels in black, organic-rich wetland sludge [15]. Owen and Otton [2] provide a comprehensive review that also includes older peat studies. They indicate that, apart from sorption, the complexation of the uranyl cation (UO_2^{2+}) by carboxyl groups, and the formation of insoluble uranyl humates, contributes significantly to the retention of U in peat [16,17]. Generally high pH-values, large atomic weights and high valences, favor metal retention by humic and fulvic acids, enabling the latter to strongly concentrate heavy metals from natural waters even at low concentrations [2,18]. Highly stable U-humate complexes may also form organic coatings ('biofilms') on otherwise largely inert coarser sediment components such as quartz pebbles and sand grains [19]. According to Crancon et al. (2010) the sorption of humic substances onto mineral surfaces creates hydrophobic mixtures that stabilize organic matter in soils. The stability of these aggregates is, in turn, dependent on soil moisture as well as the flow velocity and chemistry of the porewater [19].

Sorption onto mineral phases: Uranium (e.g., as uranyl ion UO_2^{2+}) may also adsorb directly to freshly precipitated gels of iron hydroxides (goethite: $FeO(OH)$) or iron oxide (hematite: Fe_2O_3) as well as to clay minerals displaying a negative charge of their large reactive surface area (under non-acidic conditions). Complete sorption of U onto clay and other surfaces typically occurs within a period of a few days [19]. Krepelova et al. (2006)

found that U(VI) adsorption to clay (kaolinite) under acidic conditions increases if humic acids are present due to the formation of additional binding sites for U coming from the adsorption of humic acids to kaolinite [20]. Feng et al. (2005) showed in batch experiments that peat humic acid is sorbed to different clay minerals (kaolinite and montmorillonite) via cation bridging, ligand exchange and van der Waals forces. Sorption of humic acids to mineral surfaces generally increases when the pH is lowered and the ionic strength of the solution rises [21,22]. This is in contrast to the behavior of non-organic U-species which tend to increase mobility in aquatic systems when the pH is lowered and ionic strength rises. Since clay minerals are frequently found as interspersed layers in peat deposits this may contribute to U-immobilization in peat affected by acid mine drainage that lowers the pH while simultaneously increasing the ionic strength. Interestingly, sorption of humic peat acids was much stronger to kaolinite than to montmorillonite, despite the former displaying only a fraction of the reactive surface and cation exchange capacity of the latter. The presence of multivalent cations, such as Ca^{2+} and Al^{3+}, can largely enhance organic acid adsorption to clay through cation bridges [21].

The sorption of metal ions to different mineral surfaces was found to be strongly influenced by organic acids that ubiquitously occur in most natural aquatic systems [20,21,22,23,24,25]. While U bound to inorganic species such as carbonates tends to be re-mobilized through acidification, the opposite effect occurs for U in humic complexes. In the presence of organic ligands, U sorption increases when the pH is lowered (acidification) and decreases when the pH rises [25]. That is to say, in generally acidic environments such as podzols and certain peats, the organically bound U is less mobile [19,26]. Furthermore, Ho and Miller (1985) found that the surface charge of hematite becomes negative upon adsorption of humic acid, even at low concentrations [24], potentially increasing the sorption capacity for non-organic U.

Owing to the ability to remove a wide range of contaminants, peat is frequently used as filter material for cleaning industrial waste including passive water treatment by permeable reactive barriers that remove uranium from polluted groundwater. However, the efficiency of peat to act as sorbent varies according to the composition and type of peat (degree of decomposition, fiber type and contents, ash contents, etc.) [7,12,14,27,28,29,30,31,32,33].

Precipitation of insoluble U-species: With the surplus of organic material promoting anaerobic, reducing conditions (submerged), peat may also contribute indirectly to metal fixation by triggering redox-controlled precipitation, or co-precipitation of aqueous U species [34]. The former would include the reduction of hexavalent U (U^{6+}) to its less soluble tetravalent form (U^{4+}) which could subsequently accumulate below a water table that frequently delineates the boundary between oxic and anoxic conditions [35]. However, despite thermodynamic models predicting the precipitation of tetravalent U, Schöner (2006) could not find reduced U (IV) in nature [3]. Together with the fact that the highest U levels were detected in oxidized peat horizons, this suggests that direct precipitation of reduced U species contributes little, if at all, to U accumulation in peat. While reducing conditions are often associated with the anoxic catotelm, Jacks and Norrström (2004) found that sulphate reduction increases towards the stream and preferably occurs in the surface peat, indicating the importance of degradable substance (i.e., decomposing organic matter) for the sulfate reducers [36].

Co-precipitation of U along with secondary minerals: This mechanism refers to an indirect, less ion-specific form of immobilization including co-precipitation or occlusion with secondary minerals such as iron hydroxide or calcite [3]. Since precipitation of iron hydroxides as well as calcite is strongly pH-dependent, pronounced diurnal oscillations of the pH following solar-depended photosynthesis and temperature cycles may result in distinct daily differences in the rate of U immobilization [37]. In contrast, results from a case study in South Africa showed for stream sediments strong correlations between uranium and a range of other heavy metals with the carbonate contents suggesting the existence of a non-metal-specific mechanism of U-removal (e.g., co-precipitation along with calcite) [38]. Smuts (1997), for a range of different South African peatlands, found a high positive correlation between the concentration of several heavy metals and hematite (Fe_2O_3), as well as a correlation between heavy metal concentrations (including U) and the ash content of peat [39]. In situations where high phosphate inputs are present (e.g., through the discharge of untreated sewage effluents into streams) the formation of insoluble uranyl phosphates could further contribute to U-removal from the water phase [2]. On surface and within the near-surface capillary fringe of peat increased evapotranspiration and associated water losses may result in dissolved U-complexes precipitating due to super-saturation in near surface porewater. In cases where all water evaporates precipitating salts such as carbonates, sulfates and, to a lesser degree, chlorides and nitrates often contain elevated U levels which can readily be re-dissolved through subsequent re-wetting (e.g., through rainfall).

Deposition of colloidborne U: Schöner (2006) further established a clear association of U with aluminium-rich colloids suspended in water [3]. Displaying particle diameters from approximately 5 to 200 nm colloids constitute an intermediate phase between the truly dissolved and solid states of U (the borderline between dissolved and particulate matter is commonly defined by 0.45 μm filtration). Colloids can be of inorganic and organic origin with the former consisting of clay particles, alumo-silicates and metal oxy-hydroxides of iron (goethite, hematite) or other heavy metals including U (e.g., $U(OH)_4$). Organic colloids are often less well-defined in their structure and are formed by naturally occurring, soluble organic substances (such as humin, humic and fulvic acids). These organic acids form metallo-organic complexes through binding U to their negatively charged ligands [21,22,23,25]. Investigating U migration through podzolic soils of battlefields in Kosovo, Crancon et al. (2010) estimate that a maximum of 10% of the total dissolved U (VI) species may be bound to humic colloids. Although small, this fraction was found to be responsible for the rapid and far-reaching transport of U into nearby aquatic systems [19]. Immobilization of colloidborne U may be facilitated by gravitational deposition of colloids as well as electrostatic interactions with mineral surfaces [19]. Gravitational deposition of colloids will often require prior aggregation to larger-sized particles through flocculation or coagulation. This, in turn, can be triggered by chemical processes such as changes of pH and redox conditions. In peat, such changes in porewater chemistry may be triggered by event-related hydrodynamical processes, such as the percolation of rainwater, the lateral infiltration of surface water from adjacent streams, rising groundwater levels, as well as photosynthesis- and temperature-driven diurnal and seasonal oscillations of water chemistry [37,40,41,42].

Dissolved U and other heavy metals, however, not only interact with colloids but also with particles of a larger diameter ('suspended solids'). In this context, iron and calcium-rich organic particles have been found to facilitate metal transport in a peat-karst system [43]. After being sorbed onto or incorporated into suspended particulate matter the physical (as opposed to chemical) mobility of U in the aquatic system is largely controlled by hydraulic processes governing, amongst others, the gravitational deposition and re-suspension (erosion) of sediments. Both processes are influenced by the diameter and density of the involved particles which are, however, not necessarily constant but may change due to aggregation, coagulation and flocculation [44,45].

Biological U uptake: A third group of indirect, peat-specific mechanisms relates to the habitat peat may provide for metal-accumulating bacteria and fungi. Apart from the latter aiding the decomposition of plant matter and the associated production of highly sorptive humic material that, in turn, enhances U absorption, bacteria and fungi themselves are able to concentrate U on or in their cell walls via metabolic biosorption [40]. Tsezos and Voleski (1981, 1982) report on a specimen of the common fungus (Rhizopus arrhizus) containing 18% U^{6+} (180,000 ppm) in its cell wall [46,47] cited in [2]. In addition Fe^{3+}-reducing bacteria are also able to obtain growth energy by coupling oxidation of various electron donors to the reduction of hexavalent U to its tetravalent form. Sikora and Keeney (1983) suggest that more bacteria are commonly found in mineral-rich fens (groundwater-fed) than in rain-fed bogs rendering fens generally better traps for U than bogs [48]. In addition, sulfur-reducing bacteria also contribute to geochemical U enrichment in peat by reducing sulfate to sulfide and triggering concomitant precipitation of U and other heavy metals ('bioprecipitation') [49,50,51]. No data have been found on how far calcite structures of diatoms, which are also a constituent of peat, may act as bio-sorbents/sinks for U. Since algae are common in surface water of natural peatlands their metabolic absorption of U may contribute to the U contents of peatlands once the dead algae accumulate as biomass that forms part of the peat. Based on batch experiments, Dienemann et al. (2008) report a linear relationship between U concentration in several Cladophera species and U levels in surrounding water with a maximum concentration of U accumulated in algae of 360 ppm. On average, algae were able to concentrate U in their bodies from the water phase by a factor of 100 [52]. Thus, in eutrophied, algae rich peatlands with U polluted surface water the deposition of contaminated dead algae biomass may indeed significantly contribute to elevated U levels in the peat.

Bhat et al. (2008) suggest the use of biomass of dried red algae (Catenella repens) as a (passive) biosorpent to clean U contaminated waste water at large scale reporting an U-loading capacity of 303 mg/g (i.e., over 3 g uranium per kg of dry algae mass) [53]. Since the dead algae essentially constitute organic matter, the observed adsorption kinetics may give some indication on the required contact time between U-contaminated water and peat for U sorption to take place. Results show that an initial electrostatic sorption of U to organic surface ligands rapidly takes place in the first 15 minutes accounting for some 80% of the total U sorption (at pH 4.5, some 60% at pH 2.5). This is followed by a slower second phase of U moving into the biomass through intracellular diffusion reaching equilibrium after 45 min. Maximum concentrations of 3,030 ppm U in the sorbent material were observed [53]. Assuming similar kinetics for true peat a relatively short contact time between U polluted stream water and peat likely to occur in nature would allow for substantial U removal from the water phase.

Phytoextraction of U: Aided by microbiological action at the rhizosphere as plant-soil interface peat-forming vegetation may also remove a certain amount of U from the water phase through biological uptake and subsequent incorporation into their biomass ('phytoextraction'). While some of the stored U will subsequently be released back into the water phase following decomposition of plant material, the proportion of U contained in only partially degraded fibers (such as roots) that contribute to the formation of peat still remains immobilized. For the plant community as a whole (with no harvesting or annual burning of reeds taking place) the seasonal vegetation cycle is likely to result in the establishment of a dynamic equilibrium between U-uptake and -release constituting a permanent (dynamic) U-reservoir. For the 12 German wetlands (Wismut region) investigated by Schöner (2006), phytoextraction was found to play only a marginal role in the overall U-accumulation in wetlands which suggests that most of the U is loosely sorbed onto biomass such as moss particles and not incorporated [3]. In contrast, Kochenov et al. (1965) established that plant debris contained 50% to 66% of the total U found in the peat [5]. Investigating common reeds growing in a pond of U contaminated seepage water near an uranium mill tailings deposit in the Czech Republic, Cerne et al. (2010) found that reeds may accumulate ^{238}U in roots and leaves by up to >300 times above the natural U level in the plant (some 2 μg/kg dry weight). Leaves and stems of 20 analyzed plants (Phragmites australis) displayed an average U concentration of 691 and 192 μg U/kg dry weight respectively [54]. With an observed reed density in South African peatlands of approximately 50 plants per m^2 and an average dry weight of 177 g/plant (n = 3) (own measurements) the reeds would contain approximately 2–6 mg U per m^2. For a 1 km^2-large wetland vegetated with reeds exhibiting the above U-levels this could translate into 2–6 kg of U per annum being removed from the wetland through harvesting the (full-grown) reeds. Compared to a fluvial U-load of some 3.2 t U/a possibly arriving at the Gerhard Minnebron peatland [1] this would be a negligible amount.

2.1.2. Mechanisms of U-Remobilization

While peat has frequently been reported to remove dissolved U efficiently from the water phase even at low concentrations, the degree to which U is retained is not well investigated and likely to depend on a number of factors. Apart from hydraulic aspects controlling the rate of water flow through the peat, these include the physico-chemical characteristics of the peat, chemical composition of the water (e.g., the presence of other heavy metals competing for sorption sites), redox potential (Eh), temperature and pH of the water controlling the speciation and solubility of U [3]. Twardowska and Kyziol (2003) add a kinetic dimension by reporting that the contact mode between peat and dissolved metals also impacts on the efficiency of metal removal [55]. Distinguishing between static equilibrium and dynamic non-equilibrium (i.e., peat in contact with stagnant or flowing water) they found in batch and column experiments, somewhat surprisingly, that through-flow conditions (i.e., shorter contact time) enhanced the sorption of Cd and Zn (but not of Cu) onto peat (mainly by reducing the formation of Cl-metal-complexes that largely prevent the metals from binding to organic matter).

Schöner (2006) further distinguishes between an initial trapping of U and a long-term fixation and in three wetlands found that an average of 82% of the sediment-bound U could be remobilized by sequential extractions mainly originating from instable and moderate mobile phases [3]. In view of the conditions (Eh, pH, and temperature) applied in sequential in vitro leaching procedures, which have frequently little in common with natural systems, it is uncertain how far these methods are able to, reliably, mimic natural in situ processes governing the remobilization of U in peat.

Based on the main fixation mechanisms, a number of different processes that lead to the remobilization of U from peat can be identified. Amongst others these include the oxidation of peat (redox-controlled liberation), the acidification of porewater (pH-controlled liberation of sorbed U), the formation of soluble inorganic and organic U complexes (speciation-controlled liberation), changes in the chemical composition of peat porewater; changes in the ionic strength of peat porewater and the physical and biological destruction of U-containing biomass (e.g., by fire).

Oxidation of peat: This is likely to occur when receding water tables expose formerly submerged peat to atmospheric conditions. Kochenov et al. (1967) showed that sorption of U onto peat is reversible under oxidizing conditions [5] in [2]. Apart from accelerating the decay of plant fibers where sufficient oxygen for microbial activity becomes available and the subsequent release of contained U into the water, the oxidation of sulfides, such as pyrite (FeS_2), formed under reducing conditions and the subsequent formation of sulfuric acid may acidify the porewater and thus aid the liberation of sorbed U [48,56]. The use of sulfuric acid (H_2SO_4) as leaching agent in many commercial uranium mines indicates the relatively high remobilization potential of this process. With pyrite concentrations of up to 28% reported for coastal peats in north west Germany [57], the acid generation and thus metal remobilization potential of some peat deposits, is considerable. Associated peaks of sulfates released from oxidised peat into wetland streams may, in their own right, have a considerable impact on water quality [58,59]. Since oxidizing conditions generally promote the transformation of existing reduced uranium (U^{4+}) to its much more soluble hexavalent form (U^{6+}), this may further remobilize U from peat. After extended exposure of dried peat to atmospheric oxygen some peat may also be affected by spontaneous combustion resulting in difficult to extinguish underground peat fires that in the process release accumulated U.

(External) Acidification: Being quite different from the above discussed acidification that follows the exposure of peat to atmospheric oxygen, this type of 'external' acidification of peat porewater refers to the direct influx of acidic water generated outside the peat. Sources of such water may include, amongst others, acid rain, spillages of acidic waste water and acid mine drainage (AMD) which all result in the lowering of pH in affected porewater. As the investigated peatland is located downstream of two large mining areas acid mine drainage is of particular relevance. Since lower pH-values were found to increase U remobilization from sediments and dissolution of mineral phases containing U (e.g., carbonates) the influx of AMD is likely to lower the net-attenuation of U in peat. Furthermore, at low pH, the surfaces of many mineral phases which act as potential U sorbents are positively charged resulting in the reduction of the overall sorption capacity for U.

In contrast, lower pH values were also found to increase binding strength of U-organic complexes such as humic and fulvic acids to clays and other mineral phases (goethite and hematite) somewhat counteracting the increased mobility of non-organic U species. Since organic U complexes commonly account for not more than 10% of the total U contents, with the balance of the U (i.e., 90%) being associated with inorganic species, the net effect of direct acidification of peat porewater is likely to be a decrease in U-attenuation i.e., a net-release of U from the AMD-affected peat.

Formation of soluble inorganic and organic complexes: A third major group of mechanisms to remobilize U, is the formation of soluble organic and inorganic complexes releasing U previously sorbed onto particles back into the aqueous phase. The latter includes uranyl complexes formed with anions such as chlorides (Cl^-) [55], phosphates (PO_4^{3-}) [3], sulfates (SO_4^{2-}) [60], as well as carbonates (CO_3^{2-}) and hydrogencarbonates (HCO_3^-) [54]. The ability of the sulfate and carbonate ions to form stable uranyl complexes is frequently used in commercially leaching uranium from milled ore and illustrates the relatively high U-mobilization potential of these ions. While most U-mines employ the faster acid leaching; for ore with tetravalent U, oxidizing agents need to be added prior to leaching. For ore of mainly carbonatic lithology the slower carbonate leaching is frequently used since the high buffer potential of carbonates renders the use of sulphuric acid uneconomical [61,62]. Francis et al. (1999) successfully used a carbonate-hydrogencarbonate extraction as well as citric acid to remove U from contaminated soil and found carbonates to be particularly selective for U [54]. Since all of the abovementioned ions are contained in commonly used products such as gypsum (SO_4^{2-}), lime (CO_3^{2-}), fertilizer (PO_4^{3-}) and table salt (Cl^-) they are abundant in many natural water resources. Furthermore, Krepelova et al. (2006) observed in batch experiments that the formation of aqueous uranyl-humate complexes reduces the sorption of U(VI) to clay minerals (kaolinite) in the near neutral pH range [20].

Changes of the ionic composition of peat porewater: Investigating the efficiency of two standard potabilization processes (i.e., chlorination only and a combined process consisting of coagulation, flocculation, settling, filtration and chlorination stages) specifically designed to remove U from drinking water in a Spanish community, Baeza et al. (2008) found that increasing concentrations of either hydrogencarbonate (HCO_3^-), calcium (Ca^{2+}) or magnesium (Mg^{2+}) ions reduced the U-removal efficiency from 90% to 60% (at optimal pH of 6) [63]. Originating from the chemical weathering (dissolution) of dolomite all three ions are abundant in the dolomitic water of the study area and thus of particular interest for the filter function of the investigated peat deposit. While the removal efficiency for U was significantly reduced by the three ions, it increased for Ra by 40% [63]. Since the presence of multivalent cations such as Ca^{2+} and Al^{3+} was found to increase the sorption of humic U complexes to mineral surface (through cation bridging) their reduction results in organic U complexes being released. Such reduction may, for example, be associated with the displacement of porewater by infiltrating rainwater.

Changes of the ionic strength of peat porewater: Apart from diluting the concentration of multivalent ions, the influx of rainwater can also decrease the ionic strength and thus weaken the sorption of organic U-complexes to clay and other mineral surfaces contained in the peat. Based on batch experiments, Crancon et al. (2010) propose that decreasing ionic strength associated with the infiltration of rainwater into sandy podzols contaminated with depleted uranium (DU) is responsible for the subsequent remobilization of U-bearing colloids and their rapid and far-reaching waterborne transport into the underlying aquifer and adjacent streams [19].

While rising ionic strength increases the adsorption of humic acids to clay minerals, it was found to have little direct effect on the sorption of U (VI) onto kaolinite [20].

In cases where ionic strength increases, e.g., through the influx of contaminated wastewater such as acid mine drainage or untreated sewage effluents, the associated increase in the concentration of major cations as well as trace metals reduces the immobilization of U through competitive adsorption to peat [14,64]. While the immobilization rate for a specific metal may be reduced, the increase in ionic strength appears to overall increase the total metal sorption capacity of peat [14].

Decomposition and physical destruction of U-containing organic matter: The extended exposure to atmospheric oxygen in some peat types may result in spontaneous combustion and subsequent burning of the organic material leaving U-enriched ash behind that could be flushed into the surface drainage.

The biological cycle of U-uptake and -release through biosorption and phytoextraction as discussed earlier, over the long-term, is likely to remain more or less constant, with possible pronounced seasonal variations. Impacting on biomass production as well as biological degradation and decomposition the influx of nutrients (N, P, S stimulating bacteriological activities, plant and algae growth, etc.) or toxins and contaminants (reducing biological activity) into the system, temperature changes, as well as seasonal and event-related water table fluctuations (changing redox conditions in the upper peat), may further impact the biological U-cycle.

The massive and sudden exposure of large amounts of peat to aerobic conditions associated with peat mining also falls into the category of physical destruction. How far this activity has a direct impact on the release of sorbed heavy metals into the water, is uncertain. Oxidation-related metal release may be limited since much of the extracted peat is swiftly removed from the water and stored on dry land for transportation to clients and would depend on the kinetics of U release processes.

2.1.3. Summary of the Chemical Sub-Model

The aim of the chemical model component is to determine factors and processes that control the ability of peat to retain U through filtering U containing water. This will allow determining and perhaps quantifying the impact events such as rainfall, changing water tables and the exposure of peat to acid mine drainage may have on the retention of U. In this regard, three different aspects need to be distinguished:

(i)the identification of physico-chemical processes that lead to the immobilization and remobilization of U in/from peat;

(ii)the identification of factors such as water chemistry (pH, Eh, ionic strength etc.) and hydraulic aspects that govern the occurrence and intensity of the above processes; and

(iii)the determination of consequences natural and anthropogenic events in peatlands may have for the attenuation of U, with special reference to the site specific conditions at the GMB peatland.

In the case of GMB peatland, U mainly originates from upstream gold mining activities. While currently the direct discharge of large volumes of U-containing (non-acidic) groundwater (pH 7–8) pumped from underground mine workings constitutes a major source of stream and groundwater pollution, future threats may include the influx of acidic water decanting from flooded mine voids. The associated change in the chemical nature of the U influx in future needs to be taken into account when assessing the filter function of the downstream peat under such a scenario.

(i) Processes affecting U-mobility

Case studies from peatland worldwide suggest that peat is an excellent sorbent for dissolved U and able to accumulate the radioactive heavy metal from very dilute concentrations in the water to significant concentrations in the peat. Owing to its exceptional filter capacity for heavy metals, including U, peat is, in fact, frequently used in industrial filters to clean U contaminated waste water.

Under oxidizing conditions prevalent in the upper, near-surface part of the peat column (acrotelm) U occurs as hexavalent ion U(VI) which frequently binds to oxygen to form the uranyl cation (UO_2^{2+}). Under reducing conditions, prevailing in deeper portions of the peat well below the water table, hexavalent U may be reduced to its tetravalent form U(IV) which is generally much less soluble than U(VI). The three naturally occurring isotopes of U (^{238}U, ^{234}U and ^{235}U) are regarded as behaving chemically in an identical manner.

The removal of U from the water phase is termed 'immobilization', a term that implies that solid U phases in peat are less mobile in the aquatic environment than dissolved U species moving along with the water phase. Under certain conditions, however, immobilized U may be released from solid phases of the peat back into the water column. This process is termed 'remobilization'. The retention potential of peat for U will be determined by the net balance of U immobilization and remobilization. For both processes a number of different mechanisms have been identified based mainly on findings reported in literature complimented by on-site field observations. For the immobilization of U these mechanisms include:

The sorption of U-species to different solid phases (especially organic matter as a major constituent of peat which displays an exceptional high sorption capacity for multivalent cations, such as U, as well as different clays and freshly precipitated gels of iron hydroxides with their large reactive surface area). Owing to competition by H^+ ions adsorption processes, being highly pH dependent and generally reduced in acidic conditions when competition by protons for available sorption sites rises, and simultaneously sorption sites are reduced due to the surface charge of mineral phases, such as iron hydroxide gels, turning positive. The mobilizing effect of decreasing pH values is somewhat counterbalanced by organic U-acid complexes which bind stronger to mineral phase at low pH.

The precipitation of insoluble U-species (largely confined to reducing conditions that allow U(IV) species to form and to U-phosphate complexes exhibiting a low solubility even under oxidizing conditions).

The co-precipitation of U, along with precipitating mineral phases, such as Fe-hydroxides and -oxides (goethite and hematite), as well as calcite. The rate of precipitation of these secondary minerals is highly pH-dependent and may thus show profound diurnal oscillations following photosynthesis-triggered pH changes of up to 2.5 pH units per day, as discussed by Winde et al. (2004) [34,37]. Furthermore, the precipitation of calcite is part of the calcium carbonate-carbon dioxide equilibrium which is controlled by the pH as well as the concentration of dissolved CO_2 that, in turn, depends on the atmospheric partial CO_2 pressure and water temperature. By implication, the latter two parameters also have some influence on the U mobility even though this is only indirect and perhaps affects only a small portion of the total U.

The biological uptake of U through metabolic activity of aquatic organisms commonly found in peat environments including fungi, bacteria, algae and peat forming plants (in the case of the GMB peatland these are mainly common reeds—Phragmites australis). Owing to the accumulation in tissue, such as cell walls of fungi, bacteria and algae, U may reach considerable concentration in living organisms. Once dead, the U contaminated biomass also contributes to the attenuation of U in peat if preserved in the peat. Dead biomass such as algae were also found to be effective bio-sorbents for U and are used for industrial waste water purification. The U uptake by plants is frequently referred to as 'phytoextraction' and used as a means of rehabilitating contaminated wetland soils.

The gravitational deposition of U-bearing particles including the macroscopic suspended solids (suspended sediment, clay particles, calcite and other mineral precipitates) as well as colloids formed by inorganic material (e.g., alumo-silicates, iron hydroxides, etc.). Owing to their very small diameter, colloids are usually highly mobile even within substrates of low permeability. However, under certain conditions U bearing colloids tend to form larger aggregates through flocculation and coagulation that allow for their gravitational deposition together with larger-sized suspended solids.

Regarding the re-mobilization of U from peat, five major processes have been identified:

De-sorption: Mainly triggered by changing pH values and/or other chemical parameters loosely sorbed U may be released from organic matter as well as other sorbents through desorption i.e., reversing the initial adsorption. While decreasing pH values favor U desorption from organic matter and mineral phases, increasing pH values tend to decrease binding strength of U-organic complexes to clay and other minerals.

The dissolution of U-bearing precipitates such as calcite and iron oxides and hydroxides under acidic conditions may also liberate U from solid phase in the peat. In the case of evaporative salt crusts on the peat surface and the capillary fringe, rewetting through rainfall or rising groundwater tables will readily dissolve these crusts and release the incorporated U into the water.

The remobilization of particle- and colloid-bound U: Macroscopic U bearing particles gravitationally deposited at low flow conditions may be washed back into the stream during flood events. Where piping occurs peat particles may also be eroded subterraneously and flushed into the fluvial system. Additionally, changing porewater chemistry altering ionic strength and composition, pH etc. can also trigger the mobilization of aggregated U-colloids and result in their erosion and subsequent downward transport through the peat column.

The physical destruction of U-bearing biomass: Since much of the U is sorbed to semi-decomposed organic matter the physical destruction of the latter by complete decay or fire also results in the mobilization of U. Apart from sorbed U, this also releases U that has been accumulated metabolically in plants and other organisms. Triggering events may include the seasonal or permanent lowering of the groundwater table exposing peat to atmospheric oxygen and the associated completion of bacterial decomposition. Decay of peat may also be triggered by pH values rising above a certain threshold (8.2). Drying out of the peat may also lead to spontaneous combustion that burns the peat and concentrates contained U in the remaining ash. In the GMB peatland the annual burning of reeds has a similar effect which, however, only affects the reed parts above the water table. Since burning is commonly done in winter the (potentially uraniferous) ash can be washed into the stream by the subsequent spring rains.

The formation of soluble inorganic and organic U-complexes: Depending on the ionic composition of the water, U (as uranyl ion) can form a wide range of highly soluble, often neutral or negatively charged complexes with sulfate, nitrate and carbonate, that hardly bind to any sorbent and thus are highly mobile in the peat-water system. Being contained in frequently used products such as lime (carbonate), fertilizer (nitrate) and gypsum (sulfate) all three U-complex-forming anions are often abundant in natural aquatic systems. Especially in acidic environments, U also forms organic complexes with dissolved organic matter (Dissolved organic matter e.g., humic or fulvic acids) which display a similar high mobility. However, they seldom account for more than 10% of the total U content.

(ii) Factors governing immobilization and remobilization processes

The above discussed processes are strongly controlled by two major aspects, namely water chemistry and hydraulics. The former comprises parameters, such as pH, redox potential (Eh), ionic strength, chemical composition of the water phase in the peat-water system, and mainly control the intensity of the different processes. Hydraulic factors, on the other hand, are of more fundamental importance as they determine where and for how long peat comes into contact with water. With regard to the latter, the duration of water-peat contact is a crucial parameter that indirectly influences water chemistry (e.g., the permanent saturation of organic matter with stagnant water will result in anoxic/reducing conditions) and directly affect U-mobility. The latter is due to processes such as adsorption or precipitation requiring a certain amount of time depending on their kinetics. Therefore, contact time between water and peat determines whether a process can take place, and to what extent. The mode of the water-peat contact will also influence how much of the total water volume may be affected by the im- or re-mobilization processes. Furthermore, the contact mode, e.g., whether peat is in contact with stagnant or flowing water, also determines whether and to what extent concentration gradients at the solid-aqueous interface may be maintained and equilibria be established. This will be discussed in more detail in connection with the hydraulic filter component.

In the following the focus is on summarizing the impacts on U mobility associated with water chemistry including the following parameters: pH, Eh, ionic strength, chemical composition of the water. Since some of the impacts were mentioned in connection with the different im- and re-mobilisation processes some degree of overlap can hardly be avoided.

Impacts of pH: The pH exerts control over a wide range of processes affecting U-mobility including sorption/desorption, precipitation and co-precipitation, dissolution of secondary mineral phases, decay of peat, binding strength of U-organic complexes to mineral sorbents, aggregation and dispersion of colloids, solubility of U-complexes, etc. Generally, neutral to weak alkaline conditions prevalent in the GMB peatland favor the electrostatic sorption of U to organic matter and mineral (i.e., inorganic) peat constituents such as clay minerals, iron hydroxides, etc. Therefore, lowering of the pH in such systems is likely to increase U mobility by promoting desorption. Depending on the degree of acidification, this may also lead to the dissolution of hydroxide gels and other instable mineral phases. Events that may trigger acidification include the drying out of previously waterlogged peat (indirect acidification through the oxidation of sulfides and the associated generation of sulfuric acid) as well as the input of acidic water from external sources, such as acid mine drainage, acid rain, etc. However, decreasing pH-values not only result in U-mobilization but also favor the formation of U-organic complexes and their sorption onto mineral phases such as clay. Also, the optimum for biological U-uptake frequently occurs in acidic conditions.

Impacts of redox potential (Eh): As a redox-sensitive metal, the mobility of U is strongly Eh-depended. Under oxidizing conditions U is hexavalent and occurs as U-oxide (uranyl ion) which tends to form highly soluble inorganic and organic aqueous complexes that remain dissolved in the water phase. Under reducing conditions, U is tetravalent and forms insoluble species that precipitate out of the water.

Impacts of ionic strength: For U bound to organic complexes, increasing ionic strength somewhat unexpectedly results in decreasing U mobility. This effect has also been found for colloidal U which is mobilized in acidic soils by infiltration of low ionic strength rainwater. For U sorbed to mineral phases and organic matter via electrostatic forces, the opposite effect has been reported, i.e., increasing ionic strength mobilized U, inter alia, through metallic ions competing for sorption sites. Increasing ionic strength also has the opposite effect by promoting aggregation, and subsequent gravitational deposition of (U containing) dissolved organic matter, through coagulation and flocculation.

Impacts of chemical composition of the water: The chemical composition of the water phase in the peat-water system mainly impacts on U mobility via constituents with which U can form complexes that are either soluble and thus increase its mobility or are insoluble and therefore promote U retention. This includes inorganic U complexes of which those with sulfate, carbonate, nitrate and chloride display a high solubility under most natural conditions while U complexes with phosphate tend to be much less soluble. Owing to the ubiquitous presence of dissolved organic matter in most natural waters and especially in peatlands, U is likely to form soluble complexes with humic and fulvic acids particularly at low pH. While this generally increases U mobility, in the presence of Ca^{2+} and Al^{3+} ions organic U complexes tend to bind to clay thereby immobilizing U. Furthermore, Ca^{2+}, Mg^{2+} and HCO_3^- have each been found to significantly reduce the efficiency of filtration and flocculation used in water treatment to remove U. Since these ions are naturally produced through the chemical weathering (dissolution) of dolomite ($CaMgCO_3$), all three ions are abundant in the GMB peatland that is fed by a dolomitic spring.

(iii) Natural and anthropogenic events affecting U mobility in peatlands

Based on the above processes and governing factors, a range of plausible natural and anthropogenic events are identified that are likely to impact the degree of U attenuation in peat. Natural events commonly occurring in peatlands include rainfall, flooding, droughts and associated variations of the groundwater table, as well as solar input variations (diurnal and seasonal). Anthropogenic events/impacts, with regard to the study area include the exposure of the peatland to acid mine drainage, elevated levels of mining-related U in (non-acidic) dolomitic water, the annual burning of reeds, the excavation of peat (commercial peat mining), abstraction and diversion of dolomitic water feeding the wetland into an irrigation canal, as well as occasional flooding through re-routing diverted water from the irrigation canal back into the wetland for maintenance purposes.

In terms of effects on the U mobility, these events can be grouped according to the main underlying processes that impact U attenuation, such as oxidization (including subsequent acidification), direct (external) acidification, physical destruction of U containing matter, etc. The lowering of the water table during the dry winter months, as a consequence of droughts or of the diversion/abstraction of water, for example, results in the exposure of previously waterlogged peat and organic matter to atmospheric oxygen and the subsequent oxidization of sulfides. Apart from the associated decomposition of organic matter (i.e., reducing sorption potential for U) this results in the formation of sulfuric acid and lowering of the pH, as well as possible dissolution of U-bearing mineral phases. Similar processes may also apply to peat excavated from the wetland and deposited on land. A more direct way of lowering the pH is the input of acidic water from external sources such as acid rainfall or acid mine drainage (AMD) (here termed 'external acidification'). While rainfall, as well as AMD, lowers the pH, the net effects on U mobilization may be quite different owing to differences in ionic strength and chemical composition discussed earlier.

Figure 1 provides an overview on the processes, governing factors and events, impacting the ability of peat to retain (immobilize) U and those leading to the release (remobilization) of U after the initial sorption.

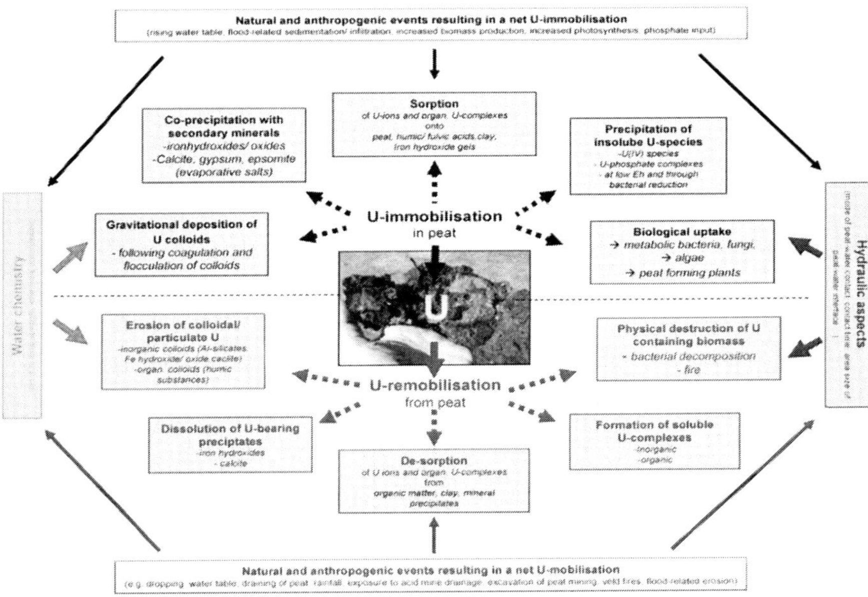

Figure 1. Overview on processes governing the immobilization and remobilization of U in/from peat including controlling factors and triggering events (chemical component of the U filter model for peat).

2.2. The Hydraulic Filter Component

In order for peat to act as a filter, the U polluted water must somehow come into contact with peat. Regarding the hydraulic properties of in situ peat deposits occurring in peatlands a significant body of literature has accumulated over the past four decades or so (e.g., [65,66,67,68,69,70,71,72,73,74,75,76,77,78,79,80,81,82,83,84,85,86,87,88, 89,90,91,92]). With respect to the hydraulic properties of peat, two distinctly different horizons are important; the acrotelm as upper part of the peat column, consisting of roots and decomposing plant material which is frequently above the water table; and the much denser lower part of the peat (catotelm) that is most of the time covered by water exhibiting anaerobe (i.e., reducing, anoxic conditions). Apart from the hydrochemical different milieus, both horizons display distinctly different hydraulic properties with the loosely structured acrotelm exhibiting a hydraulic conductivity (k_f) that on average is some 10,000 times higher than that of the underlying catotelm (1×10^{-2} vs. 1×10^{-6} m/s) [93]. Holden and Burt

(2002) indicate the range of k_f for the acrotelm somewhat wider (10^{-2}–10^{-6} m/s) but still between 4–6 orders of magnitude (10,000 to 1 million times) higher than the hydraulic conductivity of the underlying catotelm (10^{-8}–10^{-10} m/s) [74]. Regarding the latter they also point out that depth is not a significant control on hydraulic conductivity of the acrotelm since k_f-values can spatially vary by several orders of magnitude within only a few horizontal or vertical centimeters [74]. For peat-water contact to occur, the following range of different possible hydraulic modes was identified based on literature findings and field observations in the GMB wetland.

2.2.1. Contaminated Surface Water Flows Past the Peat Deposit Without Penetrating to Any Significant Degree (Skimming Over the Surface):

Owing to relatively high flow velocities, surface water exhibits when flowing over peat (e.g., during flood events) or when flowing past bank-forming peat that forms the stream channel, a comparatively short contact period exists for any specific point of the water-peat interface. However, for the peat-water contact area as a whole (e.g., the entire length of the bank-forming or flooded peat), the contact time between stream water and peat may be sufficient to allow for significant U sorption to occur. Under near-laminar flow conditions typically found in low energy wetlands, U sorption would, however, be restricted to a narrow layer of stream water directly in contact with the peat while most of the flowing water body would remain unaffected. Observing that flood events in peatland streams always occur at times of high groundwater table, Evans et al. (1999) suggest that the generation of overland and near-surface runoff is controlled by saturation of the catotelm rather than rainfall exceeding the infiltration capacity [82]. By implication, flooding of peatlands adjacent to the stream channel would result in little, if any, stream water percolating through the underlying peat and thus limiting the possible removal of U. While this is true for dissolved U, it may be different for particle-bound U which may be deposited in flooded areas through gravitational sedimentation. Since U mainly arrives in dissolved form at the GMB peatland particulate U transport is of limited importance there. In cases where pools of stagnant water remain after the flooding has subsided, evaporative precipitation of U may, however, be a possible pathway of peat's U pollution.

2.2.2. Contaminated Surface Water Infiltrates into the Peat (Downward Matrix Flow)

Downward vertical matrix flow refers to gravitational infiltration of rain or surface water that percolates slowly through the peat matrix by successively filling interstitial pores on its way to the lower lying water table. This type of diffuse flow through the peat matrix allows for relatively long contact time between polluted stream water and peat and may thus enhance U removal through sorption. However, the possible volumes of water affected are limited by the hydraulic resistance of the peat. Values for the hydraulic conductivity of peat reported in literature vary by up to ten orders of magnitude [83]. This large variation is due to a range of factors impacting the water flow through peat, including the degree of water saturation, the direction of flow (lateral vs. vertical), the composition of the peat (fiber-, clay- and ash-content), the degree of decomposition (the more decomposed, the lower the hydraulic conductivity), the compaction of the peat (deeper peat is more compressed, and thus less penetrable, than shallow peat near the surface), duration of peat's possible drying periods, and the existence of preferential pathways within the peat [4,66,76,80,84,85,86,87].

The potential filter effect associated with downward vertical water flow is limited to the relatively shallow upper unsaturated part of the peat (acrotelm) located above the mean annual water table [80]. This zone commonly comprises only a few centimeters but may drop much deeper (during extended dry periods [82]. Percolating surface water reaching the water table is assumed to change to a lateral direction of flow [88].

During high-flow conditions U polluted streams may flood adjacent peat deposits. Owing to the high infiltration capacity of the upper peat layer this is likely to result in stream water percolating through the areas where the acrotelm is not saturated towards the (shallow) water table. With flow velocities in the region of centimeters per day [82], such percolation would allow for sufficient contact time for U sorption to take place given that up to 80% of the dissolved U were found to sorb within the first 15 minutes [53].

A decrease in (saturated) hydraulic conductivity with depth was also found by Price and Whittington (2010), based on column experiments using Canadian Sphagnum moss peat. They report a ten times lower hydraulic conductivity already at the rather shallow depth of 25 cm compared to near-surface peat at 5 cm depth ($\sim 2 \times 10^{-4}$ vs. $\sim 2 \times 10^{-3}$ m/s respectively), mainly due to a decreasing size of interstitial pores in the lower lying peat. They also report a drastic decrease in hydraulic conductivity for the near-surface peat by five orders of magnitude after it had been drained, and explained this by merging air bubbles that increase the length of the water flow path through the peat matrix [80]. This is supported by observations of a depressed rate of water table recovery after a prolonged period of low water tables suggesting that peat undergoes physical changes once it dried out [82]. An important consequence of this observation is that drained peat subsequently offers a significant resistance to the infiltration of surface water and would—if the same applies to reed derived peat at GMB—limit infiltration-related U removal after extended dry periods.

2.2.3. Rapid, Downward Concentrated Flow through Peat (Non-Matrix Macropore- and Pipe-Flow)

While the peat matrix generally presents a porous medium that allows for Darcy-type of diffuse water flow, this may be different where preferential pathways exists. In the acrotelm (the upper, aerobic part of the peat that supports plant growth [4]) these preferential pathways may include root channels, animal burrows as well as macropores, while pipes and ash layers may act as preferential flowpaths in the underlying, more dense peat substrate (catotelm). In cases where extensive piping affects peat deposits (e.g., in sloped areas) subsurface water flow through the peat may be rapid with short contact times allowing for little filtering to take place [69,74,82,89,90,91]. Accounting for approximately 30% of the runoff production within the acrotelm, concentrated (i.e., non-diffuse) flow through macropores is an important flow mechanism in near-surface peat. Although macropores only account for less than 1% of the peat volume Holden and Burt (2002) estimate that they facilitate around 80% of the near-surface water flux in peat [74]. Investigating water flow in pipes developed in shallow peat of the Measnant catchment (Wales) by analyzing over 200 storm events with associated flow rates in stream channels and pipes measured at 17 separate gauging stations, Jones (1979) found that subsurface pipeflow may account for up to 49% of the stormflow and 46% of the baseflow, indicating the significance of the process [74,92] (cited in [82]). While the contribution of pipeflow may be of lesser importance in generally flat topography of the karst fen at Gerhard Minnebron (owing to the lack of sloped terrain), pipes may still form in the near-stream bank-forming peat due to subterraneous water flow following the groundwater table dipping towards the stream [93]. While this would restrict the occurrence of pipes to the near surface, Holden and Burt (2002) observed in upland blanket peats of the U.K. that pipes also developed in deeper peat layers well below the groundwater table, preferentially forming at the interface between peat and the underlying mineral substrate. With diameters of 3 to 70 cm and length of up to 150 m pipeflow in deep (water saturated) peat was found to contribute 10% to 30% to the stream flow [74].

2.2.4. Lateral Water Flow through the Peat (Through-Flow)

Where horizontal layers of different hydraulic conductivity exists (e.g., layers of ash, clay, gravel, and bedrock), a preferential lateral water flow through these layers ('throughflow' [88], 'lateral seepage' [73]) is likely to occur. With limited contact of the U containing water to the highly sorptive peat matrix, this layer-bound type of throughflow would result in limited removal of U from the water phase. However, also the peat itself, according to Reeve et al. (2001), "usually contain(s) complex sequences of strata that have highly variable botanical, physical and hydraulic properties" [94] and may thus allow for different rates of lateral flow through the peat matrix enhancing U attenuation.

Van Loon et al. (2009) determined that exfiltrating groundwater from an adjacent (not subjacent) aquifer entering the peat via upward vertical flow gets redistributed laterally once it reaches the loosely structured, and thus highly permeable, root zone of the near-surface peat (saturated k_f-value: 500 m/d = 6.8 × 10^{-3} m/s) [88]. Apart from the hydraulic conductivity (i.e., resistance to flow) the velocity of the laterally flowing porewater is controlled by the hydraulic gradient (head) driving the flow, and for layer-bound flow, the dip of the layer.

According to Reeve et al. (2001), hydrodynamic dispersion associated with the lateral water flow through peat may also drive vertical solute fluxes (e.g., from underlying mineral soils at the bottom of the peat) towards the surface, even in the absence of an upward directed groundwater flow [94]. Reeve et al. (2000) suggest that in peatlands where vertical flow of exfiltrating groundwater is limited by mineral soils of low permeability, the lateral flow of water dominates the hydrology in the upper portion of the peat column [95].

Lateral flow or throughflow in peat may also be induced by surface water infiltrating from adjacent stream channels or ponds/lakes into the peat in response to rising surface water levels. This process is reversible and the infiltrated water, temporarily stored in the peat, may be released back into the stream/lake once the water level in the lakes/streams subsides again. Depending on the focal point of investigation, this process is either referred to as 'bank storage' (if the stream flow is the main focus) or 'flow reversal' if the porewater movement is of interest [77,96].

Owing to pumping regimes by deep level gold mines utilizing day-night difference in electricity tariffs for minimizing their considerable pumping costs, water levels in mining-affected streams often show pronounced diurnal fluctuations of water levels. Associated consequences for the immobilization of U in bank sediments affected by infiltrating water are discussed in Winde and van der Walt (2004) [81].

Investigating the transport of phosphorous (P) in an altered peatland in northern Israel, Sade et al. (2010) found that much of the water from a canal running through the peatland infiltrated into adjacent peat via rapid lateral flow in cracks that developed through frequent drying and rewetting in the upper peat and underlying marl ($k_{sat.}$: 1.7 × 10^{-3} m/s] [97]. Based on differences in flow velocity and chemical conditions between the cracks (rapid flow, oxidizing conditions, low solute concentrations) and the surrounding matrix (slow flow, reducing conditions, high solute concentrations) they propose a conceptual 'dual-domain model' for waterborne P-transport. In this model a thin layer of precipitated $Fe(OH)_3$ covering the walls of the crack acts as a trap for P. Formed by reduced Fe^{2+} that migrates from the matrix pores into the oxidizing cracks such layer could, in theory, also act as a barrier/trap for dissolved U. On the other hand, U may be re-released through the erosion of particles from the Fe-hydroxide layer in the form of particulate U. The higher solute concentrations in the matrix porewater compared to water free-flowing in the canal and in cracks are caused by chemical gradients that drive diffusion of molecules from the cracks to the pores after hydraulic pressure differences between the crack and the matrix that cause advective influx of solutes have equilibrated. The overall water flux is controlled by the (highly variable) hydraulic gradient between the canal water level and the peat [97].

2.2.5. Upward (Non Gradient) Water Flow through Overlying Peat

The upward flow of water through the peat may be associated with a range of different processes involving exfiltrating groundwater, peat porewater, as well as surface water from adjacent water bodies such as streams, ponds and lakes. Hydraulic processes leading to the vertical movement of water against gravitation include the following:

The flow of exfiltrating groundwater from an underlying confined aquifer (e.g., entering the peat via subaquatic springs) [98]. Groundwater that, under artesian pressure, moves upward through the overlying 3–5 m thick peat deposit to the surface may perhaps occur at the GMB peatland owing to subaquatic discharges from underlying, highly karstified dolomite. However, on site field observations at the GMB peatland suggest that groundwater discharge is localized and confined to subaquatic springs, above which no peat accumulates. This results in well-like, often round holes in the peat layer within which groundwater wells up directly from the dolomitic bedrock to surface. In this situation, it appears unlikely that groundwater also infiltrates into the surrounding peat that offers significant hydraulic resistance. While the groundwater may not filter vertically through the peat it subsequently moves laterally through the near-surface layer draining towards the stream in the centre of the peatland. During this lateral flow U may potentially be removed.

Exfiltrating groundwater entering the peat from a higher-lying, adjacent aquifer sloping towards the peat (i.e., a valley situation) may also move vertically through the peat matrix if the piezometric level is above the peat surface [88]. Owing to the generally flat topography at GMB such water flux would probably be of low intensity. It could be triggered by the higher lying water level in the adjacent river (the Mooi River) that runs parallel to the GMB stream. In which case, disproportionately high, localized groundwater responses to rainfall may occur in the capillary fringe leading to the formation of groundwater mounts (also known as 'groundwater ridging') [71,77].

Lateral water flow: In peat areas not permanently fed by groundwater high evapotranspiration losses during the growing season were found to result in falling water tables that temporarily redirect originally lateral water flow to the peat surface [88].

Capillary flow: In the near-surface portion of the peat column capillary effects are responsible for vertical, non-gradient flow of groundwater [80]. Reflecting the varying intensity of evaporation and transpiration such capillary water flux results in diurnal water table oscillations of several centimeters per day [82,99]. In column experiments with Canadian Sphagnun peat, Price and Whittington (2010) found that peat porewater, owing to capillary action, moves upwards from deeper storage reservoirs containing some 47 vol.% of water to maintain evaporation from the surface over a 34 day period without exhausting the water contents [80]. While many hydraulic properties of Sphagnum peat are not comparable to reed-derived peat found at the GMB wetland, upward movement of porewater, especially during the extended dry periods in winter, is likely to occur there too. An indication of this is, amongst others, the continuous growth of evaporative salt crusts on surface during winter when the groundwater table is low and no rain occurs.

Groundwater ridging: In sloped areas at the edge of the peatland, capillary effects in conjunction with the infiltration of rainwater are suggested to cause disproportionally high increases of the water table known as 'ground water ridging' which subsequently lead to the discharge of pre-event water to adjacent streams [15,26,77,100,101]. Other non-Darcy type of vertical, up-gradient water movement may be associated with fluctuating groundwater tables caused by seasonal (i.e., changes in the climatic net-water balance), diurnal (e.g., solar cycle driven oscillation of evapotranspiration, plant growth, etc.) or event-related processes, such rainfall, floods and droughts [15].

2.2.6. Significance of the Hydraulic Processes for the Chemical Model Component

Since the concept of U mobility explored in this paper is based on peat-water systems, the conditions under which water comes into contact with peat are of fundamental importance for understanding U attenuation in peatlands. This aspect is covered by the hydraulic component of the conceptual model on the U-retention in peat, and focuses on different possible modes of how peat-water contact occurs. This includes, for example, the relatively slow diffuse flow through the interstitial pore space of the peat (matrix flow), as well as more rapid and concentrated flow over peat and through macropores. Consequences of the different hydraulic modes mainly relate to the associated contact time between peat and water and the size of the contact area. While the latter is relatively large in matrix flow, it is much lower in rapid flow through pipes and macropores. Generally it is assumed that interactions between dissolved U in the water phase and solid peat components, are more pronounced the larger the area of interaction and the longer the peat-water contact time. The latter is of importance with regard to the (often unknown) kinetics of U immobilization and remobilization processes. It may, however, also cease to be relevant if, for example, an equilibrium state is reached that effectively stops the adsorption of U regardless how long peat may be exposed to U polluted water. Such a scenario is likely to be confined to stagnant or very slow moving peat-water systems. In contrast, in systems where water moves comparatively rapid, saturation of the near-interface water layer may be prevented by the ongoing exchange of water. Preventing that ionic saturation acts as a diffusion limiting factor, flowing water may allow for continued U adsorption onto peat or 'leaching' of U from peat. The peat-water contact time has also indirect consequences for the U mobility (i.e., not only with regard to kinetics of certain reaction) since extended water-saturation of organic material, for example, results in anaerobic, reducing conditions that may contribute to U immobilization and vice versa. In many instances immobilizing processes, such as precipitation or co-precipitation, are more rapid than remobilization e.g., through the dissolution of precipitates, desorption, etc. Therefore, systems with short water-peat contact time may tend to favor a net retention of U.

2.2.7. Summary of the Hydraulic Model Component

Based on literature and observations in the studied peatland, five different peat-water contact modes have been identified significantly varying in associated contact time and area of possible peat-water interaction. Owing to pronounced differences in hydraulic conductivity water movement through the upper, loosely structured root zone of the peat (acrotelm) is up to five orders of magnitude faster than through deeper lying, permanently waterlogged peat. However, owing to horizontal layers of sand, ash and clay, more rapid water fluxes may also occur in deeper parts of the peat column either through these layers (in the case of sand or ash) or at the peat-layer interface (e.g., clay). Moreover, rapid flow also occurs in macropores in the near-surface root zone of the peat and possibly also through pipes which may form in the upper, as well as in the lower parts, of the peat column. Whether piping does indeed occur in the GMB peatland is at this stage of the investigation still uncertain.

Sources of water in the GMB peatland include:

- -stream water infiltrating from the stream channel into the bank peat;
- -exfiltrating groundwater moving laterally from shallow aquifers into peat at the edges of the wetland;
- -surface runoff from the surrounding catchment entering the margin of the peatland;
- -dolomitic groundwater exfiltrating under artesian pressure from the underlying karst aquifer; and
- -rainwater directly infiltrating into the acrotelm.

During flood events stream water may also affect peat in areas far from the stream.

In addition to rapid concentrated flow, either horizontally in the stream channel or vertically through macropores and perhaps pipes, a certain degree of slow matrix flow is likely to occur especially in the loosely structured near-surface peat. Apart from gradient-driven downward and vertical flow, this may also include temporal (horizontal) infiltration and exfiltration of surface water following changing stream water levels. In deeper peat vertical (upward) matrix flow may occur as a result of diffusion following evapotranspiration losses on the peat surface. The observed formation of evaporative salt crusts, covering large areas of the peat surface, further points to ground- and or pore-water ascending to surface through capillary forces. The hydraulic properties of the peat and different modes of peat-water contact possibly occurring at the GMB peatland are depicted in Figure 2.

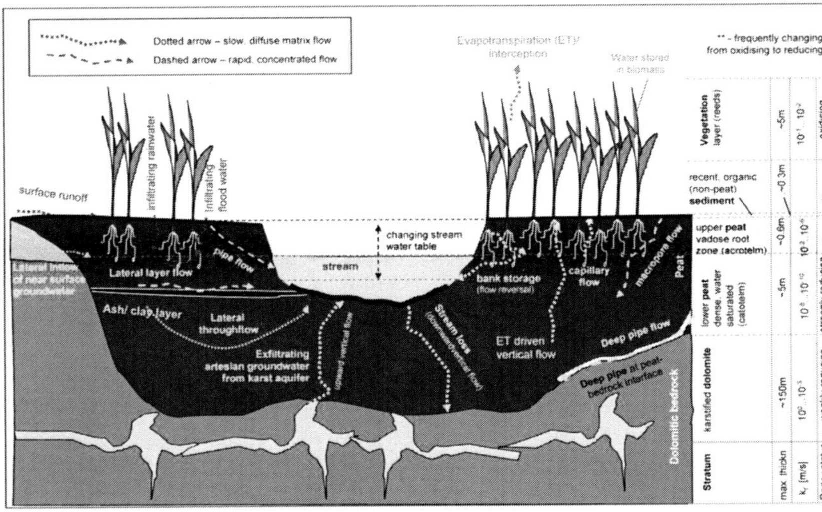

Figure 2. Overview on possible hydraulic processes in the GMB Peatland depicting different modes of surface—and groundwater moving through peat (not to scale) (hydraulic component of the U filter model for peat).

3. COURSE OF PROPOSED FIELD AND LABORATORY INVESTIGATIONS

Based on the above discussed conceptual model, the following 3-phase approach was adopted to obtain site-specific input data into the conceptual model, through a range of different field and laboratory investigations outlined below.

The first phase of the investigation was aimed at characterizing and quantifying the hydraulic filter component. In essence that means determining the volume of water that flows through the peatland. Owing to a lack of sufficient flow data for the wetland existing gauging records from the Department for Water Affairs (DWA) for the surrounding region were analyzed in order to establish the possible contribution of the wetland as a whole to the regional system. This was followed by two field campaigns to measure the flow rates at sites near, and in, the wetland for which no flow data were available and to obtain, spatially, more detailed flow estimates. In the final step, the wetland area was screened by measuring water quality parameters (temperature, EC, pH) along horizontal and vertical gradients to detect possible inflow sites of surface- and/or groundwater into the wetland. After establishing a first order approximation of the total volume of water entering and leaving the wetland, and a semi-quantitative apportionment to the different identified sources, attempts were made to determine different peat-water contact modes as described in the hydraulic model component. For this purpose, the hydrodynamics of peat porewater were monitored using quasi-continuous measurements of water levels, and quality parameters, by datalogger controlled in situ probes placed in the unmined section of the remaining natural peat.

In the second phase of the investigation, the main objective was to characterize and quantify the chemical filter component, i.e., to determine how far peat in the study area under local conditions with regard to chemical water composition, U-speziations and many other site-specific factors controlling the sorption capacity of peat, is indeed able to effectively remove dissolved U from the water column and to retain it. In a first step, U-levels in surface- and groundwater, as well as in sediments and peat samples from the wetland and its surroundings, were analyzed. Based on this, the extent of U-pollution and possibly associated accumulation of U in the wetland sediments and peat was assessed as a first proxy for quantifying the peat filter capacity for U. In a second step of this phase, batch experiments were conducted using sampled peat and two types of typical mine water to determine the U-removal efficiency of peat as well as the remobilization potential.

In the final (third) phase the above results are synoptically integrated into a preliminary, conceptual model, describing the ability of the remaining peat deposit at GMB to act as filter for U-polluted water under current conditions, as well as under a possible post-mine closure scenario, where acidic mine water may flow into the peatland from a vast system of flooded underground mine voids.

4. SUMMARY AND CONCLUSIONS

Numerous case studies reported in the literature indicate that peat, under natural conditions, acts as a very efficient filter for dissolved U and is able to accumulate the radioactive heavy metal from extremely low (natural background) concentrations in surface and groundwater to levels that may exceed those found in commercially mined, low grade, U ore.

To assess how far local peat at the GMB peatland, under the specific conditions of the study area, may indeed be able to remove U from mining polluted water and thereby protect a downstream municipality, a generic conceptual model on processes and factors governing U attenuation in peat was developed. This model consists of two linked sub-models, namely a chemical filter component addressing the ability of peat to filter and retain U, and a hydraulic filter component that refers to the mode in which U polluted water may come in contact with peat. Based on a significant body of literature, an overview on mechanisms responsible for the removal (immobilization) and subsequent release (remobilization) of U in peat has been compiled that informs the chemical filter component. In a similar manner, the general characteristics of the hydraulic model component were derived.

With regard to the different water-peat contact modes, the associated contact time and the size of the involved reactive surface area of the peat-water interface are of particular importance. They determine whether, and if so, to what extent the various chemical processes governing U mobility in the peat-water system may take place, and what portion of the total throughflow may be affected. In addition, the chemical sub-model also identifies factors such as pH, Eh, ionic strength and others, that determine the occurrence and intensity of the various processes. Finally, based on the conceptual model, a range of plausible natural and anthropogenic events have been discussed that may affect U attenuation in the GMB peatland.

The chemical sub-model contains 10 different processes that govern the attenuation of U peat, all of which are at least to some degree controlled by factors relating to water chemistry. The attenuation of U in peat is ultimately controlled by the net-balance of two different types of processes, namely immobilization and remobilization. The former refers to the removal of dissolved (mobile) U from the water phase either through transforming aqueous U into solid (immobile) phases (phase transition) that then accumulate in peat, or through the incorporation of dissolved U onto existing solid peat constituents by sorption or biological uptake. Immobilization based on phase transitions include the precipitation of insoluble U-species, the co-precipitation of U along with other precipitating secondary minerals, such as calcite and iron hydroxide as well as the gravitational deposition of aggregated U bearing colloids. The latter is somewhat ambiguous as colloids represent a state between solid and truly dissolved U. The remobilization of U from peat also comprises five different processes which either result in the release of U ions or soluble U-complexes back into the aqueous phase (desorption; dissolution of mineral U-phases and formation of inorganic and organic U complexes) or the release of particulate/colloidal U (physical destruction of U-containing organic matter through fire for example and the remobilization of U-colloids/erosion of deposited U bearing particles). Due to the impact of chemical parameters, such as changing pH, redox-potential, ionic strength and chemical composition of the water, these processes can be triggered by a range of natural and anthropogenic events including changing water levels, rainfall, influx of acidic mine water, annual burning of reeds, excavation of peat through commercial mining, etc.

In order to verify the applicability of the developed generic model to the local conditions at the GMB peatland a combination of different, site-specific approaches has been designed. For the chemical component, these include an analysis of U concentration levels and geochemical enrichment factors in selected sediment-water systems in and around the wetland, as well as a batch experiment for quantifying the removal efficiency and remobilization rate for two typical U polluted mine waters found in the study area. A similar approach was chosen for the hydraulic component. Based on a range of identified possible modes of water flow through or over peat, different types of field investigations were designed to characterize and, if possible, quantify the mode and rate of how water comes into contact with the peat and other wetland sediments as potential U filters. This included a statistical analysis of historical flow data, a screening survey of water temperature and EC-levels to detect possible underground inflow of groundwater and quasi-continuous observations of porewater dynamics using electronic in situ probes. The results of these investigations are discussed in Part III for the hydraulic component and Part IV for the chemical component of the proposed U-filter model for peat [102,103].

ACKNOWLEDGEMENTS

The study is funded by the Department of Water Affairs and Forestry of South Africa (project no. 213-2006) which is gratefully acknowledged. I also wish to thank two anonymous reviewers for their constructive comments that helped to improve the paper.

REFERENCES

1. Winde, F.; Erasmus, E. Peatlands as filters for polluted mine water?—A case study from an uranium-contaminated karst system in South Africa. Part I: Hydrogeological setting and U-fluxes. Water 2010, 3, 291–322.

2. Owen, D.E.; Otton, J.K. Mountain wetlands: Efficient uranium filters—potential impacts. Ecol. Eng. 1995, 5, 77–93.

3. Schöner, A. Hydrogeochemische Prozesse der Uranfixierung in natürlichen Wetlands und deren Anwendbarkeit in der 'passiven' Wasserbehandlung (Hydrogeochemical Processes of Uranium Fixation in Natural Wetlands and Their Applicability in Passive' Water Treatment). Dissertation, Chemisch-Geowissenschaftliche Fakultät, Friedrich-Schiller-Universität Jena, Jena, Germany, 2006; p. 373.

4. Smuts, W.J. Characteristics of South African Peats and Their Potential Exploitation. PhD Thesis, Faculty of Science, University of Pretoria, Pretoria, South Africa, 1997; p. 222.

5. Kochenov, A.V.; Zinevyev, V.V.; Lovaleva, S.A. Some features of the accumulation of uranium in peat bogs. Geochem. Int. 1965, 2, 65–70.

6. Lopatkina, A.P. Conditions of accumulation of uranium in peat. Geochem. Int. 1967, 4, 577–588.

7. Szalay, A. Accumulation of uranium and other micrometals in coal and organic shales and the role of humic acids in these geochemical enrichments. Ark. Mineral. Geol. 1974, 5, 23–36.

8. Moore, G.W. Extraction of uranium from aqueous solution by coal and some other materials. Econ. Geol. 1954, 49, 652–658.

9. Coggins, A.M.; Jennings, S.G.; Ebinghaus, R. Accumulation rates of the heavy metals lead, mercury and cadmium in ombrotrophic peatlands in the west of Ireland. Atmosph. Env. 2005, 40, 260–278.

10. Farmer, J.G.; Anderson, P.; Cloy, J.M.; Graham, M.C.; MacKenzie, A.B.; Cook, G.T. Historical accumulation rates of mercury in four scottish ombrotrophic peat bogs over the past 2,000 years. Sci. Total Environ. 2009, 407, 5578–5588.

11. Gosset, T.; Trancart, J.L.; Thevenot, D.R. Batch metal removal by peat: Kinetics and thermodynamics. Water Res. 1986, 20, 21–26.

12. Denham, M.; Looney, B. Recommended Amendment Mixture for in-Situ Treatment of Waste, Management Unit Groundwater, Ashtabula Closure Project; Report to the US Department of Energy

no. 2004, WSRC-TR-2004-00185; Savannah River Site: Aiken, SC, USA, 2004.

13. Nakashima, S.; Disnar, J.R.; Perruchot, A.; Trichet, J. Experimental study of mechanisms of fixation and reduction of uranium by sedimentary organic matter under diagenetic or hydrothermal conditions. Geochim et Cosmochim Acta 1984, 48, 2321–2329.

14. Brown, P.A.; Gill, S.A.; Allen, S.J. Metal removal from wastewater using peat. Water Res. 2000, 34, 3907–3916.

15. Winde, F. Urankontamination von Fließgewässern—Prozessdynamik, Mechanismen und Steuerfaktoren. Untersuchungen zum Transport von gelöstem Uran in bergbaulich gestörten Landschaften unterschiedlicher Klimate. (Uranium Contamination of Streams—Process Dynamics, Mechanisms and Governing Factors. Investigations into the Transfer of Dissolved Uranium in Mining-Impacted Landscapes of Different Climatic Conditions); Habilitationsschrift; Chemisch-Geowissenschaftliche Fakultät, Friedrich-Schiller-Universität Jena: Jena, Germany, 2003; p. 375.

16. Titayeva, N.A. Association of radium and uranium with peat. Geochem. Int. 1967, 4, 1168–1174.

17. Idiz, E.F.; Carlisle, D.; Kaplan, I.R. Interaction between organic matter and trace metals in a uranium rich bog, Kern County, California, USA. Appl. Geochem. 1986, 1, 573–590.

18. Ibarra, J.V.; Osacar, J.; Gavilan, J.M. Retention of metallic cations by lignites and humic acids. Fuel 1979, 58, 827–830.

19. Crancon, P.; Pili, E.; Charlet, L. Uranium facilitated transport by water-dispersible colloids in field and soil columns. Sci. Total Environ. 2010, 408, 2118–2128.

20. Krepelova, A.; Sachs, S.; Berhard, G. Uranium(VI) sorption onto kaolinite in the presence and absence of humic acid. Radiochim. Acta. 2006, 94, 825–833.

21. Feng, X.; Simpson, A.J.; Simpson, M.J. Chemical and mineralogical controls on humic acid sorption to clay mineral surfaces. Organ Geochemistr. 2005, 36, 1553–1566.

22. Weng, L.; Van Riemsdijk, W.H.; Hiemstra, T. Adsorption of humic acids onto goethite: Effects of molar mass, pH and ionic strength. J. Colloid Interface Sci. 2007, 314, 107–118.

23. Murphy, R.J.; Lenhart, J.J.; Honeyman, B.D. The sorption of thorium(IV) and uranium(VI) to hematite in the presence of natural organic matter. Colloid. Surface. Physicochem. Eng. Aspect. 1999, 157, 47–62.

24. Ho, C.H.; Miller, N.H. Effect of humic acids on uranium uptake by hematite particles. J. Colloid Interface Sci. 1985, 106, 281–288.

25. Lenhart, J.J.; Honeyman, B.D. Uranium(VI) sorption to hematite in the presence of humic acid. Geochim et Cosmochim Acta 1999, 63, 2891–2901.

26. Bradley, C. Transient Modelling of Water-Table Variation in a Floodplain Wetland, Narborough Bog, Leicestershire. J. Hydrol. 1996, 185, 87–114.

27. Cohen, A.D.; Rollins, M.S.; Zunic, W.M.; Durig, J.R. Effects of chemical and physical differences in peat and their ability to extract hydrocarbons from water. Water Res. 1991, 25, 1047–1060.

28. Couillard, D. The use of peat in wastewater treatment. Water Res. 1994, 28, 1261–1274.

29. Kao, C.M.; Lei, S.E. Using peat biobarrier to remediate PCE/TCE contaminated aquifers. Water Res. 2000, 34, 835–845.

30. Ringquist, L.; Holmgren, A.; Oeborn, I. Poorly humified peat as an adsorbent for metals in wastewater. Water Res. 2001, 36, 2394–2404.

31. Ringquist, L.; Oeborn, I. Copper and zinc adsorption onto poorly humified sphagnum and carex peat. Water Res. 2001, 36, 2233–2242.

32. Rasmussen, G.; Fremmersvik, G.; Olsen, R.A. Treatment of creosote-contaminated groundwater in a peat-sand permeable barrier—A columns study. J. Hazard. Materials 2002, B93, 285–306.

33. Al Faqi, L.; Johnson, P.D.; Allen, S.J. Evaluation of a new peat-based sorbent for metals capture. Bioresour. Technol. 2007, 99, 1394–1402.

34. Boonchayaanant, B.; Nayak, D.; Du, X.; Criddle, C.S. Uranium reduction and resistance to reoxidation under iron-reducing and sulfate-reducing conditions. Water Res. 2009, 43, 4652–4664.

35. Nedwell, D.B.; Watson, A. Production, oxidation and emission in a U.K. ombotrophic peat bog: Influence of SO42− from acid rain. Soil Biol. Biochem. 1995, 27, 893–903.

36. Jacks, G.; Norrström, A.C. Hydrochemistry and hydrology of forest riparian wetlands. Forest Ecol. Managem. 2004, 196, 187–197.

37. Winde, F.; Wade, P.; van der Walt, I.J. Gold tailings as a source of waterborne uranium contamination of streams—The Koekemoerspruit (Klerksdorp goldfield, South Africa) as a case study. Part III: Fluctuations of stream chemistry and their impacts on uranium mobility. Water SA 2004, 30, 233–240.

38. Winde, F.; Wade, P.; van der Walt, I.J. Gold tailings as a source of waterborne uranium contamination of streams—The Koekemoerspruit (Klerksdorp goldfield, South Africa) as a case study. Part I: Uranium migration along the aqueous pathway. Water SA 2004, 30, 219–226.

39. Bhat, S.V.; Melo, J.S.; Chaugule, B.B.; D'Souza, S.F. Biosorption characteristics of uranium(VI) from aqueous medium onto Canella repens, a red alga. J. Hzard. Mat. 2008, 158, 628–635.

40. Simonsen, J.F.; Harremoes, P. Oxygen and pH fluctuations in rivers. Water Res. 1978, 12, 477–489.

41. Jones, C.A.; Nimick, D.A.; McCleskey, R.B. Relative effect of temperature and pH on cycling of dissolved trace elements in prickly pear creek, Montana. Water Air Soil Poll. 2004, 153, 95–113.

42. Domínguez-Villar, D.; Arteaga, C.; García-Giménez, R.; Smith, E.A.; Pedraza, J. Diurnal and seasonal water variations of temperature, pH, redox potential and conductivity in gnammas (Weathering Pits): Implications for chemical weathering. Catena 2008, 72, 37–48.

43. Mavrocordatos, D.; Mondi-Couture, C.; Atteia, O.; Leppard, G.G.; Perret, D. Formation of a distinct class of Fe-Ca(-Corg)-rich

particles in a complex peat-karst system. J. Hydrol. 2000, 237, 234–247.

44. Winde, F. Untersuchungen zur Genese, Schwermetallkontamination und hochwassergebundenen Verlagerung rezenter Gerinnebettsedimente in Nebenvorflutern der Halleschen Saaleaue (Investigations into formation, heavy metal contamination and flood-induced re-suspension of recent stream channel sediments in side arms of the River Saale near Halle). In Late Quaternary and Recent Earth Surface Systems in Europe; Aurada, D., Billwitz, K., Lampe, E., Eds.; Martin-Luther-Universität Halle-Wittenberg: Halle, Germany, 1998; pp. 105–122, Supplementband der Zeitschrift für Geomorphologie.

45. Maggi, F. Biological flocculation of suspended particles in nutrient-rich aqueous ecosystems. J. Hydrol. 2009, 376, 116–125.

46. Tsezos, M.; Voleski, B. The mechanism of uranium biosorption by Rhizopus Arrhizus. Biotechnol. Bioengin. 1982, 24, 385–401.

47. Tsezos, M.; Voleski, B. Biosorption of uranium and thorium. Biotechnol. Bioengin. 1981, 23, 583–604.

48. Sikora, L.J.; Keeney, D.R. Further aspects of soil chemistry under anaerobic conditions. In Ecosystems of the World; Elsevier: New York, NY, USA, 1983.

49. McNevin, D.; Barford, J.; Hage, J. Adsorption and biological degradation of ammonium and sulfide on peat. Water Res. 1999, 33, 1449–1459.

50. Morris, K.; Raiswell, R. Biochemical cycles and remobilisation of actinide elements. In Interactions of Microorganisms with Radionuclids, 1st ed.; Keith-Roach, M.J., Livens, F.R., Eds.; Elsevier Science Ltd.: New York, NY, USA, 2002; Chapter 4.

51. Van Roy, S.; Vanbroekhoven, K.; Dejonghe, W.; Diels, L. Immobilization of heavy metals in the saturated zone by sorption and in Situ bioprecipitation processes. Hydrometall 2006, 83, 195–203.

52. Dienemann, C.; Dienemann, H.; Dudel, E.G.; Schurig, C. Uranium fixation by Cladophera spec. In Uranium, Mining and

Hydrogeology; Merkel, B.J., Hache-Berger, A., Eds.; Springer Verlag: Berlin and Heidelberg, Germany, 2008; pp. 103–110.

53. Bhat, S.V.; Melo, J.S.; Chaugule, B.B.; D'Souza, S.F. Biosorption characteristics of uranium(VI) from aqueous medium onto Canella repens, a red alga. J. Hzard. Mat. 2008, 158, 628–635.

54. Cerne, M.; Smodis, B.; Strok, M. Uptake of radionuclides by a common reed (Phragmites australis Cav. Trin. Ex Steud.) grown in the vicinity of the former uranium mine at Zirovski vrh. Nucl. Eng. Des. 2010.

55. Twardowska, I.; Kyziol, J. Soprtion of metals onto natural organic matter as a function of complexation and adsorbent-adsorbate contact mode. Environ. Int. 2003, 28, 783–791.

56. Tipping, E.; Smith, E.J.; Lawlor, A.J.; Hughes, S.; Stevens, P.A. Predicting the release of metals from ombotrophic peat due to drought-induced acidification. Environ. Poll. 2003, 123, 239–253.

57. Dellwig, O.; Böttcher, M.E.; Lipinski, M.; Brumsack, H.J. Trace metals in holocene coastal peats and their relation to pyrite formation (NW Germany). Chem. Geol. 2002, 182, 423–442.

58. Mandernack, K.W.; Lynch, L.; Krouse, H.R.; Morgan, M.D. Sulfur cycling in wetland peat of the New Jersey pinelands and its effect on stream water chemistry. Geochim et Cosmochim 2000, 64, 3949–3964.

59. Eimers, C.M.; Dillon, P.J.; Schiff, S.L.; Jeffries, D.S. The effects of drying and re-wetting and increased temperature on sulphate release from upland and wetland material. Soil Biol. Biochem. 2003, 35, 1663–1673.

60. Wade, P.W.; Woodbourne, S.; Morris, W.M.; Vos, P.; Jarvis, N.V. Tier 1 Risk Assessment of Radionuclids in Selected Sediments of the Mooi River; WRC-Project, K5/1095; Water Research Commission: Pretoria, South Africa, 2002.

61. Francis, C.W.; Timpson, M.E.; Wilson, J.H. Bench- and pilot-scale studies relating to the removal of uranium from uranium-contaminated soils using carbonate and citrite lixivants. J. Hazard. Mater. 1999, 66, 67–87.

62. Schulze, G. Bestandsaufnahme und Charakterisierung der dtofflichen Auswirkungen des Uranerzbergbaus und der Uranaufbereitung (Standort Seelingstädt) am Beispiel des Wasserpfades. (Stock-Taking and Characterisation of Substance-Related Impacts of Uranium Mining and Milling (Site Seelingstädt) using the Aquatic Pathway as an Example.). Veröff. Museum Gera, Naturwiss. Reihe 1993, 20, 40–73.

63. Baeza, A.; Salas, A.; Legarda, F. Determining factors in the elimination of uranium and radium from groundwaters during standard potabilization process. Sci. Total Environ. 2008, 406, 24–34.

64. Quin, F.; Wen, B.; Shan, X.Q.; Xie, Y.N.; Liu, T.; Zhang, S.Z.; Khan, S.U. Mechanisms of competitive adsorption of Pb, Cu, and Cd on peat. Environ. Pollut. 2006, 144, 669–680.

65. Boelter, D.H. Hydraulic conductivity of peats. Soil Sci. 1965, 100, 227–231.

66. Ingram, H.A.P.; Rycroft, D.W.; Williams, D.J.A. Anomalous transmission of water through certain peats. J. Hydrol. 1974, 22, 213–218.

67. Dooge, J. The water balance of bogs and fens. In Hydrology of Marsh-Ridden Areas: Proceedings of the Minsk Symposium; UNESCO: Paris, France, 1975; pp. 233–271.

68. Dasberg, S.; Neuman, S.P. Peat hydrology in the hula basin, Israel: I. Properties of peat. J. Hydrol. 1977, 32, 219–239.

69. Chanson, D.B.; Siegel, D.I. Hydraulic conductivity and related physical properties of peat, lost river peatland, Northern Minnesota. Soil Sci. 1986, 142, 91–99.

70. Boeye, D.; Verheyen, R.F. The hydrological balance of a groundwater discharge fen. J. Hydrol. 1992, 137, 149–163.

71. Gilman, K. Hydrology and Wetland Conservation; John Wiley & Sons, Ltd.: New York, NY, USA, 1994.

72. Fraser, C.J.D.; Roulet, N.T.; Lafleur, M. Groundwater flow patterns in a large peatland. J. Hydrol. 2001, 246, 142–154.

73. Bragg, O.M. Hydrology of peat-forming wetlands in Scotland. Sci. Total Environ. 2002, 294, 111–129.

74. Holden, J.; Burt, T.P. Piping and pipeflow in a deep peat catchment. Catena 2002, 48, 164–199.

75. Holden, J. Going beyond the acrotelm-catotelm model in blanket peat catchments. Geophys. Res. Abstr. 2003, 5, 01233.

76. Bradley, C.; van Den Berg, J.A. Infiltration mechanisms in a herbaceous peat: Results of a controlled infiltration experiment. Hydrol. Sci. J. 2005, 50, 713–725.

77. Bradley, C.; Brown, A.G. Shallow groundwater modeling and the over-bank contribution to a small floodplain bog. In Geomorphology and Groundwater, 1st ed.; Brown, A.G., Ed.; Wiley: Chichester, UK, 1995; pp. 37–52.

78. Daniels, S.M.; Agnew, C.T.; Allot, T.E.H.; Evans, M.G. Water table variability and runoff generation in an eroded peatland, South Pennines, UK. J. Hydrol. 2008, 362, 214–226.

79. Holden, J. Flow through macropores of different size classes in blanket peat. J. Hydrol. 2009, 364, 342–348.

80. Price, J.S.; Whittington, P.N. Water flow in Sphagnum hummock: Mesocosm measurements and modelling. J. Hydrol. 2010, 381, 333–340.

81. Winde, F.; van der Walt, I.J. Gold tailings as a source of waterborne uranium contamination of streams—the Koekemoerspruit (Klerksdorp goldfield, South Africa) as a case study. Part II of III: Dynamics of groundwater-stream interactions. Water SA 2004, 30, 227–232.

82. Evans, M.G.; Burt, T.P.; Holden, J.; Adamson, J.K. Runoff generation and water table fluctuations in blanket peat: Evidence from UK data spanning the dry summer of 1995. J. Hydrol. 1999, 221, 141–160.

83. Wise, W.R.; Annable, M.D.; Walser, J.A.E.; Switt, R.S.; Shaw, D.T. A wetland-aquifer interaction test. J. Hydrol. 2000, 227, 257–272.

84. Rycroft, D.W.; Williams, D.J.A.; Ingram, H.A.P. The transmission of water through peat: I. Review. J. Ecol. 1975, 63, 535–556.

85. Siegel, D.I. A Review of the Recharge-Discharge Function of Wetlands. In The Ecology and Management of Wetlands: Ecology of Wetlands; Hook, D.D., Ed.; Croom Helm: London, UK, 1988; pp. 59–67.

86. Price, J.S.; Schlotzhauer, S.M. Importance of shrinkage and compression in determining water storage changes in peat: The case of a mined peatland. Hydrol. Proc. 1999, 13, 2591–2601.

87. Reeve, A.S.; Siegel, D.I.; Glaser, P.H. simulating vertical flow in large peatlands. J. Hydrol. 2000, 227, 201–217.

88. Van Loon, A.H.; Schot, P.P.; Griffioen, J.; Bierkens, M.F.P.; Wassen, M.J. Throughflow as a determining factor for habitat contiguity in a near-natural fen. J. Hydrol. 2009, 379, 30–40.

89. Siegel, D.I.; Reeve, A.S.; Glaser, P.H.; Romanowicz, E.A. Climate-driven flushing of pore water in peatlands. Nature 1995, 374, 531–533.

90. Hemond, H.F.; Goldman, J.C. On non-darcian water flow in peat. J. Ecol. 1985, 73, 579–584.

91. Blodau, C.; Moore, T.R. Macro porosity affects water movement and pore water sampling in peat soils. Soil Sci. 2002, 167, 98–109.

92. Jones, J.A.A. Extending the Hewlett model of stream runoff generation. Area 1979, 11, 110–114.

93. Jones, J.A.A.; International Geographical Union, Chair Commission on Water Sustainability, retired professor, University of Wales, Aberystwyth. personal communication, July 2009.

94. Reeve, A.S.; Siegel, D.I.; Glaser, P.H. Simulating dispersive mixing in large peatlands. J. Hydrol. 2001, 242, 103–114.

95. Reeve, A.S.; Siegel, D.I.; Glaser, P.H. Simulating vertical flow in large peatlands. J. Hydrol. 2000, 227, 201–217.

96. Fraser, C.J.D.; Roulet, N.T.; Lafleur, M. Groundwater flow patterns in a large peatland. J. Hydrol. 2001, 246, 142–154.

97. Sade, R.; Litaor, M.I.; Shenker, M. Evaluation of groundwater and phosphorous transport in fractured altered wetland soils. J. Hydrol. 2010, 393, 133–142.

98. Lowry, C.S.; Fratta, D.; Anderson, M.P. Ground penetrating radar and spring formation in a groundwater dominated peat wetland. J. Hydrol. 2009, 373, 68–79.

99. Bauer, P.; Thabeng, G.; Stauffer, F.; Kinzelbach, W. Estimation of the evapotranspiration rate from diurnal groundwater fluctuations in the Okavango Delta, Botswana. J. Hydrol. 2004, 288, 344–355.

100. Bradley, C.; Brown, A.G. Shallow Groundwater Modeling and the Overbank Contribution to a Small Floodplain Bog. In Geomorphology and Groundwater; Brown, A.G., Ed.; Wiley: Wiltshire, UK, 1995; pp. 37–52.

101. Cloke, H.L.; Anderson, M.G.; McDonnell, J.J.; Renaud, J.-P. Using numerical modelling to evaluate the capillary fringe groundwater ridging hypothesis of streamflow generation. J. Hydrol. 2006, 316, 141–162.

102. Winde, F. Peatlands as filters for polluted mine water?—A case study from an uranium-contaminated karst system in South Africa. Part III: Quantifying the hydraulic filter component. Water 2010, 3, 356–390.

103. Winde, F. Peatlands as filters for polluted mine water?—A case study from an uranium-contaminated karst system in South Africa. Part IV: Quantifying the chemical filter component. Water 2010, 3, 391–423.

INDEX

K

L

M

O

P

R

S

U

V

W